Computer Simulation in Materials Science

Papers presented at the
1986 ASM Materials Science Seminar

4-5 October 1986
Lake Buena Vista, Florida

Sponsored by the Science Seminar
Committee of the Materials Science
Division of ASM International

Edited by
R. J. Arsenault
University of Maryland
J. R. Beeler, Jr.
North Carolina State University
D. M. Esterling
George Washington University

ASM INTERNATIONAL
Metals Park, Ohio 44073

Library of Congress Catalog Card Number: 87-70937

ISBN: 0-87170-296-7

SAN: 204-7586

Editorial and production coordination by
Carnes Publication Services, Inc.

PRINTED IN THE UNITED STATES OF AMERICA

Preface

The intent of this seminar was to present in a hierarchical format the currently most advanced techniques in Computational Materials Science, encompassing both the Atomistic Model and the Continuum Model aspects of this field, as well as to show how these two aspects complement each other in the consideration of a complex problem such as forging. To fully understand materials behavior, it is necessary to progress along a sequence of atomistic, microstructural, and continuum model aspects of materials behavior. Methods for the synthesis of these models in treating a complex problem have not been fully developed. There are major gaps in the theory and practice of this synthesis. A central theme of forging was suggested to the authors to maintain a fixed technological center as well as to exemplify important gaps in the synthesis of atomistic and continuum models in one particular area of application.

Forging induces high-strain, high-strain-rate phenomena at elevated temperatures. It poses questions in Continuum Mechanics—What are the proper constitutive relations?—and in Materials Science—How does one describe the mechanisms for high-strain, high-strain-rate plastic deformation, and how does one describe atomic interactions in a highly deformed crystal?

The original plan was that each paper would be a tutorial event. The authors were encouraged by the organizing committee to correspond with each other as they developed their respective papers. As a consequence of this correspondence, variations ranging from a predominantly tutorial to a predominantly descriptive character developed in some of the assignments. This cooperative effort eliminated needless overlap among several papers and lent a general informational improvement to the seminar as a whole.

The timing of the seminar was fortunate. Several practical methods for estimating electronic density effects and N-body interaction effects in computer simulation of solids were developed and published during the three years preceding the seminar. This was a boon for the atomistic model simulation aspects of the seminar. In addition, recent and important advances also had been made in framing constitutive relations and in applying the finite element to die design by simulating billet behavior during a forging process. This enriched the continuum model simulation aspects of the seminar.

Some of the gaps which seem to exist in a synthesis of atomistic model and continuum model simulation techniques are phantoms. These phantoms arise because, at the atomistic level, the crystal defect interaction rules given by the atomistic model approach are qualitatively different from those given by the continuum model approach. The rules given by the atomistic model approach are the correct rules. The phantoms vanish when this fact is understood.

Another source of controversy centers on the fact that existing practical models for atomic interactions are primitive; they, at best, only partially account for the vital *directional* aspects of the electron density distribution. On the positive side, however, they are sufficiently accurate to explain some of the gross features of hydrogen embrittlement and the surface properties of metals.

The book begins with Lahoti's general overview of forging. His chapter provides general background knowledge of forging and emphasizes the areas in which computer simulation has been used in designing dies, preforms, and the forging process. After this introduction, the topical progression of the seminar is: (1) atomistic models, (2) continuum models, and (3) the synthesis of atomistic and continuum models in the design of a forging process.

The atomistic model portion opens with Johnson's chapter on the development of practical ways to describe interactions among atoms in a condensed phase. He deals with both pairwise, central-force models and N-body interaction models. Beeler describes the procedure for writing a molecular dynamics computer program, illustrated with flow diagrams and program listings, for both pairwise and N-body interaction models. His chapter also briefly describes Monte Carlo applications. The application of the molecular dynamics method to computing the thermodynamic properties of a metallic solid is demonstrated by Straub, with emphasis on entropy calculations. His chapter shows how otherwise unobtainable estimates of both equilibrium and nonequilibrium thermodynamics properties of metals can be obtained from molecular dynamics simulations. Wolf presents the results of his atomistic model simulations of grain boundaries. He deals largely with the key physical attributes which make a boundary stable, covering both ionic and metallic systems. Simulation of the effect of hydrogen on crack propagation and helium on crack nucleation in nickel is covered by Baskes. He uses the embedded atom method atomic interaction model (an N-body model) rather than a central-force model in his simulations.

Esterling and Arsenault provide a bridge between the atomistic and continuum model portion of the seminar with an account of their simulation work on thermally activated dislocation motion in both a pure metal and a metal containing solute atoms and inclusions. This involves a discussion of how to determine the intrinsic lattice resistance to dislocation motion and continuum model simulation of dislocation motion in a field of obstacles. Moyer shows how to set up a finite-element-method program to compute stress and strain. His chapter also includes practical advice on selection of element shape, how to design a test case, and optimal techniques for equation solving. Wilson reviews his work on friction modeling in both a forging and an extrusion context. He demonstrates that friction modeling is a vital concern in forging and that there are many research opportunities in this field. Thomas explains the development of constitutive equations for high-temperature deformation, for use in finite-element-method simulations of forging. He also shows experimental results which verify the constitutive relations he has developed.

The seminar closes with Gegel's general summary of how atomistic and continuum model results have been synthesized in technologically useful forging simulations. He points out that optimized forging processes have been observed to lie near the steady-state boundary on a temperature–strain rate map. This locus correlates with the maximum rate of internal entropy production in the work piece. Atomistic model simulations, dealing with entropy production and energy flow during plastic deformation, are needed to further improve the synthesis of atomistic and continuum models in forging simulations.

We wish to thank the authors for their thought-provoking and informative papers. These papers summarize the state of the art for Computational Materials Science in a cogent and concise manner. Those participants in the seminar who did not give formal presentations also made vital contributions to the seminar. They refined and augmented the formal presentations by asking probing questions and by describing their own insights and research results during the discussion sessions.

R. J. Arsenault
University of Maryland

J. R. Beeler, Jr.
North Carolina State University

D. M. Esterling
George Washington University

Contents

1

Review of Forging: Past History, Recent Advances*

G. D. LAHOTI

The Timken Company
Technology Center
Canton, OH 44706

and

S. L. SEMIATIN

Battelle Memorial Institute
505 King Avenue
Columbus, OH 43201

Abstract

Forging is the oldest and most basic process for forming metals. New knowledge in the deformation properties of metals and the increasing application of computer-aided technology underlie increased conservation of materials and productivity in the process. Thus, interaction of science and technology is making forging production more economical.

Introduction

Forging, the compressive deformation of materials between dies, is probably the oldest method used by man to work metals into useful shapes. Prehistoric man used it to form his crude hunting tools and weapons. In the Middle Ages, smiths (blacksmiths, goldsmiths, silversmiths, etc.), working in one- or two-man shops, forged tools and weapons as well as other products, such as coins, utensils, and nails. During the Industrial Revolution, the size of forging shops increased, and the size and complexity of forged items increased as well. In modern times, forged metallic articles can be seen everywhere, being used in fields as diverse as transportation, construction, agriculture, and plumbing. The list of forged products is equally varied, ranging

*The bulk of the information contained in this paper was published by the authors in the August 1981 issue of *Scientific American* under the title "The Forging of Metals."

from simple bolts, gears, and locks, to large and detailed forgings used in aircraft landing gear and steam-driven equipment for electric-power generation.

Forging is a metalworking operation involving a complex interaction of the metal and the dies and equipment used to deform it. Hence, it is not surprising that advances in forging technology through history have entailed breakthroughs in both metallurgy and equipment design. Perhaps the earliest recorded use of forging involved the simple hammering of gold and copper in various regions in the Middle East around 8000 B.C. The forming of these metals was crude, relatively speaking, since the art of refining by smelting had not been discovered, and the ability to forge the metal (that is, the forgeability or workability) was limited by the presence of impurities remaining after separation of the metal from the ore. With the advent of copper smelting (4000 B.C.), a useful method of purifying metals through chemical reactions in the liquid state was discovered. This heralded the beginning of the Copper Age and the increased utilization of metals in general and forging in particular. During this period, metalworkers also found that hammering resulted in desirable increases in strength (due to "work hardening" of the metal). This desire also spurred the search for alloys that were inherently strong, and led to the utilization of alloys of copper and tin (the Bronze Age, circa 2500 B.C.) and iron and carbon (the Iron Age, 1200 B.C.).

The Iron Age and the forging of iron alloys was relatively late in its coming because, prior to that, no one had been able to achieve the high temperatures needed to melt and refine iron ore. In addition, it was found that iron alloys were easily forged only when the temperature was greater than three-fifths of the absolute melting point. This so-called hot forging or hot working resulted in the formation of "black scale" (iron oxides), from which blacksmiths got their name.

The development of the tilt hammer in the thirteenth century A.D. was a boon to blacksmiths since it relieved them of much of the labor needed to forge metals. This device employed water power to raise a lever arm on one end of which was a hammering tool. After raising the hammer, the smith allowed it to fall, producing the forging blow. This relatively simple concept remained in use for several hundred years and was used primarily for forging of bars and plates (Fig. 1). With the development of processes for making large quantities of iron and steel during the Industrial Revolution (1750–1850) and the increased demand for consumer and machinery-type products during this period, a need arose for equipment with larger capacity. To meet this need, high-speed steam hammers (employing steam power to raise the hammer) and hydraulic forging presses (in which the forging load is supplied by hydraulic pressure) were invented in the mid-nineteenth century to make products ranging from firearms to locomotive parts (Fig. 2 and 3).

Fig. 1. Hot forging of steel in the Middle Ages is depicted by Georgius Agricola in *De Re Metallica* (1556). The steel is either forged by hand (top of figure) or in a water-powered hammer (bottom of figure). In either case, the steel is quenched subsequent to forging in order to impart maximum strength to the metal.

In the last hundred years, new types of forging equipment and new materials with special properties and applications have been developed. The utilization of these materials has greatly benefited from new knowledge (much of it acquired in the last 25 years) of the effects of temperature and deformation speed on the basic working characteristics of metals and the effects of die design and lubrication on metal flow and load requirements during forging. In turn, the application of this knowledge to actual forging operations has been facilitated by the development of sophisticated mathematical analyses of forging processes, which are best carried out on high-speed digital computers.

Fig. 2. Forging hammer is being used to hot forge a bar of steel. The ram, or hammer, falls by force of gravity and with the assistance of compressed air. Compressed air is also used to raise the ram.

Characteristics of Metals Used in Forging

It is now commonly accepted that the reason for the ability to impose permanent or plastic deformation on metals lies in the nature of the metallic bond. This bond consists of sharing of the outer electrons of the metal atoms and is nondirectional. The nondirectionality allows close packing of metal atoms in regular three-dimensional crystalline structures. The presence of imperfections in these structures, known as *dislocations*, enables shearing of the lattice on close-packed planes along close-packed directions (Fig. 4). Thus, it is not surprising that the relative workability of different metals depends a great deal on the ease with which dislocations can be made to move without introducing gross fracture (or other kinds of defects) into the material being worked.

Fig. 3. Hydraulic press at Alcoa's Cleveland Works is one of the two largest (445 MN [50,000 tons]) in the United States. Used to forge a variety of products, the press is driven by hydraulic pumps at the top of the press. A constant oil pressure is maintained by using the accumulators to the left of the press. (*Photograph courtesy of Mesta Machine Company.*)

The ease with which dislocations can be made to move without fracture of the metal is a function of the structural features, the temperature of the workpiece, and the deformation rate. Structural features include crystal structure (cubic or hexagonal packing of atoms, for example), presence of alloying elements, and presence of other dislocations. Probably the most important of these features, though, is the grain size. Grains are the building blocks of metal aggregates. Each grain has a different orientation of its crystal lattice. When the grain size is large relative to the overall size of the piece of metal being worked (known as the workpiece), cracks may initiate and propagate easily along the boundaries separating grains. This problem is of particular concern when ingot or cast structures are forged (Fig. 5).

Cast structures are found in metals produced by solidification of molten metal in ingot molds. During solidification, metal freezes first at the cold ingot walls and coarse, columnar grains, which may be several centimeters long, grow inward. These structures are very difficult to forge because the interfaces between the grains are planes of weakness. In addition, impurities

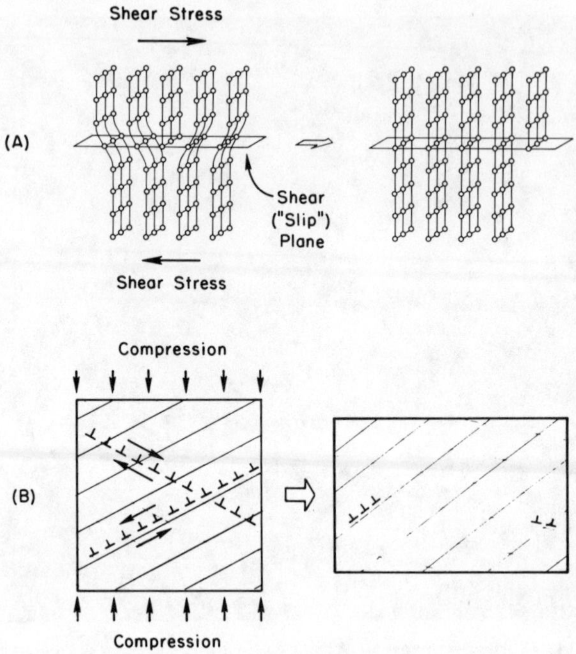

Fig. 4. Permanent, or plastic, deformation of metals occurs by the motion of dislocations through the crystal lattice in (A). The dislocations, or line defects in the stacking arrangement of metal atoms, allow easy shearing of the lattice on close-packed planes ("slip" planes) along close-packed directions. When an actual deformation is imposed on a metal aggregate, each crystal deforms along several sets of planes and directions ("slip systems"). As shown in (B), this involves motion of dislocations (denoted by ⊥) on each set of planes and a simultaneous rigid-body rotation of the crystal lattice.

(inclusions, dirt) segregate to the center and top of the ingot, also producing weak areas of low workability there. Lastly, the chemical elements in alloys are not distributed uniformly. For these reasons, the temperature range over which the ingot structure can be forged is very limited. Typically, these structures must be hot worked (i.e., at temperatures greater than three-fifths of the absolute melting point of the alloy). The upper temperature limit in this range is usually just below the melting point of the metal. The melting point of as-cast alloys is usually below that of alloys in the worked or wrought state because of the chemical inhomogeneities and the presence of low-melting-point compounds, which are usually found at the grain boundaries. Forging too close to the melting point of these compounds may lead to grain-boundary cracking when heat developed during deformation

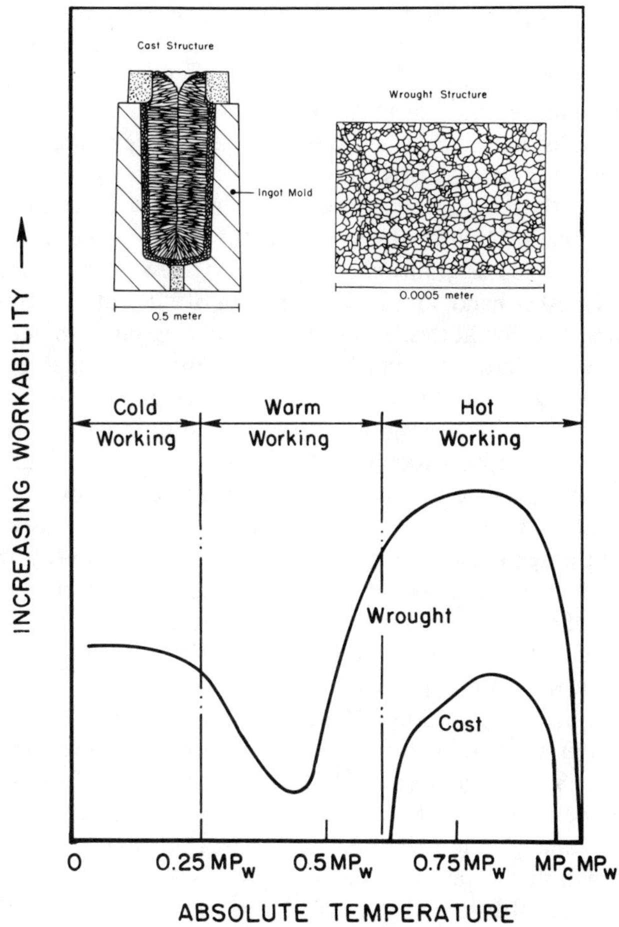

Fig. 5. Workability, or forgeability, of metals depends greatly on the size of the crystals, or grains, in the aggregate. Ingot-cast materials, which have large grains and are usually chemically heterogeneous, are typically hard to work. They also have a narrow working temperature range, the upper limit of which (the melting point of the cast material, MP_c) is below that of the wrought structure. Wrought structures have fine grain sizes and may be worked at a variety of temperatures between absolute zero and the melting point of the wrought material, MP_w.

raises the temperature and causes incipient melting of these compounds. This fracture mechanism is called *hot shortness*, and it is avoided by forging at deformation rates sufficiently slow to allow the heat developed by deformation to be dissipated into the dies.

The lower temperature limit for hot forging cast structures is the temperature at which there is some but not a great deal of dislocation multiplication. Dislocation multiplication, or creation of new dislocations, occurs during deformation when dislocations intersect each other or when they come against obstacles such as grain boundaries. Some dislocation multiplication is helpful in that it helps build a reservoir of stored energy which, in combination with thermal energy (supplied during hot deformation or during post-deformation heat treatment), is useful in breaking down the boundaries between the coarse grains and in generating new, smaller recrystallized grains. On the other hand, an excessive amount of dislocation multiplication, which occurs at temperatures below the hot working range, hinders dislocation motion in the lattice too much and thus deformation may be forced to occur at the weak grain boundaries, leading to cracking.

Breaking down the coarse-grained structure during forging results in a chemically homogeneous wrought structure with much finer grain size (usually several hundredths of a millimeter) and far superior workability (Fig. 5). It is with metals of this structure that actual forged products are manufactured. Even though most wrought metals can be worked over a wide range of temperatures and deformation rates, care must yet be exercised in selecting forging temperature. Hot forging requires minimum working loads and imparts a maximum in workability, but it is difficult to control the dimensional tolerances of hot forged metal products because of nonuniform thermal contraction upon cooling. Hence, metals are often forged at room temperature, at which most metals lie in the cold working regime (absolute zero to approximately one-fourth the absolute melting point of the metal). At these temperatures, most metals still retain a large amount of workability, which is somewhat less than that at hot working temperatures, but still far greater than that obtained at the intermediate temperatures between one-fourth and three-fifths the absolute melting point, the warm working temperature regime.

Selection of Forging Conditions—Loads

Besides dimensional control and attainable part precision, the main considerations in establishing the forging conditions are the forging load needed to carry out deformation (and, therefore, the capacity of the equipment used to do the forging) and the workability of the metal. Secondary considerations, mainly economic, are production rate, die wear, and die life.

Estimates of forging loads for different metals are often obtained by compressing cylindrical samples between well-lubricated flat parallel dies, a deformation mode which simulates the simplest kind of forging operation, known as *upsetting*. From this widely used uniaxial compression test, the deformation resistance (or "flow strength") in terms of stress (load/area) is plotted as a function of some measure of deformation, usually strain (logarithm of the ratio of the initial to final sample heights) or percentage reduction in height. The stress-strain curves so obtained (Fig. 6) show the effect of amount of deformation, rate of deformation, and test temperature on overall deformation resistance levels, and are essential input to mathemati-

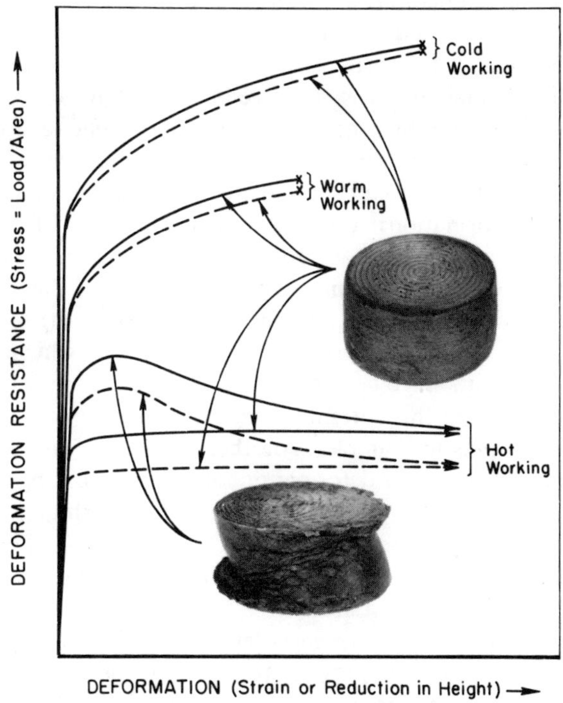

Fig. 6. The deformation resistance of wrought materials is often measured by uniaxial compression tests on cylindrical samples. Plots of this resistance in terms of stress as a function of deformation also indicate the effects of temperature and deformation rate. Increasing temperature tends to lower the deformation resistance while increasing rate (high rates are solid lines and low rates are broken lines) increases it, particularly at hot working temperatures. After testing, compression samples are usually cylindrical (photograph) except when flow softening, decreases of stress with increasing strain found sometimes at hot working temperatures, occurs. This flow softening may lead to grossly nonuniform flow during compression testing (photograph) as well as sometimes during forging. Workability limits may also be studied in the compression test. The x's indicate the relative workability in cold and warm working regimes, while that in hot working is typically much greater than in either cold or warm working.

cal modeling of forging processes. At cold working temperatures, these stresses are high primarily because of the difficulty of moving dislocations through the lattice and rapid dislocation multiplication. This dislocation multiplication leads to large amounts of work hardening, or increasing stress with strain. The workability limits at cold working temperatures are modest. The stress levels are somewhat lower at warm working temperatures, but there is still a large amount of work hardening, and workability is low.

At hot working temperatures, the stress levels are the lowest because thermal energy causes large fluctuations of atoms around their equilibrium positions. This enables dislocations to move through the lattice more easily. Moreover, the thermal energy is effective in "annealing" out excess dislocations during deformation by processes known as *dynamic recovery* (in which individual dislocations are annihilated) and *dynamic recrystallization* (in which masses of dislocations are annihiliated when new grain boundaries are formed during deformation). For this reason, the ease with which dislocations move remains fairly constant during deformation, and the stress-strain curves exhibit no work hardening. In extreme cases, the structure may change so drastically during hot working that decreasing stress-strain curves may even be observed. The microscopic mechanisms by which this work, or flow softening, occurs have been documented by Jonas and Luton,[(1)] who also demonstrated that this softening can cause grossly nonuniform deformation in the compression samples. Another important feature of hot working is the noticeable effect of deformation rate on the working loads. In cold working, this effect is small, a tenfold increase in deformation rate causing a load increase of perhaps several percent for most metals. For hot worked metals, this same tenfold increase may cause load increases of upwards of 50 percent. These load increases can be a serious consideration in equipment selection. Workability is usually greatest at hot working temperatures, as can be shown by results from compression tests. If there is little or no flow softening, strains that are obtainable can be many times those obtainable in cold or warm working. More will be said later about the important workability considerations.

For an actual forging, the forging load and the forging pressure depend markedly on the friction at the die-material interface as well as the deformation resistance measured in uniaxial compression tests. Under perfectly lubricated conditions, the pressure required to forge a cylindrical sample is uniform and is equal to the flow strength of the material. When friction is present at the interface, the forging pressure distribution takes the shape of a tent, with minimum pressure at the outer rim (equal to the flow strength of the material) and maximum pressure occurring at the center of the sample. Associated with this pressure distribution are radial compressive stresses in the sample (whose magnitude increases inwardly), which develop from the constraining effects of friction. Depending upon the magnitude of friction and deformation, the maximum pressure in forging can be as high as ten times the flow strength of the material. This tent-shape distribution also applies to more complex shapes and has been termed the "friction hill" in forging terminology. Nadai,[2] one of the founders of the modern theory of deformation, postulated an analogy between the "friction hill" and the "sand hill" developed by pouring sand over one end of the forging (Fig. 7). The volume of sand in the hill seems to have some correlation to the required forging load.

Friction conditions at the forging interface are governed by lubrication and, in warm and hot forging, the temperature gradient between cold dies and hot material that can lead to chilling of the material and "locking" of the metal touching the dies. Lubricants are employed to reduce the friction and to control or improve the uniformity of metal flow. If a nonlubricated cylinder is compressed between clean, flat dies, it will barrel (that is, deform into the shape of a barrel) since the metal at the die-workpiece interface does not flow as easily as at the mid-height of the cylinder. Good lubricants eliminate this effect. Furthermore, lubricants at the interface reduce die wear and keep the workpiece from sticking to the dies.

The most commonly used lubricants in cold forging are soaps (over zinc phosphates). In terms of everyday experience, many of us have discovered the effectiveness of soap as a lubricant when trying to start sheet-metal screws into untapped holes in sheet-metal panels. A small amount of bar soap makes the job much easier. Soap provides excellent forging lubrication at room temperature and can be removed easily from forged parts without affecting their appearance.

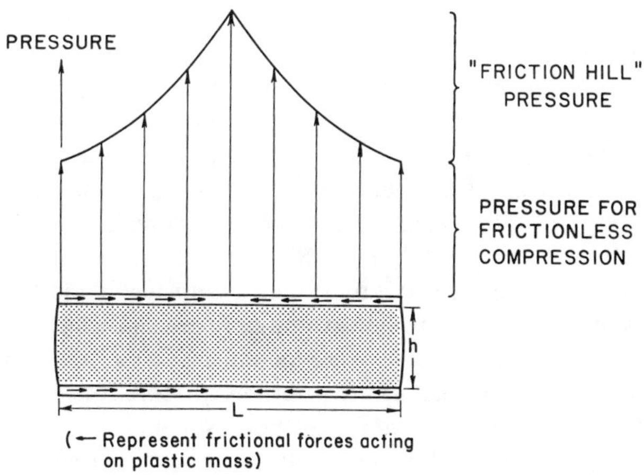

Fig. 7. Compression of a plastic mass between flat, parallel plates.

In warm and hot forging practice, graphite, in various dispersions and preparations, is perhaps the most widely used lubricant. The latest trend in the forging industry is to use water-based graphite lubricants since they are nonfuming and nontoxic. Forging lubricants are often evaluated by conducting a "ring test," in which a ring-shaped sample of known geometry is deformed between two lubricated flat dies to a known reduction in height. The change in the inside diameter of the ring specimen is very sensitive to the friction at the interface; it increases if the amount of friction is small and decreases if the amount is large. For this reason, the percentage of change in the inside diameter of the ring sample is used to rate the effectiveness of lubricants.

When simple geometries are forged, load estimates are relatively easy to obtain using measured flow strength and friction data. On the other hand, load estimates for a complex part are difficult to obtain. They depend strongly on the starting shape or preform of the workpiece, the die design, and the degree of die-material contact. The degree of contact is relatively low in upset or open-die forging and high in impression or closed-die forging. In addition, manufacture of complex parts typically involves several forging steps. Designing the number and configuration of intermediate shapes (preforms or blockers), the dies, and the forging conditions for the different steps is perhaps the most difficult task of the forging engineer. A forging such as the connecting rod for an automobile engine may require three stages to obtain the finished shape (Fig. 8). In forge shops, experienced design engineers usually follow empirical guidelines in reaching their decisions. Factors such as the complexity of the forging, the desired tolerance, the material, and the lot size influence the designer's decision. Often modeling materials such as wax, plasticine, lead, and clay compositions are used to study metal flow in small-scale model dies and to estimate the required forging load and energy under actual forging conditions. These materials are easy to deform and the modeling tools can be made from some inexpensive material. Sometimes a test specimen is sectioned into halves (usually along a diameter in round parts) and glued together after a grid pattern is inscribed on one of the flat faces. This specimen is then deformed in the model dies to obtain detailed information on metal flow, such as regions in which deformation may localize undesirably (Fig. 9). Furthermore, the shapes of the grid lines may be analyzed to determine regions in which undesirable stress, or loading, states may develop. For example, regions in which geometrical effects may lead to tensile stresses ("secondary tensile stresses") even under the nominally compressive loading are determined. More recently, computer-aided techniques have been utilized for the purposes of designing forging preforms and dies, and to simulate metal flow behavior (Fig. 10).

For designing dies for complex three-dimensional parts, such as the struc-

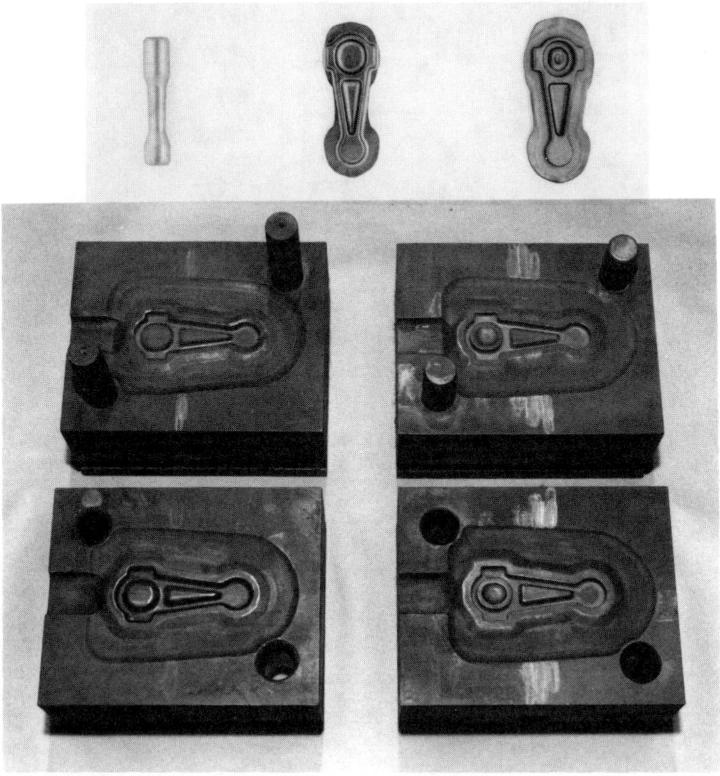

Fig. 8. Forging of automobile engine connecting rods involves several stages, each requiring a preform (or starting shape) and set of dies. Initially, a round bar is forged to obtain the preform shown on the left in the upper photograph. This preform is then forged in the blocker dies (left-hand set, lower photograph) to obtain the shape shown in the top middle photograph. After trimming the excess metal, or flash, this shape is finish forged in the other set of dies in the lower photograph to obtain the finish forging from which flash must be removed also.

tural forging in an aircraft wing, it may be necessary to consider metal flow patterns across several transverse cross sections. For each cross section, the designer develops a forging envelope (to provide clean-up allowance), determines the parting line, provides the inside and outside draft (for easy removal of the forged part from the dies), decides the corner and fillet radii (to avoid laps or folds over the corners), and determines the flash thickness (Fig. 11). Proper design of flash thickness is important in order to obtain filling of the die cavity without exerting excessive forces on the tooling and

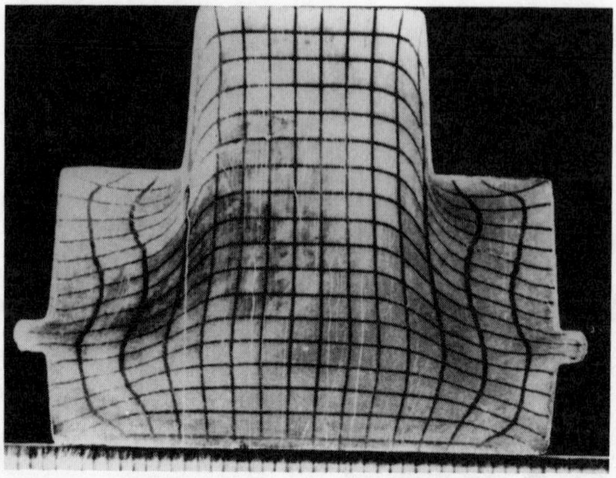

Fig. 9. Deformed grid pattern in an experimental forging. Square grid pattern was inscribed on the billet and then the billet was deformed in a set of forging dies.

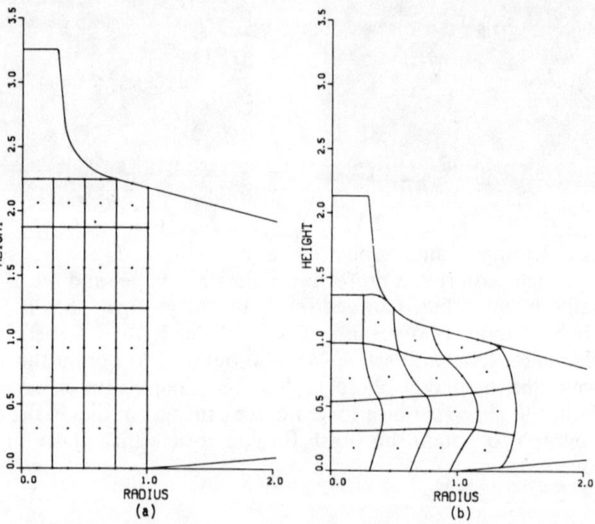

Fig. 10. Computer-aided simulation of metal flow in spike forging. Initial rectangular grid pattern (a) is input into the computer program, which predicts the distorted grid pattern (b).

achieving this with a minimum loss of metal as the result of trimming of the flash. In certain precision forging operations, such as forging of turbine blades, the die designers have to consider factors such as thermal shrinkage and distortion of the forging and elastic deflection of the tooling.

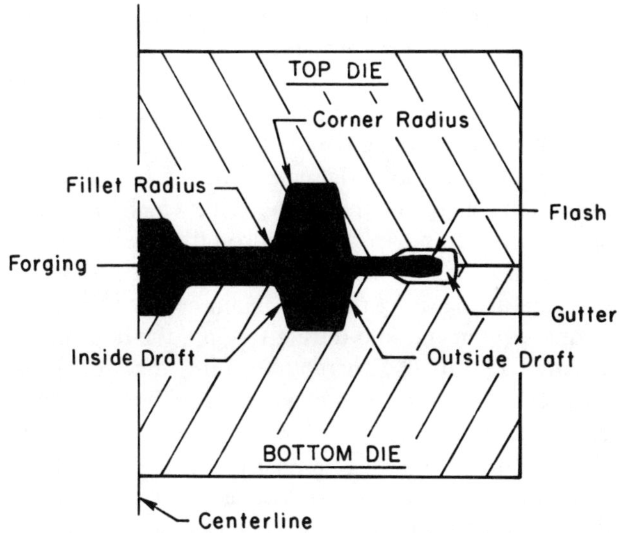

Fig. 11. Design of forging dies involves determination of fillet and corner radii (to enhance metal flow) and draft angles (to allow easy part removal from dies) as well as provision for areas into which excess metal, or flash, can move (gutters).

Once the tools or energy required to carry out the deformation are determined, the die material and forging equipment are selected. Die materials range from common-alloy and high-strength tool steels to superalloys (which can withstand high pressures at high temperatures). When the forging pressures and temperatures are moderate and the contact time is small, common alloy steel dies are adequate. Sometimes the alloy steel dies are coated with hard, wear-resistant superalloys to obtain good die life at a reasonable cost. In many cold forging operations, tungsten carbide die inserts (small dies placed in larger, supportive containers) provide excellent die life. At the other end of the spectrum, in exotic isothermal, hot forging operations, where the die and workpiece temperatures are nearly equal, the entire die assembly is made from special materials (such as IN-100 [a nickel alloy] or TZM-molybdenum), and it is heated to as high as 1000 °C under controlled atmosphere to avoid oxidation of the die material. Such a forging operation is used for precision forging of alloys of metals such as titanium, nickel, and beryllium.

Finally, the forging engineer must select the proper equipment on which a particular part will be forged. Although this selection is based primarily on load capacity, the engineer must also consider machine characteristics such as speed, which may affect workability. There are basically three types of forging equipment—namely, the drop hammer (either gravity-drop hammer or power-assisted-drop hammer), the mechanical press, and the hydraulic press. The basic difference between the three types of equipment is their forging speeds and how they store energy. Forging hammers (Fig. 2) are the fastest of all, with speeds ranging from 3 to 9 m/s (120 to 360 in./s). They store energy in the form of potential energy of the raised weight, which is converted into kinetic energy of the falling ram, which in turn is dissipated in deforming the metal. In power-assisted-drop hammers, steam or air pressure is applied against the ram during the downstroke and acts in conjunction with gravity in accelerating the ram. Hammers are the least expensive and most versatile equipment, and during forging, the ram strikes the workpiece on the anvil repeatedly, thereby forming the workpiece in steps. Because of their characteristics, hammers are rated not in terms of the load they can supply, but rather the energy they can deliver. Large hammers have capacities of about 680,000 N-m (500,000 ft-lb). Hammer forging is very noisy and often produces significant vibrations in the foundation. However, it is the mainstay of small forge shops. The energy in a mechanical press (Fig. 12) is stored in its flywheel and the ram is actuated by some mechanical system, such as a crank or a cam. Since mechanical presses have accurately controlled strokes, they are often used for close-tolerance forgings. The velocity of the ram in mechanical presses is less than in hammers (0.06 to 1.5 m/s, or 2.4 to 60 in./s). Large mechanical presses can produce loads in excess of 90 MN (10,000 tons). Hydraulic presses (Fig. 3) are the slowest of all (0.06 to 0.30 m/s, or 2.4 to 12 in./s) since they are operated by large pistons driven by a high-pressure fluid, which stores the energy required to deform the metal. Some of the largest presses in the world are hydraulic presses (the two largest hydraulic presses in the United States are rated at 445 MN [50,000 tons] each).

Selection of Forging Conditions—Workability

Workability During Cold Forging

The second major consideration in determining forging conditions is workability. As mentioned before, the fracture modes in forged metals are very dependent on temperature. When metals are cold forged, fracture usually occurs as cracking on the free surfaces (that is, surfaces other than those touching the dies). These cracks are a result of bulging or barreling of the free surfaces, which in turn leads to the development of secondary tensile stresses here. The effect of deformation rate on workability in these instances is small except when it affects the lubricant performance.

Although the exact sequence by which bulging occurs depends on all of the forging and material parameters, gross trends can be inferred by examining

Fig. 12. Mechanical forging press in a TRW production facility is being used to hot forge titanium turbine engine blades. In presses of this type, energy is stored in a flywheel (on the right of the press) and transferred to the ram by a crank or cam. Mechanical presses typically have much faster ram speeds than hydraulic presses and are often used in conventional hot forging of temperature-sensitive metals such as titanium alloys. (*Photograph courtesy of TRW, Inc.*)

surface fracture in upsetting of cylindrical samples of various height-to-diameter ratios (h/d) between flat dies with various lubrication conditions. From tests of this sort, it is found that the amount of bulging for a given height reduction, and thus the driving force for cracking, increases with decreasing aspect ratio (smaller h/d's) and poorer lubrication (more friction). Viewed differently, the obtainable reduction without fracture decreases with decreasing aspect ratios and poorer lubrication. This trend may have been expected qualitatively. However, Lee and Kuhn[3] found that if the surface strains at the initiation of surface cracking are measured (using grids printed on samples prior to deformation) and plotted, a fracture locus in the form of a straight line is obtained for a variety of carbon steels and stainless steels (Fig. 13). When the strain history at the eventual crack site is plotted in this diagram, it is seen why more overall deformation can be imposed in well-lubricated upsetting of samples with large aspect ratios. Further research with this kind of test has shown this fracture locus to be capable of describing the behavior of a wide range of metals. In all cases, the slope of

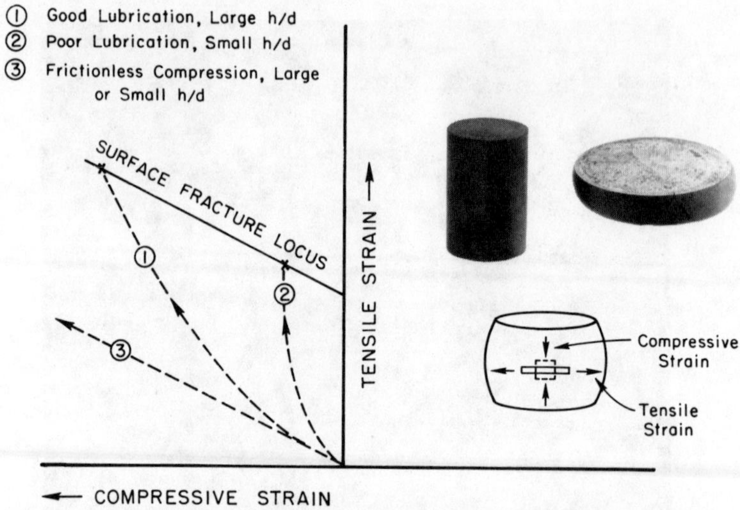

Fig. 13. Workability in cold forging is determined largely by the resistance of the metal to fracture on the free surface (photograph showing undeformed and deformed fractured samples). This free-surface fracture is affected by geometry and lubrication conditions and may be studied by upsetting cylindrical specimens of various aspects ratios (h/d) under various lubrication conditions. The strain paths of material elements on the equatorial-free surface are indicated by the broken lines. Fracture loci, or plots of the free-surface strains at failure, serve as a workability measure and are commonly used in forging design.

the locus is $-1/2$, the ratio of the tensile and compressive strains on the free surface of a sample compressed without friction and thus without bulging. Only the intercept on the tensile strain axis varies from one metal to another. Because of this, only one upset test need be conducted for a given metal to determine its cold workability relative to that of other metals.

Other kinds of fractures and defects occur in cold forging but are either less common or only negligibly affect the service properties of the part and thus are overlooked. One of these defects is porosity, or holes. Arising from secondary tensile stresses generated internally in the workpiece, the holes may range in size from microscopic, only a fraction of the grain size, to large and macroscopic, many times the grain size. These large holes are called *central bursts* (Fig. 14). If small enough, the microscopic holes usually do not affect service properties of the part. The large central bursts generally lead to gross fracture of the workpiece during forging, and thus products containing them are rarely employed for their intended use. The small holes initiate around dispersed microscopic particles in the metal structure which are

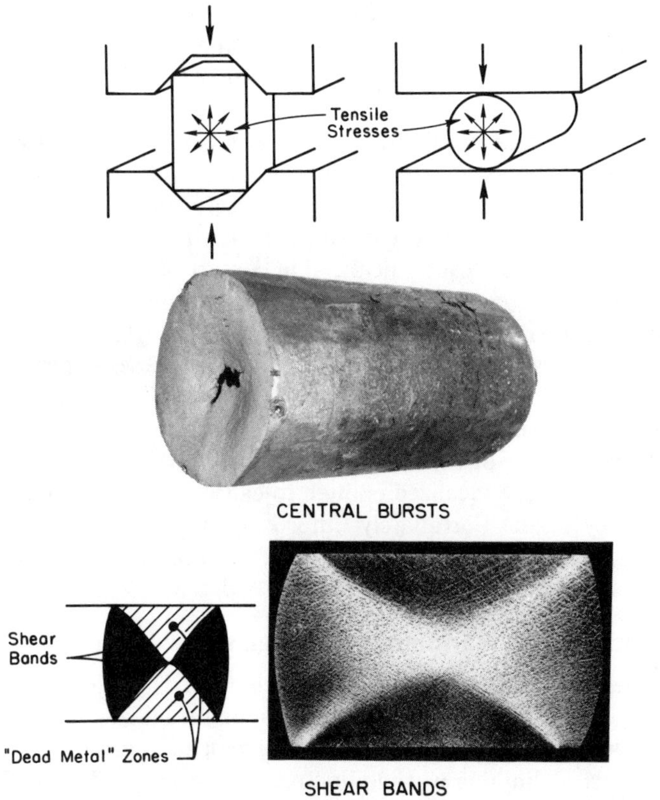

Fig. 14. Less-common defects in cold forging are central bursts and shear bands. Central bursts develop as a result of high internal secondary tensile stresses developed by poor choice of preform and/or die design and the presence of structurally unsound areas in the starting material. Shear bands result when friction constrains the deformation of the metal against the dies, setting up "dead metal" zones. Deformation is higher in the regions separating these dead metal zones from the rest of the workpiece. The photographs show a stainless steel bar forged between flat dies (causing central bursts) and an etched cross section of a carbon steel cylinder compressed between unlubricated dies (causing dead metal zones and shear bands).

produced during processing to increase service strength. In contrast, the large holes develop around structurally weak areas such as regions with a large amount of foreign matter (such as dirt carried over from the ore melting and refining process). The central burst defect is also seen in breaking down the cast structures of ingots in which foreign matter segregates to the center of the ingot. To avoid the defect, preform or die design may be changed (such as by using vee-shaped dies instead of flat dies to forge round shapes) in order to eliminate secondary tensile stresses. Another alternative is a change in the melting and refining processes to lower the levels of foreign materials which are the initiation sites of the central bursts.

Another defect occasionally found in cold forging is known as *shear bands* (Fig. 14). Shear bands are regions of localized shear deformation which develop between areas of widely different deformation levels within the workpiece. One of the main causes of the nonuniform internal deformation is friction. Friction may cause the metal in contact with the dies to "lock-up" by raising the load required for its deformation considerably beyond that required for deformation of the remainder of the workpiece. Called "dead metal zones," these regions may assume various shapes depending on the geometry and friction conditions involved. In cold forging, the shear bands separating the dead metal zone from the rest of the workpiece are usually quite diffuse because of the influence of work hardening, and only in rare cases do they become localized enough to cause gross fracture in the areas in which they occur. Fortunately, improvements in lubrication or minor changes in preform and die design are successful in alleviating the defects when they do occur.

Workability During Warm and Hot Forging

The development of internal holes or shear bands, which is normally of small consequence in cold forging, plays an important role in workability in warm and hot forging. Holes or cavities developed at grain boundaries are the major factors limiting workability in warm forging (Fig. 15). There are two major causes of these holes. One is the tendency of secondary tensile stresses to open up holes, as mentioned before. These holes are initiated by the action of dislocations moving through the lattice and impinging on the grain boundaries. If there are microscopic particles at the grain boundaries, the dislocations may induce a shearing action upon them, opening weak interfaces between the particles and the grain boundaries themselves. The secondary tensile stresses cause the holes to grow and, if the imposed deformation is large enough, the holes impinge on each other and cause fracture.

Common at high warm working and low hot working temperatures, the second type of hole that may be developed occurs where three grains meet in the metal, that is, at the triple junctions. At these temperatures, shearing along the grain boundaries becomes thermally activated and, because of moderately high stress levels, may lead to stress concentrations and holes at the triple junctions. The lattice deformation at these temperatures is not enough to relax the stress concentration. Thus, cracks traversing the entire length of the grain boundaries may initiate at the triple junctions, eventually causing gross fracture of the workpiece.

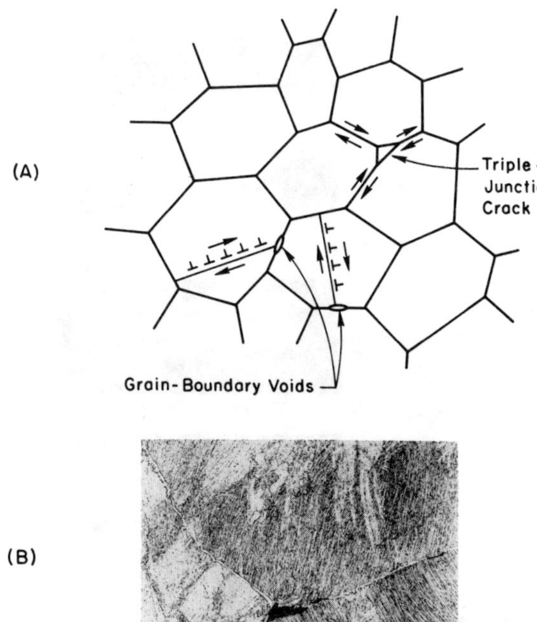

Fig. 15. Triple-junction cracking and grain boundary void formation limit workability at warm working and moderate hot working temperatures. Triple-junction cracks result from thermal activation of grain-boundary shearing. Grain-boundary voids are initiated by the action of dislocations impinging on particles found at the grain boundaries. The photograph (magnification 250×) shows examples of these defects in a titanium alloy. The three grain boundaries are delineated by the thin white ribbons which converge at the triple-junction crack.

Often, simply increasing the forging temperature to the hot working regime solves workability problems such as cavitation and triple-junction cracking. However, care must yet be exercised even at moderate and high hot working temperatures because the working stress at these temperatures is typically very sensitive to temperature (with large decreases in flow strength with increasing temperature). The effect is particularly important in forging of aerospace alloys such as the titanium alloys used in jet engine and airframe applications. When the stress is very sensitive to temperature, shear bands and shear cracks along the shear bands may develop during forging (Fig. 16).[4]

The primary source of shear bands in hot forging is heat transfer between the dies and the workpiece. This occurs because the die temperature is maintained only several hundred degrees Celsius above room temperature in

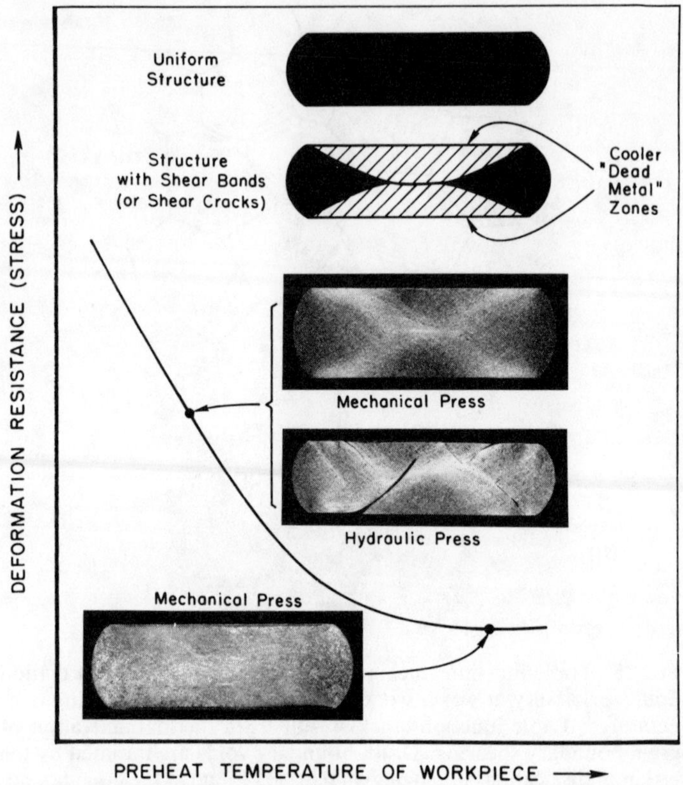

Fig. 16. Shear bands in conventional hot forging result when cold dies are used to deform hot metal. When the preform is preheated to a temperature at which the deformation resistance is not too sensitive to temperature and the deformation is carried out rapidly (as in a mechanical press), deformation is uniform. In contrast, when the preheat temperature is in a range in which the deformation resistance is very sensitive to temperature, dead metal zones and shear bands are easily formed. The severity of shear bands in these cases depends on heat transfer (and hence the magnitude of the thermal gradients set up in the workpiece) and is affected greatly by the speed of deformation. Fast working in a mechanical press leads to modest shear bands and slow working in a hydraulic press leads to shear bands and shear cracks. Photographs are metallographic cross sections of hot forged titanium-alloy bars.

conventional hot forging, whereas the workpiece may be preheated to temperatures of around 1000 °C. When the dies come in contact with the metal during forging, heat transfer causes the metal touching the dies to chill, thus raising its working stress and the load required to deform it above that for the rest of the workpiece. Dead metal zones in contact with the dies are thus formed in much the same way as those formed as the result of friction in cold forging. With the formation of dead metal zones, shear bands are generated, separating them from the still-deforming bulk. If a very large deformation is imposed, the shear deformation between the dead metal zone and the bulk may lead to cracking. Alternatively, cracking may occur along the shear bands even with modest amounts of overall reduction if enough heat transfer has occurred to lower the average temperature of the workpiece to a low, warm working temperature of inherently low workability. Shear bands may cause problems even if they do not lead to cracking during forging since they may be sites of easy crack initiation or propagation when parts are put into service.

Several simple techniques exist to minimize the occurrence of shear bands in conventional hot forging. Perhaps the simplest is to raise the preheat temperature of the workpiece to one where the stress is not too temperature dependent. This method allows for some thermal gradients to be set up during forging without generating regions of widely different working stress. This method is not always feasible, though, because of the tendency of metals to oxidize more readily at high temperatures or to undergo grain growth at high temperatures (which makes them less workable). Another obvious way to minimize shear bands is to minimize heat transfer. This may be accomplished by using lubricants (which act as thermal insulators between the dies and the workpiece) or by reducing the time during which heat transfer may occur. Contact time can be reduced using equipment with faster rams such as mechanical presses or hammers. As mentioned before, equipment of this sort is capable of imposing deformation at rates up to 100 times those of hydraulic presses (Fig. 3). Besides alleviating the occurrence of shear bands, high-speed equipment may also result in lower hot forging loads. This effect may seem paradoxical in view of the previous comment that increased deformation rates usually give rise to increased deformation resistance. The paradox is resolved by realizing that heat transfer and chilling typically play a more important role than deformation rate in determining deformation resistance in conventional hot forging.

Yet another way of minimizing heat transfer and one that is receiving significant attention is the technique of isothermal hot forging. Using special devices, the dies are heated to the same (or nearly the same) temperature as that of the workpiece. Without heat transfer, parts can be effectively hot forged in slow hydraulic presses. Because of the slow rates, working loads are smaller than in conventional hot forging done using mechanical presses or hammers. In addition, the working temperature range (that is, the workpiece preheat temperature) is greatly expanded compared with that for conventional hot forging.

Typically, only occasional grain-boundary void formation and shear bands are seen in isothermal, hot forging. As shown by Gegel, Nadiv, and Raj,[5] the source of the voids in isothermal forging is large secondary tensile stresses that can be eliminated by changes in die or preform design. Semiatin and Lahoti[6] demonstrated that shear bands may also be generated for certain materials in isothermal forging (Fig. 17). In these cases, they are a result of the flow softening effect sometimes encountered at hot working temperatures (Fig. 6) rather than the formation of dead metal zones found

(a)

(b)

(c)

(d)

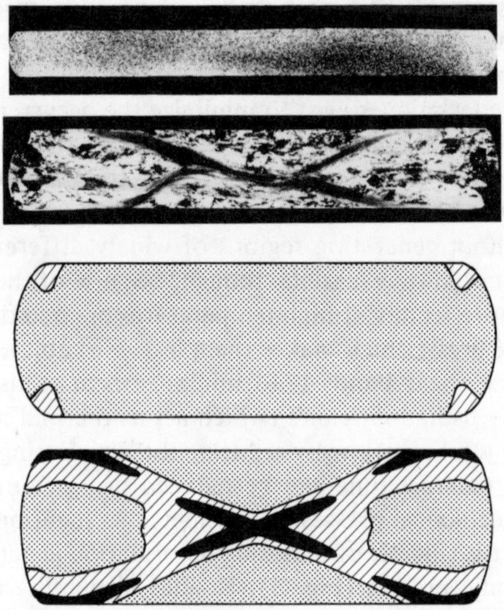

Fig. 17. Isothermal hot forging may lead to shear bands under certain conditions. (a) and (b) are cross sections of isothermally hot forged titanium bars thermal-mechanically processed prior to forging to have different properties. Deformation is uniform in (a), but shear bands are developed in (b). The compression stress-strain curve for (a) had a flow-stress plateau and that for (b) exhibited extensive flow softening, which is the source of shear bands in isothermal hot forging. These observations were quantified by computer simulation of the deformation process. Contour plots of relative deformation rate in (c) and (d) were generated using forging conditions and material properties similar to those of the bars in (a) and (b), respectively. In these plots, high numbers indicate large deformation rates and low numbers modest deformation rates. Simulation in (c) shows modest and relatively uniform deformation rates similar to observation in (a). Simulation in (d) shows large deformation rates and large gradients in deformation rate, the pattern of which is analogous to the flow localization observed in (b).

in cold working and conventional hot forging. Within the shear band, deformation is higher than it is on either side of it. The localization of strain in this instance is analogous to the localization of deformation that occurs when metals are *pulled* or *stretched* beyond a certain extent, giving rise to local thinning or necking. In tensions, localization begins when deformation can proceed under continuously decreasing load. Similarly, when the material properties are such that the stress-strain plot has a maximum (as with hot worked metals that exhibit flow softening), localization may occur in compressive forming modes such as forging. The degree of localization of deformation within shear bands in isothermal hot forging depends a great deal on forging geometry and the rate at which the material softens with deformation. Because the degree of softening varies with temperature and imposed deformation rate, forging conditions can often be found to avoid the shear bands. Preform and die designs that tend to constrain the deformation to two dimensions ("plane strain" deformation) are more likely to give rise to shear bands that allow three-dimensional deformation. It is difficult to predict the detailed effects of geometry on constraint levels, however. In this regard, computer simulation becomes very useful in predicting the occurrence of shear bands.

Application of Mathematical and Computer Modeling in Forging

As discussed earlier, often the preform and die designs are too complicated to enable the development of simple and general methods for predicting working loads, metal flow, and occurrence of defects in forging. Until recently, overall forging process design and die design were done by experienced people using trial-and-error methods. These procedures are not only expensive but also time consuming, often requiring major experimental setups and complex measurements. For this reason, an ever-increasing amount of research is being performed to develop general mathematical and computer techniques capable of simulating the forging operation under arbitrary conditions.

In the past, the mathematical modeling of forging processes has been done by one of the several approximate methods, such as the slab method (approximate stress analysis), the slip-line method (the method of characteristics), or the upper-bound method (a method utilizing an energy principle). Although these studies contributed significantly toward understanding the mechanics of deformation in forging, they lacked generality and often did not provide accurate estimates of required forces and energy. In the last 15 years, another method, the finite-element method, has been successfully applied to solve a wide range of forging problems. In this method of analysis, a cross section of the deforming material is divided into a two-dimensional network of discrete elements (finite elements), and the deformation at selected points (nodes) in each element is determined by the application of some variational principle. Thus, by analyzing an aggregate of elements one at a time, the deformation pattern in a complex shape can be determined. The accuracy of the predictions increases with the number of elements used. However, the application of the finite-element method to practical forging problems has been limited because of the requirements of a large computer and an experienced user. Recently, a user-oriented general-purpose finite-element computer program was developed at Battelle by Oh et al.[(7)] This program is capable of solving forging problems with complex geometries and does not require an excessive amount of computer time. In addition, this type of forging simulation provides detailed mappings of the stress, strain, and temperature fields within the workpiece and the dies. This computer program was used by Lahoti and Semiatin to predict shear bands in isothermal forging (Fig. 16). This is the first time that it has been possible to predict local deformation patterns in forging by using advanced computerized techniques. The same program was used by Mallas et al.[(8)] to predict the stress field in the forging of titanium jet-engine compressor disks. They were able to correlate the regions of high hydrostatic tensile stress to regions in which holes were observed experimentally.

Computers are also being used to design the forging dies. These designs take into account the corrections needed as the result of thermal shrinkage of hot metal (in hot forging), elastic deflection of tooling, and the required clean-up allowances. Computer models can predict the stresses in the critical areas of the dies, thereby forewarning the designer of potential problems related to die breakage. As just mentioned, the designer can also predict, for example, where deformation may localize or holes open up as the result of high secondary tensile stresses and thus make beneficial changes in preform design to eliminate these and other problems. With the advent of interactive graphics programming, a designer can quickly see the results graphically on a computer cathode ray tube (CRT) display and interact with the computer to make the necessary changes (Fig. 18). This type of interaction is almost essential in design since there are no unique solutions to a given problem. Most recently, the work of designers and die makers has been made somewhat easier by the introduction of stand-alone computer-aided design and drafting systems. These systems are independent computers with advanced software for geometrical design and display. They not only assist the engineer in designing the dies on the screen, but when the design procedure is completed, they also produce a complete set of engineering drawings or a magnetic tape which can be sent to the plant for automated machining of the dies. This technology is currently being used at an increasing rate by the more advanced forging shops, especially for aerospace components.

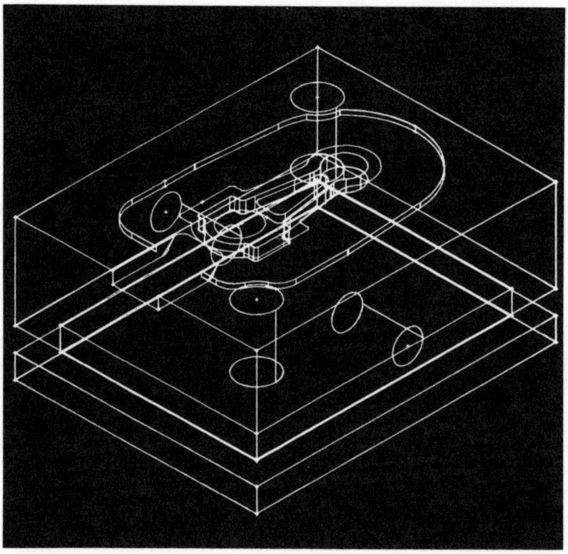

Fig. 18. Computer-aided design of dies for an automobile engine connecting rod is illustrated in this figure, which has been drafted by the computer.

With the advent of forging design and simulation on computers, scientists and engineers have obtained a powerful tool. Increasingly, this tool will enable them to design preforms and dies better, and eliminate the old trial-and-error methods of actually machining and testing several sets of preforms and dies to obtain a desired part. Also, they will be able to predict the detailed metal flow and occurrence of defects in forging, allowing them to change designs to avoid defects and to prevent bad parts from reaching service. Hidden costs in forged products will be eliminated, and productivity in the forging industry will be increased.

References

1. J. J. Jonas and M. J. Luton, "Flow Softening at Elevated Temperatures," in *Advances in Deformation Processing*, edited by J. J. Burke and V. Weiss, Plenum Press, New York (1978).
2. A. Nadai, *Theory of Flow and Fracture of Solids*, McGraw-Hill, New York (1950).
3. P. W. Lee and H. Kuhn, "Fracture in Cold Upset Forging—A Criterion and Model," *Metallurgical Transactions, 4*, 969 (1973).
4. S. L. Semiatin and G. D. Lahoti, "The Occurrence of Shear Bands in Nonisothermal, Hot Forging," *Metallurgical Transactions A, 14A*, 105 (1983).
5. H. Gegel, S. Nadiv, and R. Raj, "Dynamic Effects on Flow and Fracture During Isothermal Forging of a Titanium Alloy," *Scripta Metallurgica, 14*, 241 (1980).

6. S. L. Semiatin and G. D. Lahoti, "The Occurrence of Shear Bands in Isothermal, Hot Forging," *Metallurgical Transactions A, 13A*, 275 (1982).
7. S. I. Oh, G. D. Lahoti, and T. Altan, "Application of a Rigid-Plastic Finite-Element Method to Some Metal Forming Operations," *J. of Mechanical Working Technology, 6,* #4, 277 (1982).
8. J. C. Malas, H. L. Gegel, S. I. Oh, and G. D. Lahoti, "Metallurgical Validation of a Finite-Element Program for the Modeling of Isothermal Forging Processes," in *Experimental Verification of Process Models*, American Society for Metals, Metals Park, Ohio (1983).

Bibliography

1. *De Re Metallica.* Georgius Agricola. Translation of first Latin Edition (1556) by H. C. and L. H. Hoover. Published by the Mining Magazine, Salisbury House, 1912.
2. "The Nature of Metals." A. H. Cottrell in *Scientific American*, Vol. 217, No. 3, pp. 90-100, September 1967.
3. "The Deformation of Metals at High Temperatures." Hugh J. McQueen and W. J. McGregor Tegart in *Scientific American*, Vol. 232, No. 4, pages 116-125, April 1975.
4. *Technical Notes on Forging*. Joseph W. Spretnak. Forging Industry Educational and Research Foundation, 1977.
5. *Fundamentals of Forging Practice*. A. M. Sabroff et al. Reinhold Book Corporation, 1968.
6. *Forging Equipment, Materials, and Practices*. T. Altan, Editor, The Metals and Ceramics Information Center (MCIC), Handbook No. MCIC-HB-03, 1973.

2

Interatomic Potential Development in Materials Science

R. A. Johnson

University of Virginia
Materials Science Department
School of Engineering
Charlottesville, VA 22901

The development of interatomic potentials is part of a larger project of designing a mathematical model which can be used to perform computer simulations—experiments, really—for various properties of solids which are ultimately controlled by interatomic interactions. The design of a model and the interatomic potentials used in the model are so interconnected that no attempt will be made to discuss them separately. The purpose of this chapter is therefore more to introduce the ideas and mode of thinking associated with atomistic modeling of materials than to discuss the intricacies of interatomic potentials per se. There have been numerous reviews and conference proceedings covering this field, and references for additional reading are given[1-5] at the end of the chapter.

The general requirements for a model are that it be both accurate enough to permit a broad range of experiments with a reasonable expectation that the results will be physically meaningful, and simple enough to do so within your computer budget. For atomistic calculations, a mathematical model is simply an equation for the energy of the material as a function of the positions of all of the atoms—given the atomic arrangement, the energy can be computed. In practice, it is normally only the difference in energy between two sets of atomic positions, or configurations, that is of interest, which leads to considerable simplifications in the calculations.

The atoms in a gas spend most of their time in isolation: each atom is only close to other atoms in the gas or at the walls of the container a small fraction of the time. The atoms in a liquid or a solid (condensed matter) are in continual interaction. The direct goal in the development of an interatomic potential in condensed matter is to devise a simple scheme to approximate this very complex situation that can be used in energy calculations. Although there has been significant research in this field for both ionic and covalent materials, the present review is directed towards metals.

Historically, atomistic computer modeling has been carried out from the earliest availability of computers for scientific research.[6-10] Indeed, this field is a continuation of precomputer calculations dating from 1942,[11,12] which were themselves based on earlier studies of the interatomic forces in copper[13] and on analytic forms for two-particle interactions given by Mie,[14] Lennard-Jones,[15] Morse,[16] and Born-Mayer.[17]

Molecular Potentials

Perhaps the easiest way to obtain some insight into interatomic potentials is to think through the two-particle case, illustrated in Fig. 1. Two hydrogen atoms do not interact when they are widely separated, a condition defined as the energy E = 0 state. As the atoms come closer and start to overlap, there is a repulsive interaction indicated by the dashed curved as a result of the atoms occupying the same space. However, there is also an

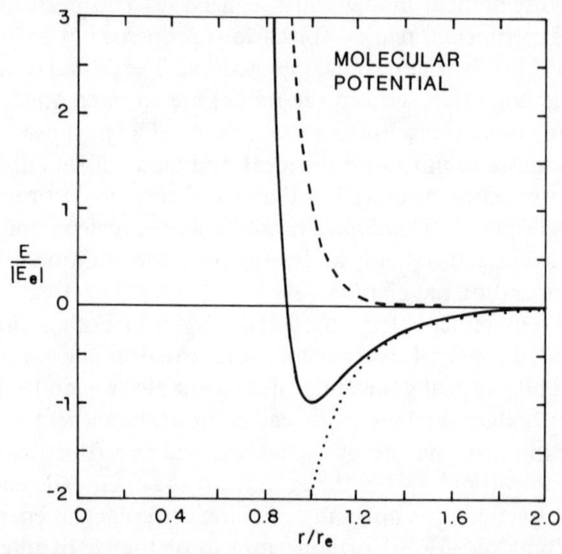

Fig. 1. The dashed curve represents the repulsion of two atoms due to overlap, that is, both trying to occupy the same space, and the dotted curve shows, for example, attraction arising from covalent binding, the quantum-mechanical sharing of electrons of opposite spin. The composite curve retains the sharp repulsion at close separation, but has an equilibrium distance at which attractive and repulsive forces just balance, and a long-range tail which asymptotically approaches zero.

attractive interaction, in this case called *covalent bonding* and shown by the dotted curve, which arises because the atoms share their electrons. The overall interaction is based on quantum mechanics and has no simple intuitive explanation, but the net effect, the solid curve in Fig. 1, is a sharp repulsion at close separations, an equilibrium distance at which the attractive and repulsive forces just balance, and a tailing off to zero as the separation becomes greater yet. The shape is similar to the parabola of an ideal spring near the minimum of the potential. For this simple case, the interatomic potential is the energy function.

An analytic form for this general shape, first suggested by Mie[14] and applied by Lennard-Jones,[15] includes a repulsive term A/r^n and an attractive term $-B/r^m$ where A and B are positive and n is greater than m. Most commonly $n = 12$ and $m = 6$, with the $m = 6$ attractive term assumed to arise from induced dipole-dipole interactions (van der Waals forces). A similar form proposed by Morse[16] includes an exponential repulsion Ae^{-ar} and attraction $-Be^{-br}$. The quantum-mechanical problem for the vibrational spectra of diatomic molecules could be solved analytically with the Morse potential, and both these forms have had considerable success.

Models and Potentials for Solids

Potentials

The concept of a potential in a solid is not as straightforward. Here we start by imagining a very large number of metal atoms, say copper, in a face-centered cubic arrangement, but with a lattice constant so large that they do not interact at all. Each copper atom has the same set of atomic energy levels for its electrons, and this state is taken as the zero in energy. As in the molecular case, the atoms are brought closer together in this thought experiment so that they start to interact. The precise atomic energy levels spread out into bands and the electrons occupy the lowest energies available in the bands. The details of the theory for this metallic bonding are incredibly complex, but the net effect of the energy per atom as a function of the lattice constant is a curve with the same general shape as a Morse or a Lennard-Jones potential. Note that this is a type of energy function, but not the one we want and not an interatomic potential in this case.

To construct an interatomic potential model of a solid, now think of connecting each atom with all the other atoms in the solid with, for example, a Lennard-Jones 6-12 potential. The energy per atom E is then

$$E = \frac{1}{2} \sum_{i=1}^{N-1} \phi(r_i) \tag{1}$$

where

$$\phi(r) = \frac{A}{r^{12}} - \frac{B}{r^6} \tag{2}$$

N is the number of atoms in the solid, and the 1/2 is required to avoid double counting (the energy for each bond between two atoms can be thought of as being equally divided between them).

The parameters A and B are as yet undetermined, so the model is incomplete. Two bits of physical information about a particular solid can be used to obtain numerical values for these parameters. If the model is intended to simulate copper, the minimum in the E versus a must occur at a = 3.615 A, the lattice constant for copper. This will give the model the correct spacing. The compressibility K is inversely proportional to the curvature of E evaluated at a, and if the experimental value of K for copper, 1/K = 0.86 eV/A^3, is also used as an input, A and B are uniquely determined and the model will have the correct response to hydrostatic pressure. Mathematically these two conditions are:

$$\left.\frac{dE}{da}\right|_e = 0 \tag{3}$$

and

$$\left.\frac{d^2E}{da^2}\right|_e = \frac{9a}{4K} \tag{4}$$

with e indicating evaluation at equilibrium. The resulting values are $64A/a^{12} = 0.21$ eV and $8B/a^6 = 0.35$ eV. The potential is shown in Fig. 2.

The model is now complete. You can imagine a large set of balls all interconnected with appropriate Lennard-Jones type springs, and you can test the model for other known physical properties of copper and carry out "experiments" with the model. The value of E at the equilibrium lattice constant is the negative of the cohesive, or sublimation energy:

$$E|_e = -E_c \tag{5}$$

Unfortunately, it turns out that the cohesive energy given by the model is smaller than the experimental value by almost a factor of three: the model value is $E_c = 1.29$ eV while the experimental result is 3.54 eV. We could also check the resistance to shear in our model, and again the values would have the correct general magnitude but would not be in particularly good agreement with experimental results. The vibrational properties of the model can similarly be compared with experiment, with corresponding results.

It is possible to obtain a better fit by using more parameters in the potential. One would think that we could match the cohesive energy exactly if, for example, n was not taken as 12 but was allowed to vary. However, as long

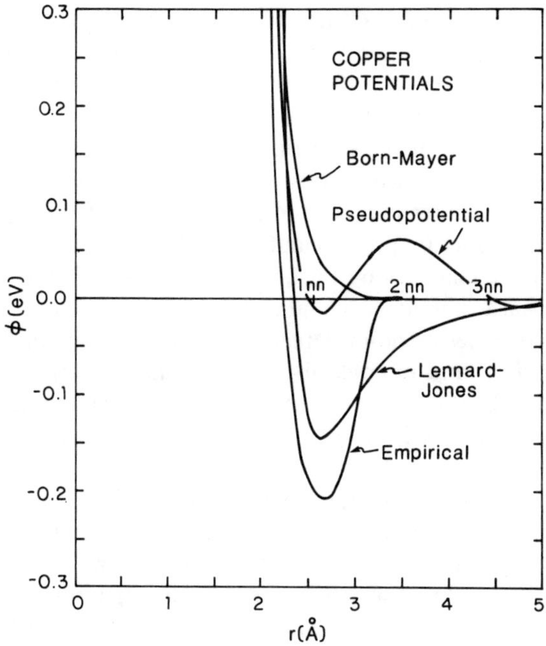

Fig. 2. Four interatomic potentials for copper are demonstrated. The Lennard-Jones 6-12 potential is derived in the present chapter, the Born-Mayer repulsive potential has been used for radiation-damage calculations,[(10)] and the empirical potential[(29)] and the interatomic potential based on pseudopotential theory[(33)] for point-defect calculations.

as m is held at 6, there is no choice of n which will permit us to exactly match the lattice constant, the compressibility, and the cohesive energy of copper. In this case, if the product mn is 25.82, then all three inputs are precisely fitted. As discussed in greater detail in the next section, other known properties of copper are better simulated by using n = 12 and m = 6 and neglecting the cohesive energy. The details given in this example are of no significance since the Lennard-Jones potential is not presently being used for metals (it does work well for rare gas solids), but they were presented to illustrate a simple model.

Models

This model is based on a "two-body" potential and is perhaps the easiest type to visualize. In addition to the above analytic calculations associated with the model, many additional properties of the metal can be calculated numerically. For example, if the model is cleaved in two, the atoms at the newly formed surface are no longer in equilibrium. Using the computer, all the atoms are allowed to relax (that is, adjust their position somewhat) so that there is no net force on any atom. If done properly, this will give a minimum in the energy of the cleaved sample, and a comparison of the energy with the uncleaved case will yield the surface energy. The calculations can then be carried out for various directions of the cleaving plane, and the surface energy can be determined as a function of orientation. In a similar vein, grain-boundary structures can be studied. The state of the art for calculations of the actual fracture process and for grain-boundary modeling is presented in following chapters by Baskes[(18)] and Wolf,[(19)] respectively.

Many of the earliest model calculations for metals were directed at understanding the properties of point defects in the lattice structure—that is, the stability, migration, and interactions between vacant lattice sites and self-interstitial atoms. The structure of lattice dislocations and the effect of other defects on their mobility have also been investigated, as reported in later chapters by Esterling[(20)] and Arsenault.[(21)]

Calculations for the energy and relaxations associated with various possible atomic configurations for self-interstitial atoms illustrate the experimental nature of computer simulation. A run is prepared with input parameters and an atomic arrangement which is close to some metastable configuration under study. These details are submitted to the computer with instructions to apply them to the model. At some later time one gets back the computer output, which usually requires significant analysis. Not uncommonly, something went haywire and the results are gibberish. Maybe the run was close, but slipped off the desired symmetry. One tries to figure out what went wrong and modify the run, or to add some constraints to the model for this case. For instance, several atoms might be held fixed in a particular arrangement and the run attempted again.

I have been discussing the simplest kind of calculation—a straightforward energy minimization usually called a *static calculation*. Temperature can be put into the model and the equations of motion for all the atoms solved so that atom positions are followed as a function of time. Such dynamic calculations are introduced by Beeler[(22)] and Straub[(23)] and are used in many of the subsequent chapters.

Two additional factors affecting models are the range of the potential and boundary conditions. To obtain the Lennard-Jones parameters given above, a summation had to be carried out over all neighboring atoms. In practice, the summation converges for a particular number of significant figures, but the number of terms in the summation can be very large: 10^5 atoms for accuracy of 5 significant figures. It is no problem to do this once to determine the model parameters, but it is unacceptable to carry out such large sums during computer experiments. This difficulty is discussed in greater detail in the next section.

To keep the computation time reasonable, the atoms in a small part of the crystal are normally chosen as representative. These atoms, commonly in the range of 100 to 1000, are given a full three degrees of freedom of motion or relaxation. The surrounding atoms may simply be held fixed at their perfect lattice positions (which is not as bad an approximation as it might seem), may be allowed to relax according to some elastic continuum model or some pressure which maintains equilibrium, or periodic boundaries may be used. In a dislocation calculation, for example, the set of atoms relaxed is periodic along the dislocation, but is perhaps matched to elastic solutions at some radial distance perpendicular to the dislocation.

Semiempirical Potentials

Long-Range

Lennard-Jones and Morse potentials are traditionally called *semiempirical* and, as mentioned above, require summations over many neighboring atoms for accuracy. An important advantage of these potentials is that they yield a stable model with no additional factors or approximations. Thus, while they present major computational difficulties, the definition of the model is precise. For example, there is no ambiguity at the surface of the sample as occurs in some models discussed below. A limitation in fitting experimental data is that there is an exact relationship between the compressibility and the average modulus of rigidity (shear constant) in any model based solely on two-body potentials, which is known as the *Cauchy relation.* The elastic properties of most metals are not well approximated by the Cauchy relation, and so cannot be matched by any two-body potential model, no matter how complex the potential is and how many adjustable parameters there are.

To avoid the long summations, the range of the potentials has often been shortened by imposing a cutoff distance. If the potential is simply truncated, computational problems usually arise because the energy function is no longer a continuous function of the positions of the atoms. Thus, the potentials are given an adjustment so that their value is zero at the cutoff distance. With this modification, the equilibrium conditions are changed so that new potential parameters must be found. To attain equilibrium with a purely two-body interaction model, the range of the potential must be such that either the minimum occurs at the nearest-neighbor distance and the cutoff distance is less than second-neighbor distance, or be great enough that there are both repulsive and attractive interactions in equilibrium.

The most widely used potentials in this classification are the Morse potentials with the parameters given by Girifalco and Weizer,[(24)] although Lennard-Jones potentials show the same pattern in results. The Lennard-Jones potential derived above has not actually been used in calculations, but it would give a reasonable value for vacancy formation and migration energies and for basic self-interstitial properties. If the cohesive energy were used instead of matching to the compressibility, the vacancy formation energy given by the model would be about three times too great. This mismatch between the cohesive energy and the vacancy formation energy occurs with all purely two-body potential models. Other difficulties also occur with these models: the divacancy migration energy is predicted to be much smaller than the single vacancy value, and surface relaxations are always outward, for example, in disagreement with experiment. Thus, although the simplicity of purely two-body models is appealing, they cannot be used to accurately simulate metals.

Short-Range

Models based on two-body Born-Mayer repulsions, Ae^{-ar}, are also included within the semiempirical classification. Since the two-body forces are repulsive at all distances, an additional factor must be incorporated in the model to provide stability. For calculations in bulk material, the atoms outside the set being relaxed are often held in their perfect lattice positions, or the atoms near the surface of the set are given some manner of elastic constraint. To define a self-consistent model, the additional factor in the energy is usually given as an overall volume dependence: the energy per atom is then

$$E = \frac{1}{2} \sum_{i=1}^{N-1} \phi(r_i) + CV^n \tag{6}$$

where C and n are model parameters and V is the overall volume of the sample. An advantage of this model is that the fit to the elastic properties is no longer constrained by the Cauchy relation; that is, the elastic constants can be matched exactly. A typical Born-Mayer repulsion used for copper(10) is shown in Fig. 2.

This type of model was used extensively in the late fifties and early sixties.(6,9,10,25,26) The coefficient in the exponential repulsion is sufficiently large that the interaction decays rapidly with distance. Thus, the model is essentially short-range and is computationally very efficient and easy to use. It proved to be quite successful for following the collision sequences during radiation damage, where the interactions are dominated by close interactions,(10) but had much the same problems as the other semiempirical potentials for stability calculations. A serious drawback is that there is no precise definition of the surface, so that the "volume" of the sample becomes ambiguous. Its use is therefore restricted to studies which can be approximated by taking the relaxation set of atoms to be embedded in an infinite material.

Empirical Potentials

When it became apparent that the semiempirical potentials were not providing a satisfactory model for basic defect properties in transition and noble metals, a form known as an *empirical potential* was developed.[27,28] It is similar to the Born-Mayer model in that it is short-ranged and uses a volume-dependent contribution to the energy, but the potential is not taken as an exponential repulsion. Thus, Eq 6 is still used for the energy calculation, and the elastic constants can be matched exactly. The two-body potential is commonly taken as similar in form to the long-range semiempirical potentials in that it is repulsive at close separations, has a minimum, and then approaches zero. However, it is much shorter in range, with zero value and slope at a cutoff distance which is less than second-neighbor spacing in the fcc lattice and between second- and third-neighbor spacing in the bcc lattice. For comparison with the Born-Mayer and Lennard-Jones forms, an empirical potential for copper[29] is shown in Fig. 2. The parameters for empirical potentials are often obtained from fitting the vacancy formation energy as well as the lattice constant and the elastic properties.

Calculations using empirical potentials retain the computational efficiency of the short-range semiempirical model and yield realistic values for defect properties. Although they give the right sense, they are simply too crude to yield fine detail. For example, the stable lattice structure for a given metal at a given temperature is a delicate function of the potential, and empirical potentials cannot be used to predict this. Similarly, they cannot yield a stacking-fault energy for an fcc or hcp metal. While they can be patched up to fit these properties, they acquire an ad hoc character and lose much of their simplicity in doing so. As in the Born-Mayer case, there is also arbitrariness in defining the surface for purposes of calculating the volume. However, unlike the long-range two-body models, these empirical potentials can yield sensible surface relaxations and energies.

Empirical potentials have been the workhorse of the field for years. When viewed as a first-order approximation, they do not produce nonsense and instead give considerable information about the nature of lattice defects in metals. Indeed, the intent of carrying out the calculations has been primarily to obtain insight into possibilities, with some indication of probability. As long as too much precision in the results is not expected, these models are satisfactory. Their limitations relative to surface properties and stacking-fault energies, for instance, are well understood and not misleading because of the basic simplicity of the model.

Pseudopotentials

Although empirical potentials provided a satisfactory first-order approximation for many lattice calculations, there was no satisfactory reason why, or how, one might go about improving them other than adding more complexity and fitting more data to determine the additional parameters. Electron band theory, and in particular the approach to band theory known as *pseudopotential theory*, has been used for many years to try to place interatomic potential calculations on a firmer fundamental foundation.[30,31]

The terminology here can be misleading: pseudopotentials pertain to the forces seen by conduction electrons in the determination of their quantum-mechanical wave functions and are not directly related to the interatomic potentials required for lattice calculations. The electron energy bands of some metals are surprisingly similar to the predictions of a free, or nearly free, electron model. This has led to a formulation based on a perturbation expansion for the electron wave functions in which the effective interaction between the ion cores and the conduction electrons, called a pseudopotential, is assumed to be weak. The overall lattice pseudopotential is then written as a sum of pseudopotentials arising from individuals ions.

The pseudopotentials are actually operators, but the approximation is commonly made that they are simply functions of position, in which case they are called *local pseudopotentials.* The result of these steps is that, given the positions of the atoms in a periodic array as input parameters, the energy of the system can be calculated. Furthermore, the energy can be factored into structure-independent (that is, volume-dependent) and structure-dependent components, with the majority of the cohesive energy accounted for by the volume-dependent term. And finally, the structure-dependent terms can be Fourier-transformed to yield an effective ion-ion potential. Thus, a model can be developed from first principles which involves two-body interatomic potentials and overall volume dependence of the crystal. The two-body potentials are long-range and oscillatory (they converge to Friedel oscillations[32]), as illustrated in Fig. 2 with a recently developed model for copper.[33]

This approach is not without its difficulties, however. The long range of the potentials makes it awkward to use them in efficient relaxation calculations, and, due to the oscillations, the results are dependent on the cutoff distance if they are shortened. The nature of the potential near the first-neighbor separation distance is delicately dependent on some very fine aspects of the theory so that it can be completely altered by some minor variations. The theory is not readily applicable to transition metals with significant *d*-band character, but to simple metals in which there is little overlap of the ion cores, with aluminum as the prototype[34] and to the alkali metals. The atomic arrangement is the important factor in using a model for defect calculations, so a model in which the structure-dependent terms are dominant is desirable. The effective two-body potential is quite weak in the pseudopotential model for aluminum, for example, which has three conduction electrons and a large lattice constant relative to the size of the ion cores. More-recent developments in pseudopotential theory as applied to metals are discussed in the next section.

Recent Developments

Carlsson[35] has recently emphasized that the nature of the interatomic potential which can adequately handle a given situation depends on the type of problem under consideration and on the local atomic environment. Long-range interactions such as the semiempirical forms and pseudopotentials are most applicable to cases in which the overall volume of the crystal is unchanged and there is little change in volume on a local scale, such as lattice vibrations, liquid structure factors, and bulk crystal structures. The short-range empirical potentials are applicable to situations involving "bond breaking" and significant volume change. The cohesive energy–vacancy formation energy dilemma discussed above is an example of this difference: a model based solely on two-body forces cannot accurately yield both of these energies. While this generalization is a gross simplification, it provides a starting point for discussing recent developments.

Long-Range, No Volume Dependence

The research by Machlin and co-workers on alloy phases[36] provides an example of the "constant volume" case. They find that reasonable agreement with experiment for the crystal structure and lattice constant of a broad range of alloys can be obtained with a Lennard-Jones (Mie) potential using $n = 8$ and $m = 4$. The potential parameters are functions of alloy conditions, including an electronegativity factor. The agreement with experimental elastic constants with this model is weak, but the calculated vacancy migration energy for seven fcc metals shows good agreement. Whereas vacancy formation is not a "constant volume"-type process, there is little volume change during vacancy migration.

Another example of the type of potential in this category is given by Carlsson et al.[37] While it is clear that, given a two-body potential, one can always determine a unique energy function, the converse is not obvious. They demonstrate that a two-body potential can be uniquely determined by inverting the relation between the cohesive energy function and its corresponding pair potential. This inversion scheme is then used to develop pair potentials starting with the results of quantum-mechanical calculations of cohesive energy as a function of lattice parameter, or equivalently of volume. The resulting potentials have much the same character as the long-range Lennard-Jones and Morse forms.

Long-Range, Volume-Dependent

This grouping primarily covers the two-body potentials associated with pseudopotential theory. Being in the "constant volume" category, they are most applicable to calculations with little volume change, but the two-body potentials themselves are not intended to provide a complete model. The volume dependence can be thought of as an add-on, and calculations are often constrained to have zero volume change.

Esterling and Swaroop[38] have developed an empirical fitting procedure based on long-range central interactions in which the primary physical input is the vibrational response of the model. They find that the resulting potentials yield a reasonably satisfactory match to the interatomic potentials obtained from pseudopotential theory for simple metals. These potentials are based on fitting to data which is independent of overall volume. Volume dependence based on some other physical input can be appended to this model.

There has been a series of defect calculations using pseudopotentials for simple metals[34] and noble metals.[39] As mentioned above, the approximations associated with the development of two-body interatomic interactions from pseudopotential theory are applicable to simple metals in which the ion cores do not interact and the electron behavior outside the cores is free-electron like. Thus, the extension of the model to noble metals requires an additional term in the potential to account for *d*-band interactions. Perhaps the most surprising aspect of calculations with these potentials[33] is that the results for copper are not very different from those based on a short-range empirical potential.[29] As seen in Fig. 2, the potentials have nothing in common. Also, for both types of potentials, the results decrease in agreement with experiment in going from copper to silver to gold.

Dagens has recently extended this approach to transition metals,[40] and calculations for vacancies and interstitials in nickel have been carried out.[41] As with the copper case, the results are surprisingly similar to those using empirical potentials. Harrison and co-workers[42-44] have shown that the use of the Thomas-Fermi screening approximation can eliminate the long-range oscillations in pseudopotential theory and have applied their results to noble and transition metals. No defect calculations have been carried out as yet with this approach, however. A generalized pseudopotential theory applied to transition metals has also recently been published by Moriarty[45] in which three-body forces are included. The conclusion is that three-body forces are significant and must be used for metals in the middle of a transition series, such as chromium and vanadium, but have little effect for nickel and copper at the end of the series. Three-body forces will be discussed in more detail in the summary and conclusion section.

Short-Range, Volume-Dependent

The major development in models in this category effectively has been to replace the volume dependency in Eq 6 with a more complex term which at first looks like another two-body potential:

$$E = \frac{1}{2} \sum_{i=1}^{N-1} \phi(r_i) + F(p) \tag{7}$$

$$p = \sum_{i=1}^{N-1} f(r_i) \tag{8}$$

This new term, which is used both in the embedded atom method developed by Daw and Baskes[46] and by Finnis and Sinclair,[47] can be interpreted in a variety of ways. The embedded atom derivation is based on density functional theory, and p represents the electron density at a particular atom arising from all the other atoms, taken as a linear superposition of atomic densities f(r). The function F(p) is the energy to embed that particular atom in a uniform electron gas of density p and depends only on the element. Based on tight binding theory, Finnis and Sinclair take F(p) as proportional to the square root of p, with f(r) being the sum of squares of overlap integrals.

Although not explicitly a volume contribution to the energy, this term is dependent on crowding in the lattice. From the purely empirical point of view, it can be thought of as a measure of local volume dependence. Since the model parameters are obtained by empirical fitting, results are independent of the physical interpretation.

The theory here is not intrinsically short-range, but both $\phi(r)$ and f(r) have been taken as short-range in actual calculations using this approach. Computer time is then no more than two times greater than with short-range empirical potentials, and in practice this new form would seem to be the natural successor to the empirical models. In addition, this scheme is adapted to treating surfaces, and since the embedding functions are independent of the source of the electron density, it is directly applicable to alloys. In a short period of time, quite extensive calculations have been carried out under the heading of the embedded atom method. Encouraging results have been obtained covering, for example, surfaces, impurities, lattice defects, alloy segregation at surfaces, impurity trapping at grain boundaries, liquid metals, hydrogen embrittlement, and fracture.

Summary

Whether interatomic potentials are derived with approximations from first principles or are purely empirical, their use in modeling is extended outside their original region of applicability. Usually the perfect crystal is the basis of the theory or of empirical fitting. Computer simulation can be thought of as theoretically or empirically matching one region of configuration space which is understood or for which there are data, such as the perfect crystal, and extrapolating to a different region about which less is known. There is never uniqueness: some other scheme may be found which does just as well in the known region but which gives different results in the region under study.

In principle, computer modeling of metals is seen as very straightforward. A primary input for all models is a two-body central potential, a continuous function of distance which is independent of orientation and is summed over all possible pairings in the solid. These potentials are taken as constant in most models, but calculations based on pseudopotential theory are often carried out at constant volume because the two-body interactions resulting from pseudopotential theory are functions of volume. The range of two-body potentials effectively varies from less than second-neighbor distance in the fcc lattice to thousands of atoms, and the shorter the range of the potential, the more efficient the calculations in computer time. The short-range potentials can be either purely repulsive (Born-Mayer) or have an attractive well (empirical), while the long-range potentials either have an attractive well (Lennard-Jones and Morse) or have long-range oscillations (pseudopotential theory).

The concepts become trickier when modifications are made to these basic models. The potentials can be treated as anisotropic, that is, noncentral, such that they vary with crystallographic direction. This approach is quite complex and has not met with much success.[48] Other methods are based on the inclusion of additional terms in the energy function. All such factors can be termed "many-body" contributions, and come in a variety of forms. Even noncentral two-body potentials can be put in this category because many atoms are required to define crystallographic directions. Several theoretical developments essentially are expansions in terms of two-body, three-body, four-body factors, etc.,[45,49,50] with the higher-order terms often neglected. As with the anisotropic approach, the use of direct three- and four-body terms has not had a major influence on atomistic calculations for metals.

Many-body effects can also be incorporated into a model through more-general functional dependencies. The most direct is to use a function of the overall volume in calculating the energy, as in semiempirical, empirical, and many pseudopotential models. Models based on terms which essentially are related to the local volume dependence through electron charge density or core overlap considerations have recently been developed.[46,47]

Conclusions

While models based on two-body Lennard-Jones and Morse potentials do not require any additional energy contributions, they do not provide a good simulation of metals. The many-body factor used to overcome these difficulties most commonly has been taken as a function of overall crystal volume. Short-range empirical models with this volume dependence have provided a crude but workable model for "bond breaking" calculations.

Taylor[51] has argued strongly in favor of pseudopotential over empirical models. Traditional thinking has been that pseudopotential theory is applicable only to simple metals, not noble and transition metals, and its use for practical metals has generally been restricted to aluminum, for which numerous calculations have been performed. However, recent extensions to noble[33] and transition[41] metals have been promising.

Models based on local volume dependence have recently been developed and extensive calculations have been carried out with the embedded atom method[46] which have yielded meaningful results over a broad spectrum of topics. They have an advantage over pseudopotential models in that this success has been achieved with short-range potentials; that is, the calculations are efficient in the use of computer resources.

Both the use of pseudopotentials for nonsimple metals and the development of the embedded atom method are sufficiently new that further investigation is required to understand their ramifications. Computer efficiency favors the embedded atom method, and so far there are no worrisome clouds on the horizon that indicate its results might not be physically meaningful.

Acknowledgment

Support by the U.S. Department of Energy, Office of Basic Energy Sciences, Division of Materials Sciences, Grant Number DE-FGO5-86ER45246, is gratefully acknowledged.

References

1. Calculation of the Properties of Vacancies and Interstitials, National Bureau of Standards Miscellaneous Publication 287, U.S. Government Printing Office, Washington, DC, 1966.
2. *Interatomic Potentials and Simulation of Lattice Defects*, edited by P. Gehlen, J. R. Beeler, Jr., and R. I. Jaffee, Plenum Press, N.Y., 1972.
3. I. M. Torrens, *Interatomic Potentials*, Academic Press, N.Y., 1972.
4. R. A. Johnson, *J. Physics F: Metal Physics*, Vol. 3, p. 295, 1973.
5. *Interatomic Potentials and Crystalline Defects*, edited by J. K. Lee, The Metallurgical Society of AIME, Warrendale, PA, 1981.
6. L. Tewordt, *Physical Review*, Vol. 109, p. 61, 1958.
7. A. C. Damask, G. J. Dienes, and V. G. Weizer, *Physical Review*, Vol. 113, p. 781, 1959.
8. L. A. Girifalco and V. G. Weizer, *J. Physics and Chemistry of Solids*, Vol. 12, p. 260, 1960.
9. A. Seeger and E. Mann, *J. Physics and Chemistry of Solids*, Vol. 12, p. 326, 1960.
10. J. B. Gibson, A. N. Goland, M. Milgram, and G. H. Vineyard, *Physical Review*, Vol. 120, p. 1229, 1960.

11. H. B. Huntington and F. Seitz, *Physical Review*, Vol. 61, p. 315, 1942.
12. H. B. Huntington, *Physical Review*, Vol. 61, p. 325, 1942.
13. K. Fuchs, *Proceedings of the Royal Society*, Vol. A151, p. 585, 1935.
14. G. Mie, *Annalen der Physik*, Vol. 11, p. 657, 1903.
15. J. E. Lennard-Jones, *Proceedings of the Royal Society*, Vol. A106, p. 463, 1924.
16. P. M. Morse, *Physical Review*, Vol. 34, p. 57, 1929.
17. M. Born and J. E. Mayer, *Zeitschrift fur Physik*, Vol. 75, p. 1, 1932.
18. M. I. Baskes, Chapter 6 in this volume.
19. D. Wolf, Chapter 5 in this volume.
20. D. M. Esterling, Chapter 7 in this volume.
21. R. J. Arsenault, Chapter 8 in this volume.
22. J. R. Beeler, Jr., Chapter 3 in this volume.
23. G. K. Straub, Chapter 4 in this volume.
24. L. A. Girifalco and V. G. Weizer, *Physical Review*, Vol. 114, p. 687, 1959.
25. H. B. Huntington, *Physical Review*, Vol. 91, p. 1092, 1953.
26. R. A. Johnson and E. Brown, *Physical Review*, Vol. 127, p. 446, 1962.
27. R. A. Johnson, *Physical Review*, Vol. 134, p. A1329, 1964.
28. R. A. Johnson, *Physical Review*, Vol. 145, p. 425, 1966.
29. R. A. Johnson, *Radiation Effects*, Vol. 2, p. 1, 1969.
30. W. A. Harrison, *Pseudopotentials in the Theory of Metals*, W. A. Benjamin, Inc., N.Y., 1966.
31. V. Heine and D. Weaire, in *Solid State Physics*, Vol. 24; Academic Press, N.Y., 1970, p. 249.
32. J. Friedel, *Philosophical Magazine*, Vol. 43, p. 152, 1952.
33. N. Q. Lam, L. Dagens, and N. V. Doan, *J. Physics F: Metal Physics*, Vol. 13, p. 2503, 1983.
34. L. Dagens, M. Rasolt, and R. Taylor, *Physical Review B*, Vol. 11, p. 2726, 1975.
35. A. E. Carlsson, *Physical Review B*, Vol. 32, p. 4866, 1985.
36. E. S. Machlin, in *Interatomic Potentials and Crystalline Defects*, edited by J. K. Lee, The Metallurgical Society of AIME, Warrendale, PA, 1981, p. 33.
37. A. E. Carlsson, C. D. Gelatt, Jr., and H. Ehrenreich, *Philosophical Magazine A*, Vol. 41, p. 241, 1980.
38. D. M. Esterling and A. Swaroop, *Physica Status Solidi*, Vol. (b)96, p. 401, 1979.
39. L. Dagens, *Physica Status Solidi*, Vol. (b)84, p. 311, 1977.
40. L. Dagens, *J. Physics F: Metal Physics*, Vol. 16, p. 1705, 1986.
41. N. Q. Lam and L. Dagens, *J. Physics F: Metal Physics*, Vol. 16, p. 1373, 1986.
42. W. A. Harrison, *Physical Review*, Vol. 181, p. 1036, 1969.
43. W. A. Harrison and J. M. Wills, *Physical Review B*, Vol. 25, p. 5007, 1982.
44. J. M. Wills and W. A. Harrison, *Physical Review B*, Vol. 28, p. 4363, 1983.
45. J. A. Moriarty, *Physical Review Letters*, Vol. 55, p. 1502, 1985.
46. M. S. Daw and M. I. Baskes, *Physical Review B*, Vol. 29, p. 6443, 1984.
47. M. W. Finnis and J. E. Sinclair, *Philosophical Magazine A*, Vol. 50, p. 45, 1984.
48. R. A. Johnson and W. D. Wilson, in *Interatomic Potentials and Simulation of Lattice Defects*, edited by P. Gehlen, J. R. Beeler, Jr., and R. I. Jaffee, Plenum Press, N.Y., 1972, p. 301.
49. A. E. Carlsson and N. W. Ashcroft, *Physical Review B*, Vol. 27, p. 2101, 1983.
50. M. W. Finnis, K. L. Kear, and D. G. Pettifor, *Physical Review Letters*, Vol. 52, p. 291, 1984.
51. R. Taylor, in *Interatomic Potentials and Crystalline Defects*, edited by J. K. Lee, The Metallurgical Society of AIME, Warrendale, PA, 1981, p. 71.

3

Atomistic Model Computer Experiments

JOE R. BEELER, JR.

Materials Science and Engineering Department
North Carolina State University
P.O. Box 7907
Raleigh, NC 27695-7907

1. Introduction

This chapter is dedicated to showing how to write atomistic model simulation programs in FORTRAN that can be used to simulate the properties and behavior of defects in metals and metallic alloys. It also points out how the results from such simulations can be applied in the design of forging techniques.

Forging is formally defined as the compressive deformation of material between two dies. The techniques of forging are being much altered by new knowledge about plastic deformation of metals and by increasing application of computers both in the design of forging dies and in the analysis of what happens in a piece of metal as it is being forged. In this context, new knowledge of the plastic flow process, of how to suppress fracture, and of how to enhance workability is vital to advances in forging techniques. Atomistic simulation has added new information in these areas because of its unique ability to describe structure at the atomic dimensional level and atomic movement processes. In this regard, it has provided new knowledge on the structure of screw dislocation cores, dislocation-solute atom interactions, stacking-fault structure, grain-boundary structure, interface structure, surface structure, structural aspects of the dynamics of dislocation motion, crack propagation, and the thermodynamics of solids. Atomistic model simulation of grain-boundary behavior and contributions of atomistic model simulation to the thermodynamics of solids are treated in this volume by D. Wolf and G. K. Straub, respectively. In addition, crack propagation (fracture) is treated by M. I. Baskes.

2. Atomistic Model Simulation

Computer simulation of material properties and behavior can be divided naturally into two classes: (1) continuum models and (2) atomistic models. In a continuum model, the material concerned is regarded as a continuous medium. Most continuum simulations involve the solution of partial differential equations using either the finite-difference method or the finite-element method. In contrast, in an atomistic model simulation, the material concerned is regarded as an aggregate of individual atoms, and each atom is treated on an individual basis.

The two atomistic simulation methods used most often are the molecular dynamics (MD) method and the Monte Carlo (MC) method. This chapter emphasizes the molecular dynamics method because most of the atomistic simulations that are pertinent to forging have been molecular dynamics simulations. In either case, a computational cell of individual atoms is constructed and then operated on so as to simulate the properties and/or behaviors of interest. In this chapter, the properties and behaviors of crystal defects are the center of interest. Although they use the same geometrical model of a solid, the computational operations involved in MD and MC simulations are qualitatively different. In the MD case, the computational operation consists of solving the 3N simultaneous Newtonian equations of motion for the N atoms in the computational cell. In the MC case, the operation consists of exchanging the positions of two atoms at a time in accordance with Boltzmann statistics. Used in conjunction, the MD and MC methods provide a powerful means of analyzing phase transformations induced by stress and/or temperature manipulation, as in forging.

Neither an MD nor an MC simulation can be performed without the establishment of proper boundary conditions for the computational cell. There are four principal possibilities:

1. Free-surface boundary
2. Rigid boundary
3. Flexible boundary
4. Periodic boundary

Normally, the computational cell contains 1000 to 20,000 atoms. As such it represents a very small speck of material, 600 atoms being a microfemtomole. The free-surface case pertains to a large, free molecule. Macroscopic crystals are simulated by using either rigid, flexible, or periodic boundary conditions. In the rigid boundary case, a mantle of fixed-position atoms with the crystal structure of interest is placed around the computational cell. The thickness of the mantle is made larger than the range of the interaction between atoms. The mantle represents the part of the macroscopic crystal that can interact with atoms in the computational cell. The defect under study is contained in the computational cell.

The flexible boundary approach is more realistic than the rigid boundary approach in that the boundary region is capable of small atom displacements in response to forces exerted on it by atoms at the perimeter of the computational cell. It has been shown that a rigid boundary can be used in studying point defects. However, certain types of simulations for extended defects require the use of flexible boundary conditions. One of these, for example, is a Peierls stress computation for dislocation movement. Periodic boundary conditions are often used to simulate large systems. Roughly speaking, periodic boundary conditions simply mean that atoms at the extreme right of the computational cell interact with those at the extreme left of the cell. Similarly, atoms at the top of the cell interact with those at the bottom, and atoms at the front interact with those at the back. When periodic boundary conditions are used, the cell diameter must exceed twice the interaction range between two individual atoms.

During the past four years, the capability of the periodic boundary condition approach has been vastly enlarged. As described above, the periodic boundary condition mode restricts the simulation to a computational cell with fixed volume and shape. As such, certain phase transformations cannot be simulated. Recent developments allow changes in both the volume and the shape of the computational cell as the simulation proceeds when periodic boundary conditions are used.

As discussed in Chapter 2 by R. A. Johnson, interactions between atoms can be represented either by central force pair potentials, provided a volume force recipe is added to represent the "Cauchy" pressure, or by a variation of the embedded atom approach in which the volume force is automatically incorporated in the embedding energy term of the interaction recipe. The MD subroutines given in the present chapter use a central force atom interaction model. This is done to simplify the presentation and affects only the force matrix calculation. A force routine for use with an embedded atom model is given in the Appendix to this chapter.

3. Molecular Dynamics

Molecular dynamics simulations are used to simulate the movement of individual atoms in an assembly of N interacting atoms on the basis of Newtonian mechanics. Given the initial position and velocity vectors for the atoms, the dynamical history of the assembly is generated by numerically integrating the 3N simultaneous equations of motion for the Cartesian coordinate components of the N atoms in the assembly. The 3N velocity components also are determined by the integration process. Hence, the molecular dynamics method generates a phase space-point trajectory for the N-atom system as a function of time. For the most part, three different numerical integration schemes have been used in MD simulations. These are the central-difference, predictor-corrector, and Nordsieck methods. A Nordsieck scheme is the best of the three. There is little difference between the advantages offered by the other two approaches. We will adopt the central-difference method because it is easy to use.

The essence of the central-difference method can be described adequately in one dimension with the notation X as the position coordinate, V as the velocity, and A = F/M as the acceleration, where F is the net force on an atom and M is the atom mass. The net force on a given atom is a function of the coordinates of all atoms in the assembly which can interact with the atom concerned. The one-dimensional scheme described here can be applied immediately to each of the 3N independent coordinate components for the N atoms in a three-dimensional crystal.

The central-difference method computes the atom coordinate components at integer multiples of the integration time step DT and atom velocity components at half-integer multiples of DT. This is diagrammed in Fig. 1. The known quantities on which each integration step is based are X(T), A(T), and V(T − DT/2). The net force F(T) experienced by an atom at time T is determined by the relative positions of all the other atoms that interact with it at time T, and A(T) = F(T)/M. Given A(T), the advanced time quanti-

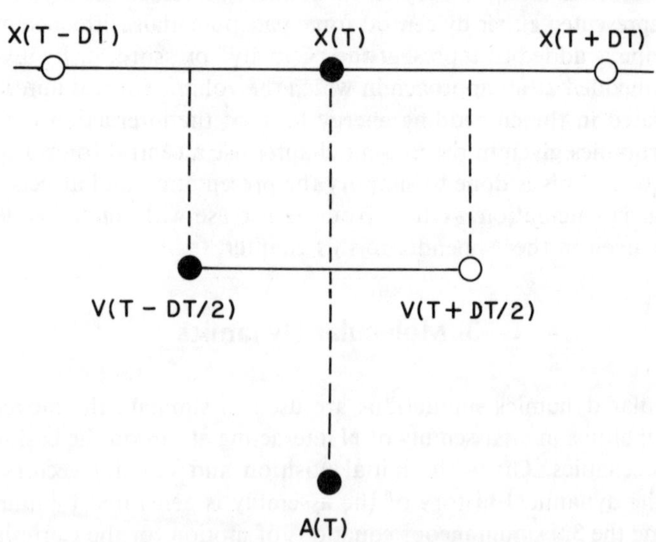

Fig. 1. The central-difference algorithm.

ties V(T + DT/2) and X(T + DT) are given by

$$V(T + DT/2) = V(T - DT/2) + A(T)DT$$
$$X(T + DT) = X(T) + V(T + DT/2)DT$$

The velocity change during the interval DT is

$$DV = (F(T)/M) * DT$$

The method assumes a constant acceleration during each time step DT, and the net force on each atom is a step function in time. This is illustrated in Fig. 2.

It turns out that an MD simulation of a small, bare crystal with either a bcc, fcc, or hcp structure can be set up easily. These crystals are stable in the sense that they retain their structural integrity when low-temperature thermal vibrations are simulated. We will introduce the integration subroutine for an MD simulation by showing how it would be applied to a small, bare crystal. This gets directly to the essence of the MD method.

The computational flow in a MD program for a bare assembly of atoms is outlined in Fig. 3. Input data (Box 1) includes the initial position and velocity of each atom, the interaction function parameters, the maximum number of time steps LAST desired for the numerical integration loop

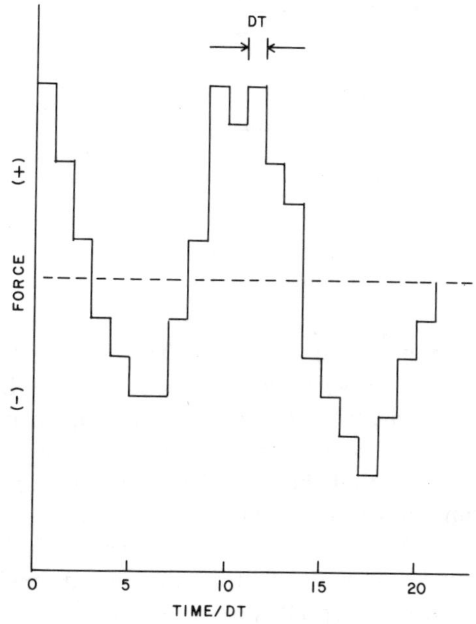

Fig. 2. The force on an atom, in a molecular dynamics simulation, is a step function.

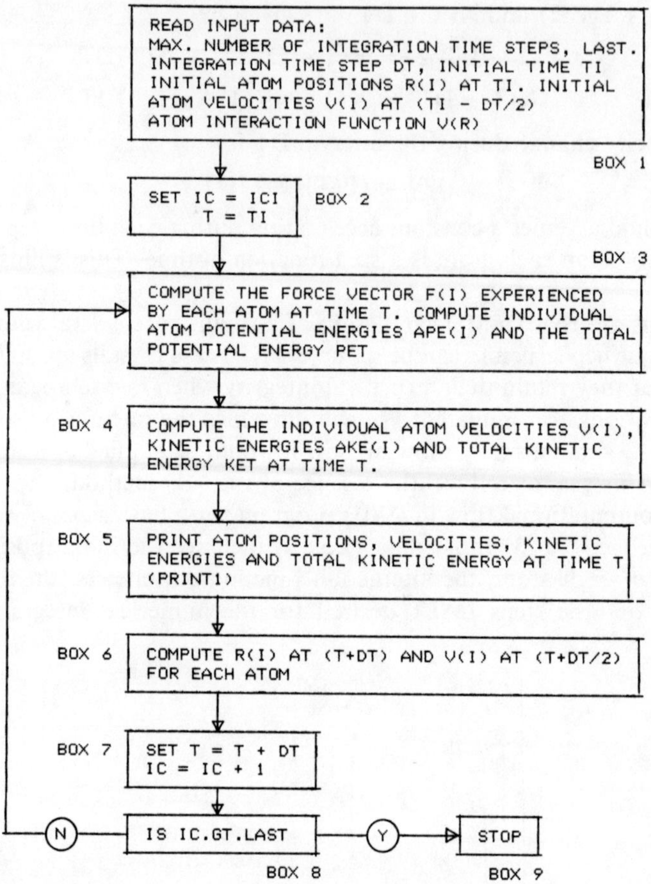

Fig. 3. Outline of the computational flow in a molecular dynamics program for a bare assembly of atoms.

(Boxes 3–8), and the integration time step DT. The initial positions are for time T = TI and the initial velocities are for time (TI − DT/2).

At the beginning of the integration loop (Box 3), the net force on each atom is computed. The velocity change experienced by each atom is then computed (Box 4). Note that the forces are computed at the beginning of each integration time step and that these forces remain unchanged throughout the time step. Each atom, therefore, follows a trajectory made up of short, straight segments. Because all atoms are moved simultaneously at the end of the integration time step (Box 6), any initial symmetry in atom positions will tend to persist when the initial velocities are zero. This fixation with one symmetry mode can be troublesome. It can be avoided by assigning each atom with a small random displacement from its initial static equilibrium position.

A convenient set of units for MD simulation is given in Table 1 to provide a specific reference for use in discussion. In Box 4 of Fig. 3, the velocity change DV would be computed using the relations

$$A(T) = 0.964847*F(T)/M$$

$$DV = A(T)*DT$$

In these relations, the proper units of force, mass, acceleration, and velocity are eV/angstrom, atomic mass unit (AMU), angstrom per tau squared, and angstrom/tau, respectively. Here, eV stands for electron volt, AMU stands for atomic mass unit, and tau is 1.0E−14 s. The kinetic energy KE of an atom with mass M and velocity V in angstroms/tau is given in eV by

$$KE = 0.518206*M*V*V$$

A listing of a simple MD FORTRAN program for a bare assembly of up to 100 atoms is given in Fig. 4. All of the key variables and constants are defined in the glossary at the beginning of the listing. This program computes the coordinates, kinetic energy, and potential energy for each atom as a function of time together with the total kinetic and potential energies for the entire assembly as a function of time. The atom pair potential energy function used in this program is a third-degree polynomial in the pair separation distance R. A way to modify this program so that an embedded atom interaction model could be used is given in the Appendix to this chapter.

3.1 Static Equilibrium Atom Position Calculations

When a defect is produced in a crystal, each atom is displaced from its initial perfect crystal position to a new static equilibrium position at which the net force on the atom again vanishes. The vector running from an atom's initial perfect crystal position to its new position, after the defect has been

Table 1. A convenient set of units for use in low-energy molecular dynamics simulations.

Quantity	Symbol	Unit	Unit magnitude
Mass	M	AMU	1.66053E−24 g
Energy	E	eV	1.60219E−12 ergs
Distance	X	A	1.0E−8 cm
Time	T	tau	1.0E−14 s
Velocity	V	A/tau	1.0E+6 cm/s
Acceleration	A	A/tau * tau	1.0E+20 cm/s * s
Force	F	eV/A	1.60219E−4 dynes

```
C***  A SIMPLE MOLECULAR DYNAMICS PROGRAM
C
C***   GLOSSARY
C
C      X, Y, Z: ATOM COORDINATES IN ANGSTROMS
C      VX, VY, VZ: VELOCITY COMPONENTS IN ANGSTROMS/TAU
C      TAU: 1.0E-14 SECONDS
C      APE(NA): ATOM POTENTIAL ENERGY IN EV
C      AKE(NA): ATOM KINETIC ENERGY IN EV
C      MASS: ATOM MASS IN ATOMIC MASS UNITS
C      A, B, C, D: COEFFICIENTS IN THE PAIR POTENTIAL:
C        V(R) = A*(R**3) + B*R*R + C*R + D EV
C      R: DISTANCE BETWEEN TWO ATOMS IN ANGSTROMS
C      DT: INTEGRATION TIME STEP IN TAU UNITS
C      LAST: NUMBER OF INTEGRATION TIME STEPS PER RUN
C      NAT: NUMBER OF ATOMS
C      NREAD: INPUT READ DEVICE NUMBER
C      NRITE: OUTPUT WRITE DEVICE NUMBER
C      KET: TOTAL KINETIC ENERGY
C      PET: TOTAL POTENTIAL ENERGY
C      FX, FY, FZ: FORCE COMPONENTS
C      AF, BF, CF: COEFFICIENTS IN FORCE POLYNOMIAL
C
       DIMENSION X(100), Y(100), Z(100), VX(100), VY(100),
      1VZ(100), FX(100), FY(100), FZ(100), APE(100), AKE(100)
       NAMELIST/INPUT1/A,B,C,D,RCUT,MASS,DT,LAST,NAT
       NAMELIST/INPUT2/X,Y,Z,VX,VY,VZ,ICI,TI
C
       REAL MASS, KE, KET, KEC
C
       NREAD = 1
       NRITE = 3
C
       READ(NREAD,INPUT1)
       READ(NREAD,INPUT2)
C
C***   INITIALIZATION
       IC = ICI
       T = TI
       NL = NAT - 1
       RCUT2 = RCUT*RCUT
       AF = 3.0*A
       BF = 2.0*B
       CF = C
       KEC = 0.518206*MASS
C
C***   CLEAR FORCE AND ENERGY TABLES
C
     5 DO 10 J = 1,NAT
       FX(J) = 0.0
       FY(J) = 0.0
       FZ(J) = 0.0
       APE(J) = 0.0
       AKE(J) = 0.0
```

Fig. 4. A simple molecular dynamics FORTRAN program for a bare assembly of up to 1000 atoms.

```
   10 CONTINUE
      PET = 0.0
      KET = 0.0
C
C***  COMPUTE FORCE ON EACH ATOM AT TIME T
C
      DO 25 I = 1,NL
      JS = I+1
      DO 20 J = JS,NAT
      DX = X(J) - X(I)
      DY = Y(J) - Y(I)
      DZ = Z(J) - Z(I)
      R2 = DX*DX + DY*DY + DZ*DZ
      IF(R2.GT.RCUT2) GO TO 20
      R = SQRT(R2)
      F = AF*(R*R) + BF*R + CF
      FR = F/R
      DFX = FR*DX
      FX(I) = FX(I) + DFX
      FX(J) = FX(J) - DFX
      DFY = FR*DY
      FY(I) = FY(I) + DFY
      FY(J) = FY(J) - DFY
      DFZ = FR*DZ
      FZ(I) = FZ(I) + DFZ
      FZ(J) = FZ(J) - DFZ
C***  COMPUTE ATOM POTENTIAL ENERGY AT TIME T
      POTA = (A*(R**3) + B*(R**2) + C*R + D)/2.0
      APE(I) = APE(I) + POTA
      APE(J) = APE(J) + POTA
      PET = PET + POTA + POTA
   20 CONTINUE
   25 CONTINUE

C***  COMPUTE ACCELERATION AT TIME T
      DO 30 I = 1,NAT
      ACX = 0.964867*FX(I)/MASS
      ACY = 0.964867*FY(I)/MASS
      ACZ = 0.964867*FZ(I)/MASS
C***  COMPUTE VELOCITY CHANGE DURING DT/2.0
      DVX = 0.5*ACX*DT
      DVY = 0.5*ACY*DT
      DVZ = 0.5*ACZ*DT
C***  COMPUTE VELOCITY AT TIME T
      VX(I) = VX(I) + DVX
      VY(I) = VY(I) + DVY
      VZ(I) = VZ(I) + DVZ
C***  COMPUTE ATOM KINETIC ENERGY AT TIME T
      AKE(I) = KEC*(VX(I)**2 + VY(I)**2 + VZ(I)**2)
C***  PRINT POSITION, VELOCITY, ENERGIES FOR TIME T
      CALL PRINT1
C***  COMPUTE VELOCITY AT TIME (T + DT/2.)
      VX(I) = VX(I) + DVX
      VY(I) = VY(I) + DVY
```

Fig. 4. continued

```
      VZ(I) = VZ(I) + DVZ
C***  COMPUTE COORDINATES AT TIME (T + DT)
      X(I) = X(I) + VX(I)*DT
      Y(I) = Y(I) + VY(I)*DT
      Z(I) = Z(I) + VZ(I)*DT
   30 CONTINUE
C
C***  STEP IC AND T
C
      IC = IC + 1
      T = T + DT
C
C***  TERMINATION TEST
C
      IF(IC.GT.LAST) GOTO 1000
      GO TO 5
C
C***  PRINT FINAL OUTPUT AND STOP
 1000 CALL PRINT2
      STOP
      END
```

Fig. 4. continued

created, is called the *displacement vector* for that atom. The collection of displacement vectors for all of the atoms is the displacement field for the defect concerned.

The simple MD program described by Fig. 3 and 4 cannot be used as is to compute the displacement field for a crystal defect, that is, the static equilibrium positions of atoms surrounding a defect. However, any MD program can be modified to compute static equilibrium atom positions by damping out the kinetic energy at appropriate times. Two methods have been used to damp out the kinetic energy: one is called the *quasidynamical method* and the other is called the *localized damping* or *microconvergence method.*

In the quasidynamical method, the program is allowed to run in the fully dynamical mode until the kinetic energy for the *entire* crystal reaches a peak value. When this peak occurs, the velocity of each atom in the crystal is set to zero but the current atom-position coordinates are retained. The calculation is then resumed, starting with zero initial velocity for each atom. Generally, the crystal kinetic energy will rise again, and when it peaks the velocity of each atom is set to zero. Each time this procedure is repeated, the kinetic energy peak is smaller than that in the preceding iteration; consequently, the atom positions converge to a set of zero net force positions for a static equilibrium configuration. A schematic of the crystal kinetic energy versus time is given in Fig. 5. This procedure works because the zero force position for an oscillating particle is also the position of maximum velocity (kinetic energy). The quasidynamical method was first proposed and used by Gibson et al.[(1)]

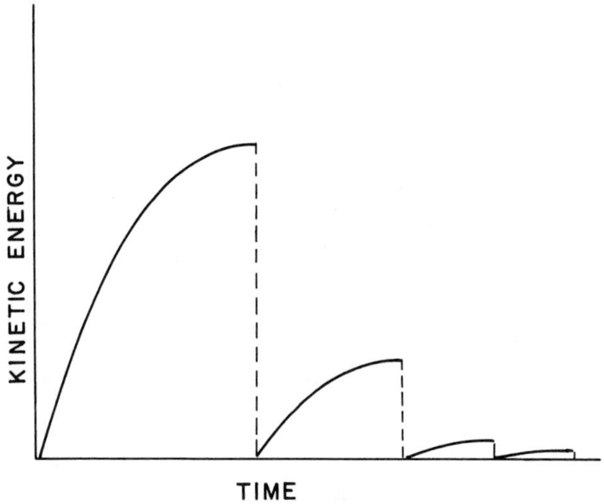

Fig. 5. The quasidynamical method.

The zero force position of an oscillating particle is also the position at which the acceleration changes sign. This fact provides an alternative way to damp out the kinetic energy in a static equilibrium atom position calculation. Figure 6 illustrates the condition that when an oscillating particle is moving toward its equilibrium position, its acceleration and velocity have the same sign, and they have opposite signs when the particle is moving away from its equilibrium position. Hence, the equilibrium position can be determined in an MD simulation by setting the velocity to zero whenever the sign of the velocity becomes different from that of the force, i.e., acceleration. Note that this scheme affects the velocity of only one atom at a time. It inhibits the motion of any atom that is moving away from its static equilibrium position without hindering the motion of atoms which are moving toward their equilibrium positions. In practice, this scheme is applied separately to each Cartesian coordinate of the velocity and acceleration vectors. As a consequence, damping occurs only for those individual velocity components of an atom that are tending to produce a deviation from the atom's static equilibrium position. This procedure was introduced by Evans.[(2)] He noticed that the quasidynamical method was not efficient when a mixture of atoms with different masses was simulated, as in the case of an alloy. This inefficiency occurs because the lighter-mass atoms tend to oscillate more rapidly than the heavier-mass atoms. Consequently, velocity damping, performed when the total kinetic energy maximum occurs, does not select the most appropriate time for damping of either the light-mass or the heavy-mass atom populations.

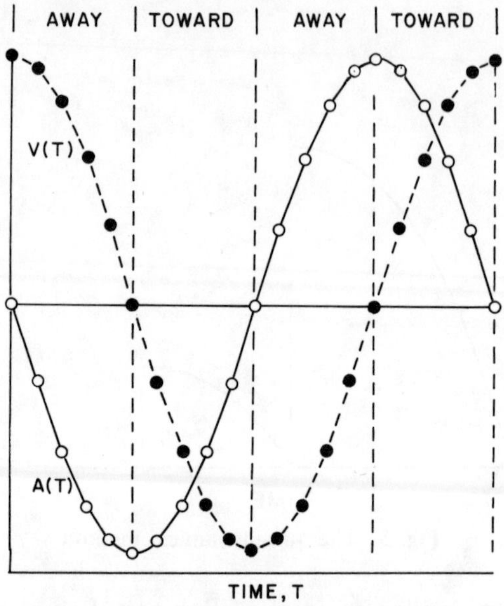

Fig. 6. The microconvergence method.

4. Computational Cell Construction

We will consider the construction of a bcc computational cell for use with rigid boundary conditions. This will be done for cases where atom interactions involve only first and second neighbors. For simplicity we will construct a cell based on (002) plane stacking. This cell, based on (002) stacking, will have NX atom sites along the X-direction edge, NY along the Y-direction edge and NZ along the Z-direction edge. The unit of length will a/2, where a is the lattice constant. This will be called a *half-lattice constant* and be denoted as hlc. It turns out that an hlc is a handy unit choice for cubic systems. The (002) stacking choice is appropriate for studying point defects but not dislocation cores, for example. In that case, for a bcc crystal, the cell should be based on (110) stacking and have (112), (110), and (111) plane faces.

An atom site map for such a cell with NX = NY = NZ = 4 atoms is given in Fig. 7. The number of (002) planes (Z-planes) in the cell is NZZ = (2*NZ − 1) = 7. There are two cases to consider: (1) When the Z-plane number NP is odd, each atom column starts with X = 0 and the Y-coordinate of the first site in the plane is Y = 0. Each column contains NX atom sites and each plane contains NY columns. (2) When NP is even, each column starts at X = 1 and the Y-coordinate of the first site in each plane is Y = 1. Each column contains NXL = (NX − 1) sites and each plane con-

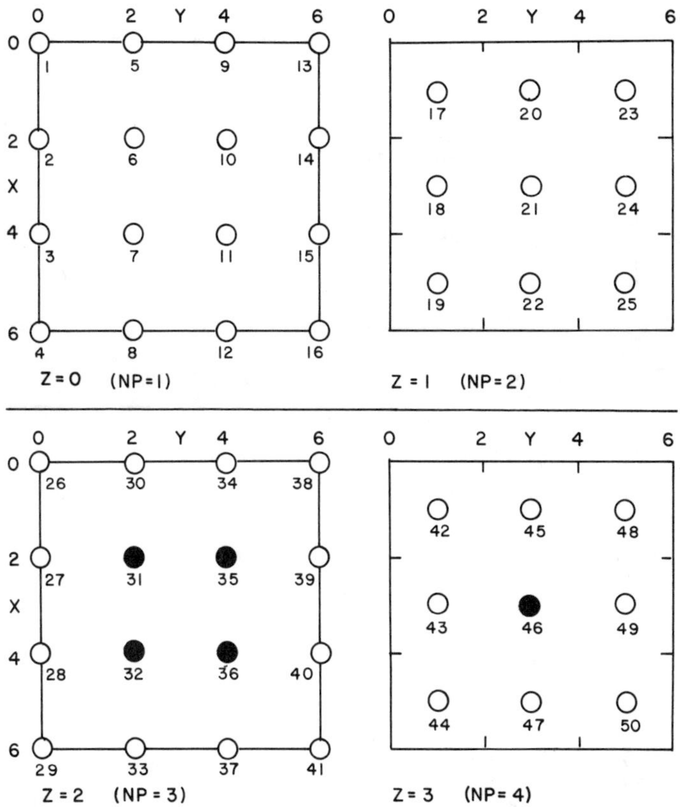

Fig. 7. An atom site map for a bcc computational cell with NX = NY = NZ = four atoms. (Figure continues on next page.)

tains NYL = (NY − 1) columns. A FORTRAN routine which generates the site-coordinate tables and site numbers for a bcc crystal based on (002) stacking is given in Fig. 8. The index JJ in this routine is used to monitor the type of Z-plane to be generated. JJ = 0 indicates that the plane number NP is even; JJ = 1 indicates that NP is odd. This index is used to institute the proper reset conditions for X, at the start of each column, and for Y, at the start of each Z-plane.

All movable atoms in the computational cell are shown as filled circles in Fig. 7 and their coordinates satisfy the conditions

$$1.5 < X < 4.5;\ 1.5 < Y < 4.5;\ 1.5 < Z < 4.5$$

The atom numbers for these movable atoms are listed in the MOVE table. A FORTRAN routine for constructing the MOVE table is given later in Fig. 11, and the results are listed in Table 2.

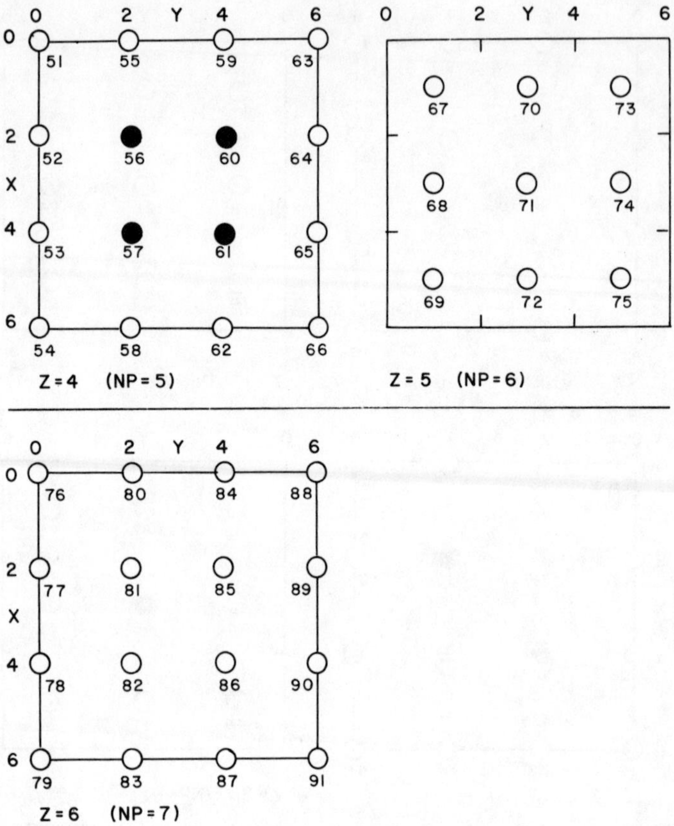

Fig. 7. continued

The interacting-neighbor table for a site in a plane with NP even is different from that for a site in a plane with NP odd. As examples, consider site NS = 46, for which NP = 4, and site NS = 35, for which NP = 3. The first and second neighbors for these two sites are listed in Table 3. The common

Table 2. Computational cell atom site number table MOVE for the cell example given in Fig. 7. KMOVE = 9

KC	MOVE(KC)	KC	MOVE(KC)
1	31	6	56
2	32	7	57
3	35	8	60
4	36	9	61
5	46		

```
      SUBROUTINE BCC
C
C     BCC CRYSTAL CONSTRUCTION: COMPUTES ATOM SITE COORDINATES
C     IN HLC FOR A RECTANGULAR PARALLELEPIPED WITH NX, NY AND
C     NZ ATOM SITES ALONG THE X-, Y- AND Z-DIRECTION EDGES,
C     RESPECTIVELY.  SITE NS = 1 HAS COORDINATES XS(1) = YS(1) =
C     ZS(1) = 0.
C
      NXX = NX
      NYY = NY
      XO = 0
      YO = 0
      ZO = 0
      JJ = 0
      NS = 0
C
      DO 5 K = 1,NZZ
      DO 3 J = 1,NYY
      DO 1 I = 1,NXX
      NS = NS + 1
      XS(NS) = XO
      YS(NS) = YO
      ZS(NS) = ZO
      XO = XO + 2
    1 CONTINUE
      YO = YO + 2
      IF(JJ.GT.0) GO TO 2
      XO = 0
      GO TO 3
    2 XO = 1
    3 CONTINUE
      ZO = ZO + 1
      IF(JJ.GT.0) GO TO 4
      JJ = 1
      XO = 1
      YO = 1
      NYY = NYL
      NXX = NXL
      GO TO 5
    4 JJ = 0
      XO = 0
      YO = 0
      NYY = NY
      NXX = NX
    5 CONTINUE
      NSITE = NS
C
      DO 10 N = 1,NSITE
      LOC(N) = N
      SITE(N) = N
      TYPE(N) = 1
      POPS(N) = 1
      NSA(N) = 0
      X(N) = FLOAT(XS(N))
      Y(N) = FLOAT(YS(N))
      Z(N) = FLOAT(ZS(N))
   10 CONTINUE
      RETURN
```

Fig. 8. A FORTRAN routine that generates the site-coordinate tables and site numbers for a bcc crystal based on (002) stacking.

Table 3. First- and second-neighbor sites of site NS = 35 (NP = 3) and site NS = 46 (NP = 4) in Fig. 7 for an odd and even Z-plane number, respectively.

	Odd NP (NP = 3) Site NS = 35	Even NP (NP = 4) Site NS = 46
First neighbors		
	NS − 15 = 20	NS − 15 = 31
	NS − 14 = 21	NS − 14 = 32
	NS − 12 = 23	NS − 11 = 35
	NS − 11 = 24	NS − 10 = 36
	NS + 10 = 45	NS + 10 = 56
	NS + 11 = 46	NS + 11 = 57
	NS + 13 = 48	NS + 14 = 60
	NS + 14 = 49	NS + 15 = 61
Second neighbors		
	NS − 25 = 10	NS − 25 = 21
	NS − 4 = 31	NS − 3 = 43
	NS − 1 = 34	NS − 1 = 45
	NS + 1 = 36	NS + 1 = 47
	NS + 4 = 39	NS + 3 = 49
	NS + 25 = 60	NS + 25 = 71

differences between NS and the site numbers of its neighbors for both even and odd NP are

DIF/−25, −15, −14, −11, −1, 1, 10, 11, 14, 25/

Differences peculiar to even NP are

DIFE/−10, −3, 3, 15/

and those peculiar to odd NP are

DIFO/−12, −4, 4, 13/

A FORTRAN routine for constructing the interacting neighbors table based on these three difference lists is given in Fig. 9.

Determination of an even or odd plane number NP (lines 1–5 in Fig. 9) works in the following way: NXY = NX∗NY is the number of sites in an odd NP plane; NXYL = (NX − 1)∗(NY −1) is the number of sites in an even NP plane. The sum NW = NXY + NXYL is the number of sites in an odd-even repeat unit along the Z-direction. KT is the number of odd-even repeat units of planes lying below the plane containing the site NS. The numerator in the expression for KT has to be (NS − 1) rather than NS in order for the routine to work properly for the last site in an even plane. JT is the number of sites lying between site NS and the last site in the KT odd-

```
C***    SUBROUTINE NABORS
C       BCC CRYSTAL INTERACTING NEIGHBORS TABLE. A ROUTINE
C       FOR CONSTRUCTING A TABLE NABS(K) OF SITE NUMBERS
C       FOR SITES WHICH ARE FIRST OR SECOND NEIGHBORS OF
C       SITE NS. AND A TABLE NABA(K) OF ATOM NUMBERS WHICH
C       ARE REFERENCED TO THESE SITES
C
        NXY = NX*NY
        NW = NXY + (NX-1)*(NY-1)
        KT = (NS - 1)/NW
        JT = (NS - KT*NW)
        IT = JT -NXY
        JJ = 0
        IF(IT.LE.0) JJ = 1
        IF(JJ.LT.1) GO TO 20
C
C       NEIGHBORS DEFINED BY DIFO
        DO 10 K = 1,4
        NABS(K) = NS + DIFO(K)
     10 CONTINUE
        GO TO 30
C
C       NEIGHBORS DEFINED BY DIFE
     20 DO 25 K = 1,4
        NABS(K) = NS + DIFE(K)
     25 CONTINUE
C
C       NEIGHBORS DEFINED BY DIF
     30 DO 35 KL = 5,14
        K = KL -4
        NABS(KL) = NS + DIF(K)
     35 CONTINUE
        NABS(15) = NS
        KMAX = 15
C       ATOM NUMBER TABLE NABA(K)
        KN = 0
        DO 40 K = 1,KMAX
        NS = NABS(K)
        IF(POPS(NS).LT.1) GO TO 40
        IF(POPS(NS).GT.1) GO TO 50
        KN = KN + 1
        NABA(KN) = LOC(NS)
        GOTO 40
     50 KN = KN+1
        NABA(KN) = LOC(NS)
        KN = KN + 1
        NABA(KN) = NSA(NS)
     40 CONTINUE
        KMAX = KN
        RETURN
```

Fig. 9. A FORTRAN routine to construct the interacting-neighbors table based on the three difference lists.

even repeat units below NS. If JT is larger than NXY, then NP must lie in an even plane since an odd plane precedes the even plane in the repeat unit. Conversely, if JT is less then or equal to NXY, NS must lie in an odd NP plane.

5. bcc Crystal, Rigid Boundary Molecular Dynamics Program

Figure 10 gives a computational flowchart for an MD program suitable for a bcc crystal computational cell with a rigid boundary region. Box 1 deals with input data. These data include all geometrical parameters needed to guide construction of a bcc crystal and the subsequent insertion of defects into this crystal. Defect insertion, in this particular example, will be limited to vacancies and self-interstitial atoms. In addition, the input data set must contain all parameters needed to define the atom-interaction model assumed for the simulation.

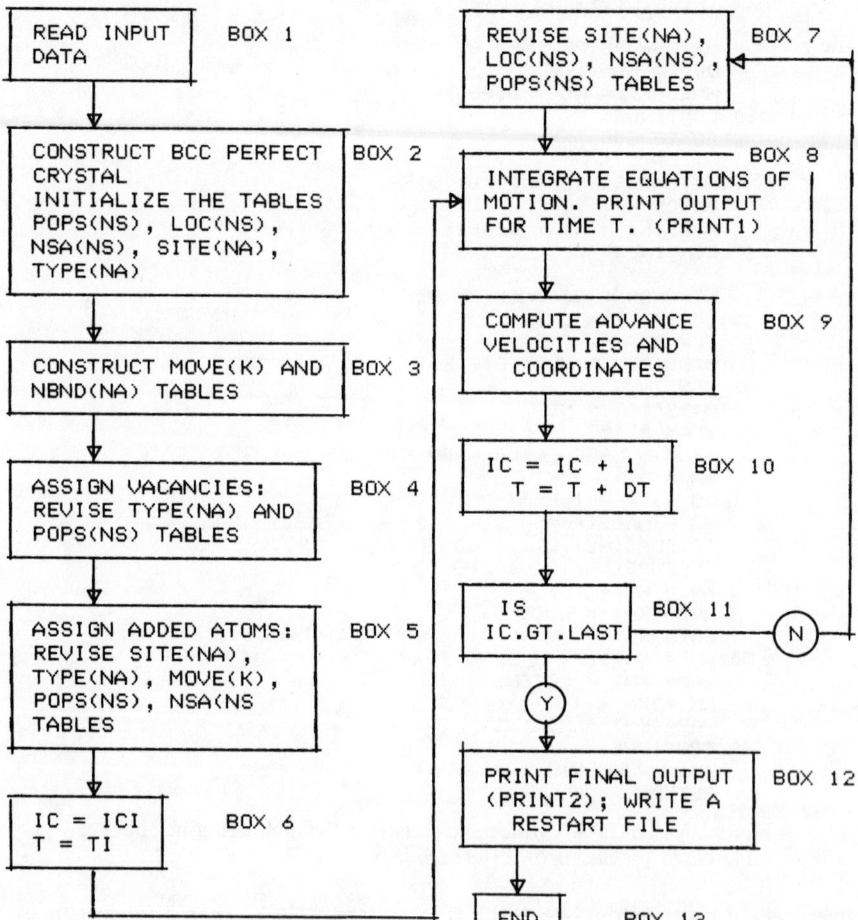

Fig. 10. A molecular dynamics program for a bcc computational cell and a rigid boundary region.

Box 2 concerns the construction of a perfect bcc crystal. As indicated in Fig. 7, the crystal volume which is constructed consists of a central movable atom computational cell (filled circle positions) nested within a rigid boundary region (open circle positions). These positions are to be regarded as perfect crystal atom sites and are numbered with the index NS as shown in Fig. 7. The subroutine which performs this construction is called Subroutine BCC. As shown in Fig. 8, the first task performed by Subroutine BCC is computation of the perfect bcc crystal atom site coordinates XS(NS), YS(NS), ZS(NS) in hlc units. These coordinates are integers. It also records the total number of perfect crystal atom sites NSITE. These perfect crystal atom site coordinates remain unchanged throughout the MD simulation. In the perfect crystal, the coordinates of atom NA located at site NS are indentical to the coordinates of site NS and, in addition, NA and NS are identical. The site coordinates are integers, whereas the atom coordinates are floating-point numbers. Hence, one has X(NS) = FLOAT(XS(NS)), Y(NA) = FLOAT(YS(NS)), Z(NA) = FLOAT(ZS(NS)) with NA = NS.

Because vacancy and self-interstitial defects can be introduced, some of the original perfect crystal sites will be vacant while others will be occupied by two atoms. Subroutine BCC also prepares five additional tables which are needed to chronicle the movement of atoms from site to site during a simulation. Atom interactions, in the program of concern here, are computed by noting what perfect crystal site NS is closest to the particular atom NA whose interactions we want to compute. This site number is given by the table SITE(NA) and is called the *reference site* for atom NA. The first- and second-neighbor sites of reference site NS are then found using the DIF, DIFE, and DIFO tables. The NABORS subroutine identifies all of the atoms {NA′} which are closest to the first and second neighbors of NS, and of NS itself, and stores them in the NAB table. The atoms stored in the NAB table are those which can interact with atom NA. This survey of atom locations is done using tables LOC(NS), NSA(NS), and POPS(NS).

Table SITE(NA) shows site number NS of the perfect crystal site which is closest to atom NA. Conversely, table LOC(NS) gives the atom number NA of the atom currently associated with site NS. POPS(NS) gives the number of atoms associated with site NS. If POPS(NS) = 2, then second atom table NSA(NS) is used to give the atom number of the second atom associated with site NS. POPS(NS) = 0 means that site NS is vacant. No more than two atoms are allowed at an atom site in the program concerned here. Two atoms is by no means an upper bound. As many as eight atoms can be allowed per site without undue complication.

The TYPE(NA) table shows what atoms were removed to create vacancies and also to designate the chemical type of the atom. TYPE(NA) = 0 means that atom NA has been removed. TYPE(NA) = 1 means that atom

NA exists in the crystal and is of the chemical type associated with index 1. Only one chemical type is considered here. Subroutine BCC assigns TYPE(NA) = 1 for each atom.

Subroutine MOVETAB constructs the table MOVE(K) of atom numbers for atoms in computational cell. These are the only atoms allowed to move in response to a net force. Since first- and second-neighbor interactions have been assumed, the perfect crystal rigid boundary region involves the outer two layers of atoms. Hence, an atom is contained in the computational cell if its coordinates satisfy the conditions

1.5 .LT. X(NA) .LT. (2∗(NX − 1) −1.5)

1.5 .LT. Y(NA) .LT. (2∗(NY − 1) −1.5)

and

1.5 .LT. Z(NA) .LT. (2∗(NZ − 1) −1.5)

The MOVETAB subroutine is listed in Fig. 11. MOVETAB also constructs the boundary table NBND(NA) which designates whether an atom NA is in the boundary region (NBND(NA) = −1) or in the computational cell (NBND(NA) = 1).

Vacancies and self-interstitials are inserted in Boxes 3 and 4, respectively, by subroutines VACAN and ADDATM. These subroutines are listed in Fig. 12. Vacancies are entered by specifying the total number of vacancies NVAX to be inserted and the atom numbers NVAC(K) of the atoms to be

```
C       SUBROUTINE MOVTAB
C       CONSTRUCTION OF THE MOVE(K) AND NBND(NA) TABLES
C
        M = 0
        DO 10 N = 1,NSITE
        XT = X(N)
        YT = Y(N)
        ZT = Z(N)
        IF((XT.GT.1.5).AND.(XT.LT.XUB)) GO TO 20
        GO TO 50
     20 IF((YT.GT.1.5).AND.(YT.LT.YUB)) GO TO 30
        GO TO 50
     30 IF((ZT.GT.1.5).AND.(ZT.LT.ZUB)) GOTO 40
        GO TO 50
     40 M = M+1
        NBND(N) = 1
        MOVE(M) = N
        GO TO 10
     50 NBND(N) = -1
     10 CONTINUE
        KMOVE = M
        RETURN
```

Fig. 11. The subroutine MOVETAB, which constructs the MOVE(K) and NBND(NA) tables.

```
C        SUBROUTINE VACAN
C
         IF(NVAX.LT.1) GO TO 20
         DO 10 M = 1,NVAX
         NA = NVAC(M)
         TYPE(NA) = 0
         POPS(NA) = 0
      10 CONTINUE
      20 RETURN

C        SUBROUTINE ADDATM
C
         IF(NADD.LT.1) GO TO 20
         NA = NSITE
         DO 10 M = 1,NADD
         NS = SADD(M)
         NB = LOC(NS)
         X(NB) = X(NB) + XDIS(M)
         Y(NB) = Y(NB) + YDIS(M)
         Z(NB) = Z(NB) + ZDIS(M)
C
         NA = NA + 1
         X(NA) = XAD(M)
         Y(NA) = YAD(M)
         Z(NA) = ZAD(M)
         NBND(NA) = 1
         TYPE(NA) = TYPAD(M)
         SITE(NA) = NS
         POPS(NS) = POPS(NS) + 1
         NSA(NS) = NA
         KMOVE =KMOVE + 1
         MOVE(KMOVE) = NA
      10 CONTINUE
         NATOM = NSITE + NADD
      20 RETURN
```

Fig. 12. The subroutines VACAN and ADDATM.

removed. POPS(NS) is set to zero for each atom site from which an atom is removed. No more than one atom can be removed from a site.

Added atoms are entered by specifying the number of added atoms NADD to be inserted, their type TYPAD(M), their site number SADD(M), and their floating-point coordinates XAD(M), YAD(M), ZAD(M), M = 1,2, . . . , NADD. If the added atom is intended to represent a split self-interstitial, the normal atom, originally at the site concerned, must be displaced from its perfect crystal position. Displacement of normal atoms at the SADD added atom sites can be accomplished by assigning displacements XDIS(M), YDIS(M), ZDIS(M). A split self-interstitial is a very strong perturbation; if the original atom is not properly displaced the simulation will fail. The atom number for the original atom is NB = LOC(NS) and NS =

SADD(M) for the M-th insertion. Adding atoms to the computational cell makes it necessary to revise the TYPE(NA), SITE(NA), POPS(NS), NSA(NS), and MOVE(K) tables. Note that the atom numbers for the added atoms are greater than NSITE.

Integration of the equations of motion proceeds in the same manner as it did for the bare assembly of atoms. Atom interactions, as mentioned previously, are computed by considering only first and second neighbors as candidates rather than all of the atoms present. Because atoms move about within the computational cell, a subroutine is needed to determine where the nearest perfect crystal site is for each atom. This search is done at the end of each integration time step DT. The subroutine which does the search is called WHERE and is listed in Fig. 13.

WHERE first scans the MOVE table to determine the number of the site NS closest to each atom NA. This must be done because atoms can move from one site to another during any time step. This updates the SITE(NA) table. WHERE also sets POPS(NS) = 0 for all sites in the computational cell. This is done to facilitate the identification of any multiply occupied sites during a second scan of the MOVE table. Having updated the SITE(NA) table, WHERE then scans the MOVE table a second time and counts the number of atoms referenced to each site NS. If POPS(NS) = 0 when a site number NS is encountered in the second scan, the revision POPS(NS) = 1 is made. If POP(NS) = 1 when a site number NS is encountered in the second scan, then NS already is occupied by one atom and the second encounter means that it is doubly occupied. Then the revision POPS(NS) = 2 is made

```
C      SUBROUTINE WHERE
C
       DO 10 K = 1,KMOVE
       NA = MOVE(K)
       CALL SITNUM
       SITE(NA) = NS
       POPS(NS) = 0
    10 CONTINUE
C
       DO 20 K = 1,KMOVE
       NA = MOVE(K)
       NS = SITE(NA)
       IF(POPS(NS) = 0) GO TO 30
       NSA(NS) = NA
       POPS(NS) = 2
       GO TO 20
C
    30 POPS(NS) = 1
       LOC(NS) = NA
    20 CONTINUE
       RETURN
```

Fig. 13. The subroutine WHERE.

and, in addition, the second occupancy table NSA is revised to show that the second atom at site NS is NSA(NS) = NA.

Subroutine SITNUM is used by subroutine WHERE to find the number NS of the site nearest to each atom NA. SITNUM is listed in the Appendix to this chapter. Given the coordinates X(NA), Y(NA), Z(NA) of atom NA, SITNUM finds the coordinates of the site nearest to atom NA using the fact that the coordinates of a bcc site are either all even or all odd integers, when expressed in hlc units, and the definition of the bcc Wigner-Seitz cell. It is a relatively long subroutine because it must select from a long list of options. It is the listing of these options that makes it long. However, the *process* of selecting the proper option is a fast process. It then uses the site coordinates to determine the site number NS.

A FORCE subroutine for a bcc computational cell with rigid boundary conditions is listed in Fig. 14. It contains only one part which is significantly different from the listing in Fig. 4 for the Simple MD Program. That part is as follows:

```
I = MOVE(K)
IF(TYPE(I.LT.1) GO TO 25              (25 CONTINUE, on I)
NS = SITE(I)
CALL NABORS
DO 20 K = 1,KMAX
J = NAB(K)
IF(TYPE(J).EQ.0) GO TO 20             (20 CONTINUE, on K)
JB = NBND(J)
IF((J.LE.I).AND.(JB.GT.0)) GO TO 20
```

Because the bcc cell can contain vacancies, a TYPE(NA) test is necessary to avoid attempting to compute interactions with a missing atom. TYPE(NA) = 0 indicates that atom NA was removed to form a vacancy. All TYPE tests in the lines above are tests for vacancies. The CALL NABORS line establishes the list of atoms which can interact with atom I = MOVE(K). There are KMAX atom numbers in this list. The last two lines in the segment listed above concern the fact that the interaction between two atoms in the computational cell is handled differently from the interaction between an atom in the computational cell with an atom in the boundary. Atom J = NAB(K) is an interacting neighbor of atom I. If J is in the computational cell, i.e., JB = NBND(J) > 0, and if J < I, then the interaction between J and I was computed and stored when atom J processed prior to atom I. Hence, when atom J is in the computational cell and J < I, atom J is passed over in the computation of interactions for atom I.

```
C***  FORCE AND ENERGY COMPUTATION FOR A BCC COMPUTATIONAL
C       CELL WITH A RIGID BOUNDARY.
C
C***   CLEAR FORCE AND ENERGY TABLES
C
    5 DO 10 J = 1,NATOM
       FX(J) = 0.0
       FY(J) = 0.0
       FZ(J) = 0.0
       APE(J) = 0.0
       AKE(J) = 0.0
   10 CONTINUE
      PET = 0.0
      KET = 0.0
C
C***  COMPUTE FORCE ON EACH ATOM AT TIME T
C
      DO 25 K = 1,KMOVE
      I = IMOVE(K)
      IF(TYPE(I).LT.1) GO TO 25
      NS = SITE(I)
      CALL NABORS
      DO 20 K = 1,KMAX
      J = NAB(K)
      IF(TYPE(J).EQ.0) GO TO 20
      IF(J.EQ.I) GO TO 20
      JB = NBND(J)
      IF((J.LE.I).AND.(JB.GT.0)) GO TO 20
      DX = X(J) - X(I)
      DY = Y(J) - Y(I)
      DZ = Z(J) - Z(I)
      R2 = DX*DX + DY*DY + DZ*DZ
      IF(R2.GT.RCUT2) GO TO 20
      R = SQRT(R2)
      F = AF*(R*R) + BF*R + CF
      FR = F/R
      DFX = FR*DX
      DFY = FR*DY
      DFZ = FR*DZ
      FX(I) = FX(I) + DFX
      FY(I) = FY(I) + DFY
      FZ(I) = FZ(I) + DFZ
      IF(JB.LT.0) GO TO 40
      FX(J) = FX(J) - DFX
      FY(J) = FY(J) - DFY
      FZ(J) = FZ(J) - DFZ
C***  COMPUTE ATOM POTENTIAL ENERGY AT TIME T
   40 POTA = (A*(R**3) + B*(R**2) + C*R + D)/2.0
      APE(I) = APE(I) + POTA
      IF(JB.LT.0) GO TO 50
      APE(J) = APE(J) + POTA
   50 PET = PET + POTA + POTA
   20 CONTINUE
   25 CONTINUE
```

Fig. 14. FORCE and ENERGY computation for a bcc computational cell with a rigid boundary. (Figure continues on next page.)

```
C***   COMPUTE ACCELERATION AT TIME T
       DO 30 IM = 1,KMOVE
       I = MOVE(IM)
       ACX = 0.964867*FX(I)/MASS
       ACY = 0.964867*FY(I)/MASS
       ACZ = 0.964867*FZ(I)/MASS
C***   COMPUTE VELOCITY CHANGE DURING DT/2.0
       DVX = 0.5*ACX*DT
       DVY = 0.5*ACY*DT
       DVZ = 0.5*ACZ*DT
C***   COMPUTE VELOCITY AT TIME T
       VX(I) = VX(I) + DVX
       VY(I) = VY(I) + DVY
       VZ(I) = VZ(I) + DVZ
C***   COMPUTE ATOM KINETIC ENERGY AT TIME T
       AKE(I) = KEC*(VX(I)**2 + VY(I)**2 + VZ(I)**2)
C***   PRINT POSITION, VELOCITY, ENERGIES FOR TIME T
       CALL PRINT1
C***   COMPUTE VELOCITY AT TIME (T + DT/2.)
       VX(I) = VX(I) + DVX
       VY(I) = VY(I) + DVY
       VZ(I) = VZ(I) + DVZ
C***   COMPUTE COORDINATES AT TIME (T + DT)
       X(I) = X(I) + VX(I)*DT
       Y(I) = Y(I) + VY(I)*DT
       Z(I) = Z(I) + VZ(I)*DT
    30 CONTINUE
```

Fig. 14. continued

6. Monte Carlo

The Metropolis algorithm[10] for generating a thermal equilibrium ensemble of configuration states, at a specified temperature, for an array of atoms is used frequently in materials science work. We will look briefly at the application of this algorithm to a binary alloy which is depicted schematically in Fig. 15. The open circles represent one type of atom and the filled circles another type. It is traditional to call these two types A and B atoms. Here we assume that the atoms must occupy sites on a triangular lattice and that we know the interaction energies E(AA), E(BB), and E(AB) for the interaction of two A atoms, two B atoms, and an AB pair, respectively.

The algorithm proceeds as a series of repetitions of the following process:

1. Select two atoms I and J from the array at random.
2. Compute the interaction energy EB(I) for atom I and the interaction energy EB(J) for atom J.
3. Interchange the positions of atoms I and J.
4. Compute the interaction energy EA(I) for atom I and interaction energy EA(J) in their new positions.
5. Determine DE = (EA(I) + EA(J)) − (EB(I) + EB(J)).

Fig. 15. A binary alloy, in which open circles represent type-A atoms and solid circles represent type-B atoms.

6. If DE is negative, retain the new configuration given by the interchange.
7. If DE is positive, compute BF = EXP(−DE/kT) where k is Boltzmann's constant and T is the absolute temperature.
8. Select a random number R from the unit interval (0,1].
9. If R is less than BF, retain the new configuration given by the interchange; otherwise reject the new configuration.

This process generates an ensemble of configuration states. The ensemble average energy provides an estimate of the average energy for the system at temperature T.

Beeler(12) describes some of the early Monte Carlo simulations which were done between 1955 and 1970 on first- and second-generation mainframe computers. These simulations concerned equation-of-state development, some explorations of the Ising model, and thermal annealing of vacancies and interstitials in an irradiated metal. The equation-of-state simulations covered pressures up to four megabars and temperatures up to 12,000 K. The Ising model work pertained to order-disorder transformations and the thermodyamics of imperfect lattice gases. These early works are of interest to someone just starting to do Monte Carlo calculations because they were performed on small systems of up to 512 atoms, and they were first-time efforts in Monte Carlo craft. Because the systems were small, most of these early computer experiments can now be duplicated on a 64K personal computer, which is equivalent, in most respects, to a mainframe computer of the 1955–70 era.

7. Sources

This section gives a brief description of a few useful information sources which give more-extensive accounts on how to construct and use atomistic model computer experiments than could be given in this chapter. Each source deals with a particular aspect of atomistic model simulation.

Computational cell construction for bcc, fcc, and hcp crystals and the use of simple periodic boundary conditions is described in detail in Chapter 2 of *Radiation Effects Computer Experiments* by Beeler.[3] The theory and procedure for molecular dynamics simulation using periodic boundary condition computational cells which can change their size and/or shape as a simulation proceeds is given by Andersen,[4] Parrinello et al.,[5] and Hoover.[6] The best flexible boundary condition model for dislocation core simulation is generally considered to be FLEX II, devised by Hoagland et al.[7] The particular embedded atom version that was the concern of this chapter was developed by Daw and Baskes.[8] A lucid and detailed account of how to set up an embedded atom model which fits a given set of experimental data is given by Finnis and Sinclair.[9] Simple Monte Carlo routines for simulating defect migration are given by Beeler.[3] Metropolis et al.[10] devised the classic algorithm so often used in Monte Carlo simulations of phase changes and to obtain equations of state. Binder's book[11] is a standard reference for Monte Carlo applications in statistical mechanics.

References

1. J. B. Gibson, A. N. Goland, M. Milgram, and G. H. Vineyard, Phys. Rev. *120*, 1229 (1960).
2. R. H. Evans, Ph.D. Thesis, Department of Electrical Engineering, North Carolina State University, 1974.
3. J. R. Beeler, Jr., *Radiation Effects Computer Experiments*, North-Holland Publishing Co., Amsterdam, 1983.
4. H. C. Andersen, J. Chem. Phys. *72*, 2384 (1980).
5. M. Parrinello, A. Rahman, and P. Vashishta, Phys. Rev. Lett. *50*, 1073 (1983).
6. W. G. Hoover, Phys. Rev. *A31*, 1695 (1985).
7. R. G. Hoagland, J. P. Hirth, and P. C. Gehlen, Phil. Mag. *34*, 413 (1976).
8. M. S. Daw and M. I. Baskes, Phys. Rev. B *29*, 6443 (1984).
9. M. W. Finnis and J. E. Sinclair, Phil. Mag. A *50*, 45 (1984).
10. N. Metropolis, A. W. Rosenbluth, M. N. Rosenbluth, A. H. Teller, and E. Teller, J. Chem. Phys., *21*, 1087 (1953).
11. K. Binder, *Applications of the Monte Carlo Method in Statistical Physics*, Springer-Verlag, Berlin (1984).
12. J. R. Beeler, Jr., in *Advances in Materials Research*, Vol. 4, edited by H. Herman, pp. 295-476, John Wiley & Sons, New York, 1970.

APPENDIX 1
Subroutine SITNUM

Given the coordinates X(NA), Y(NA), Z(NA) of atom NA in hlc, subroutine SITNUM determines the site number NS for the bcc lattice site closest to atom NA. This appendix gives a listing of SITNUM:

```
C       SUBROUTINE SITNUM
        IX = INT(X)
        IY = INT(Y)
        IZ = INT(Z)
        MX = MOD(IX,2)
        MY = MOD(IY,2)
        MZ = MOD(IZ,2)
        MSUM = MX + MY + MZ
        IF((MSUM.EQ.0).OR.(MSUM .EQ.3)) GO TO 210
        IF(MSUM.EQ.1) GO TO 310
        IF(MSUM.EQ.2) GO TO 510
  210   S = ABS(X - FLOAT(IX))
        CALL SCOMP
        IF(S.GT.1.5) GO TO 260
        GO TO 710
C
  260   IX = IX + 1
        IY = IY + 1
        IZ = IZ + 1
        CALL SCOMP
        IF(S.GT.1.5) GO TO 750
        GO TO 710
C
  310   IF(MX.EQ.1) GO TO 340
        IF(MY.EQ.1) GO TO 350
        IF(MZ.EQ.1) GO TO 360
  340   IX = IX + 1
        GO TO 370
  350   IY = IY + 1
        GO TO 370
  360   IZ = IZ + 1
  370   CALL SCOMP
        IF(S.GT.1.5) GO TO 410
        GO TO 710
C
  410   IF(MX.EQ.1) GO TO 440
        IF(MY.EQ.1) GO TO 450
        IF(MZ.EQ.1) GO TO 460
  440   IX = IX - 1
        IY = IY + 1
        IZ = IZ + 1
        GO TO 470
  450   IY = IY - 1
        IX = IX + 1
        IZ = IZ + 1
  460   IZ = IZ - 1
        IX = IX + 1
        IY = IY + 1
```

```
  470 CALL SCOMP
      IF(S.GT.1.5) GO TO 750
      GO TO 710
C
  510 IF(MX.EQ.0) GO TO 540
      IF(MY.EQ.0) GO TO 550
      IF(MZ.EQ.0) GO TO 560
  540 IX = IX + 1
      GO TO 570
  550 IY = IY + 1
      GO TO 570
  560 IZ = IZ + 1
  570 CALL SCOMP
      IF(S.GT.1.5) GO TO 610
      GO TO 710
C
  610 IF(MX.EQ.0) GO TO 640
      IF(MY.EQ.0) GO TO 650
      IF(MZ.EQ.0) GO TO 660
  640 IX = IX -1
      IY = IY + 1
      IZ = IZ + 1
      GO TO 670
  650 IY = IY - 1
      IX = IX + 1
      IZ = IZ + 1
      GO TO 670
  660 IZ = IZ - 1
      IX = IX + 1
      IY = IY + 1
      CALL SCOMP
      IF(S.GT.1.5) GO TO 750
  710 GO TO 760
C
  750 PRINT "NO SOLUTION"
      GO TO END
C
  760 NP = IZ + 1
      M = MOD(NP,2)
      IF(M.LT.1) GO TO 770
      NC = (IY/2) + 1
      NR = (IX/2) + 1
      NS = ((NP - 1)/2)*NW + (NC - 1)*NX + NR
      RETURN
  770 NC = (IY + 1)/2
      NR = (IX + 1)/2
      NS = (NP/2)*NXY + ((NP/2) - 1)*NXYL + (NC - 1)*NXL + NR
      RETURN

C     SUBROUTINE SCOMP

      S = ABS(X - FLOAT(IX))
      S = S + ABS(Y - FLOAT(IY))
      S = S + ABS(Z - FLOAT(IZ))
      RETURN
```

APPENDIX 2
Embedded Atom Method

The embedded atom method (EAM) is a simple procedure for computing the electronic contribution to cohesion in transition metals. Unlike the band-structure approach, the EAM does not require a periodic array of atoms. Consequently it can be used for disordered alloys, surfaces, cracks, dislocation cores, grain boundaries, stacking faults, and liquid- and solid-phase interfaces. The EAM is based on the Hohenberg-Kohn theorem[(1)] which states that the energy contribution of an atom in an array of interacting atoms is a function of the local electron density, due to all other atoms, at the position of the atom concerned.

During the past eight years, "practical applications," based on the Hohenberg-Kohn theorem, have been developed which enable one to closely approximate the electronic contribution to cohesion in an arbitrary configuration of atoms with solid-state and liquid-state densities, and at surfaces, both external and internal. There are several names used for this type of approach. We shall adopt the name "embedded atom method" for this general empirical approach to metallic cohesion, following Daw and Baskes,[(3)] who pioneered its systematic use in materials science and engineering.

Prior to the invention of the EAM, atom pair potential energy functions were used to describe the energetics of defects in metals. The pair potential approach proved useful in generating qualitatively correct possible local configurations of atoms in the cores of extended defects and in disordered systems. However, the pair potential approach had several drawbacks, each one following from the fact that a central force pair potential always gives C(12) = C(44) for cubic systems, where C(IJ) is an elastic constant. To circumvent this difficulty, a so-called Cauchy Pressure, P = (1/2)(C(12) − C(44)) was applied to the atomistic system to make the elastic constants of the model equal to those of the metal being simulated. Johnson[(5,6)] shows how to apply this remedy in a computer simulation of surface relaxation for fcc and bcc metals.

We will outline the use of the EAM by showing how to apply it to the tiny five-atom system defined in Fig. A1. Fig. A1a shows the complete system, which is composed of two type-A atoms and three type-B atoms; the large circles represent type-A atoms and the smaller circles type-B atoms. The electron density contribution of each atom at a particular position is estimated from the free-atom electron wave functions for the particular type of atom concerned. These can be found in the Atomic Data and Nuclear Data Tables,[(4)] for example. Having found the electron density p(1) at the position of Atom 1, in Fig. A1b, due collectively to atoms 2, 3, 4, and 5, the

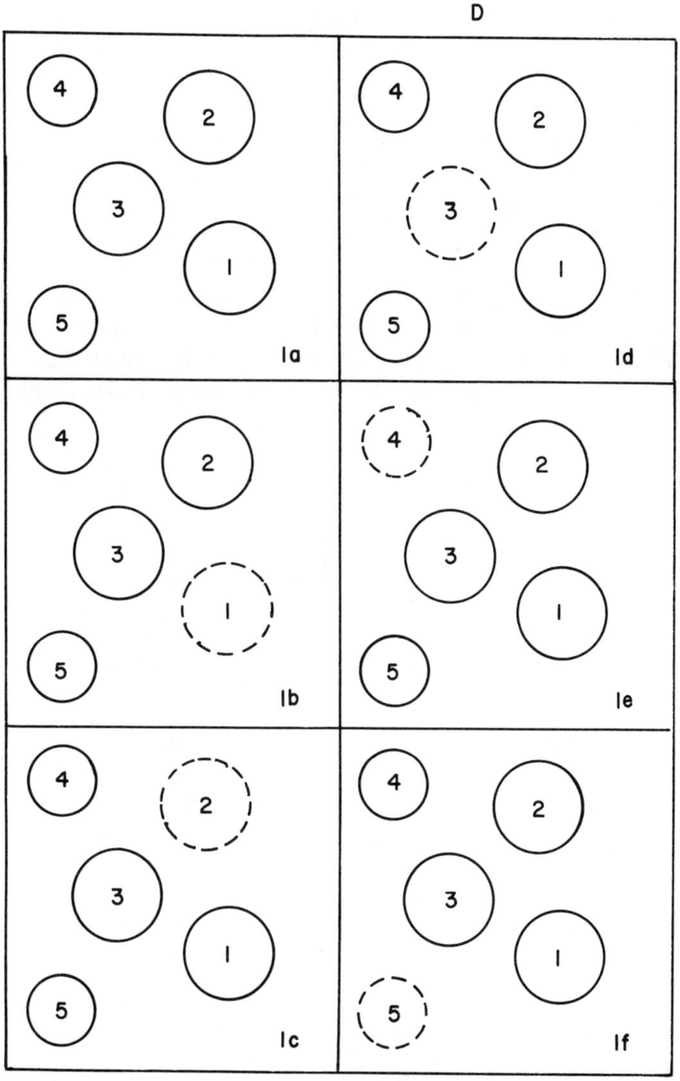

Fig. A1. The embedded atom method.

embedding energy contribution, E(1), for atom 1 would be given by the embedded atom energy function G(T,p). The index T indicates the type of atom and p is the electron density into which the atom is embedded. In this instance, T = A, since atom 1 is a type-A atom, and p = p(1), the total electron density at atom 1 due to all other atoms in the assembly. The embedding energies E(2), E(3), E(4), and E(5) would be found by repeating the procedure used to find E(1). The total embedding or N-body interaction energy EN for the assembly would be the sum of the E(n); that is,

$$EN = E(1) + E(2) + E(3) + E(4) + E(5)$$

Figure A2 shows the form of the embedding energy function G(T,p) for two rare gas atoms, hydrogen and the metal nickel. In practice, these functions are developed by fitting to experimental data. There has been limited success in deriving them from first principles.

In addition to the embedding energy, which usually represents bonding, there is a screened coulomb core-core repulsive pair potential interaction VP(ZI,ZJ,r) which depends upon the atomic numbers ZI and ZJ of the two atoms involved and their separation distance r. This contribution is a central force potential. The total pair potential interaction VP is the sum of the

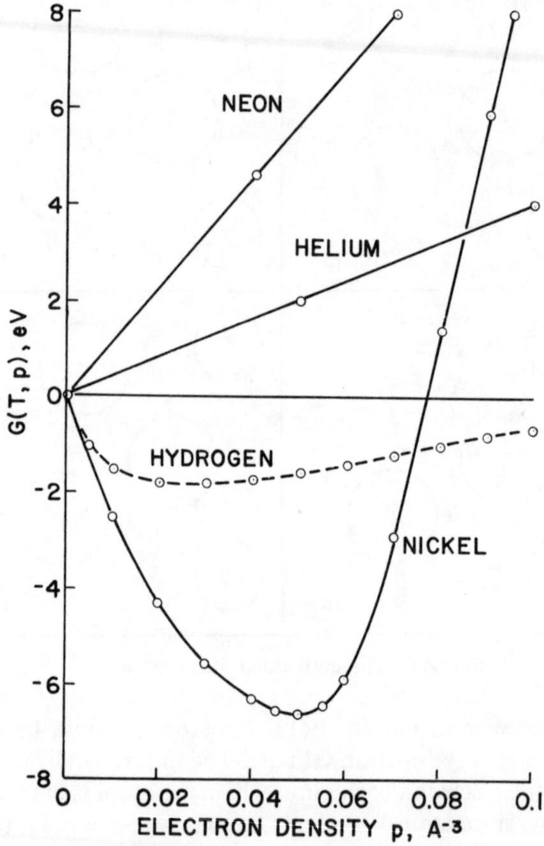

Fig. A2. The embedding energy function G(T,p) for two rare gas atoms, hydrogen and nickel.

individual atom pair interactions in the assembly. The total energy for the assembly, ET, is

$$ET = EN + VP$$

which is the sum of the N-body interaction potential energy and the pair potential interaction energy.

The X(M) component of the force F(I,X(M)) on the I-th atom, arising from EN, is given by

$$F(I,X(M)) = -SUM'[((G'(I) + G'(J)) * (dp(I)/dR) * CX(M)]$$

where SUM′ denotes summation over atoms J not equal to I, CX(M) = R(I,J,M)/R(I,J), R(I,J,M) is the M-th component of R(I,J), G′(I) = (dG/dp) at atom I, and G′(J) = (dG/dp) at atom J. The term containing G′(I) expresses the change in energy due to a change in electron density at atom site I as atom I moves. The term containing G′(J) expresses the change in energy as the result of the change in electron density at all other sites as atom I moves. In all embedded atom model calculations done thus far, the atom-atom interaction range has been restricted to first and second neighbors of atom I. Finnis and Sinclair[2] give a detailed account of how to set up an EAM model from a given set of experimental data.

References to Appendix 2

1. H. Hohenberg and W. Kohn, Phys. Rev. *136*, B864 (1964).
2. M. W. Finnis and J. E. Sinclair, Phil. Mag. A *50*, 45 (1984).
3. M. S. Daw and M. I. Baskes, Phys. Rev. B *29*, 6443 (1984).
4. E. Clementi and C. Roetti, Atomic Data and Nuclear Data Tables *14*, 177-478 (1974).
5. R. A. Johnson, Phys. Rev. *27B*, 3861 (1983).
6. R. A. Johnson, Surface Science *151*, 311 (1985).

4

Thermodynamics of Metallic Solids and Fluids From Molecular Dynamics

GALEN K. STRAUB
Los Alamos National Laboratory
Los Alamos, NM 87544

Abstract

The technique of computer simulation of the molecular dynamics in metallic systems to calculate thermodynamic properties is discussed. The nature of a metal as determined by its electronic structure is used to determine the total adiabatic potential. The effective screened ion-ion interaction can then be used in a molecular dynamics simulation. The method for the construction of a molecular dynamics ensemble, its relation to the canonical ensemble, and the definition of thermodynamic functions from the Helmholtz free energy is given. The method for the analysis of the molecular dynamics results from quasiharmonic lattice dynamics, and the decomposition in terms of harmonic and anharmonic contributions is given for solids. For fluid-phase metals, procedures for calculating the thermodynamics and determining the constant of entropy are presented. The solid-fluid phase boundary as a function of pressure and temperature is determined using the results of molecular dynamics. Throughout, examples and results for metallic sodium are used. The treatment of the transition metal electronic d-states in terms of an effective pair-wise interaction is also discussed.

1. Introduction

Computer simulation offers a very powerful technique for highly accurate calculations of the thermodynamic functions in metallic systems. We will present some of the reasons for using molecular dynamics (MD) and methodology for studying the thermodynamic properties of metals. This will include discussions on the nature of a metallic system, statistical mechanics, the construction of ensembles from MD, and the definitions of thermodynamic functions. MD is not the only way one can obtain the thermodynamic properties of a metal, but at high temperatures in the solid and in the fluid phase, MD allows a straightforward, highly accurate solution. In this chapter, we will try to elucidate the properties of an MD system and show how the fact that we are in a metal has a dominating effect on the procedures that we use. The path to thermodynamics is by way of statistical mechanics. This means that we must know how to take our MD results and interpret them as an MD ensemble. Following that, we must establish the link between an MD ensemble and a canonical ensemble, and finally, the definitions of thermodynamics in terms of a canonical ensemble.

An MD simulation consists of a system of N particles interacting with one another via some force law in a computational cell, a set of initial positions and velocities, a set of boundary conditions on the computational cell, and the time-integrated evolution of the equations of motion. We shall consider only the case of periodic boundary conditions and fixed volume for the computational cell. The total energy is assumed to be a constant of the motion. The primary product of the MD simulation is a set of positions $\{r(t)\}$ and a set of velocities $\{v(t)\}$ as a function of time. The MD system is a mechanical one that solves for the time evolution of the *classical* equations of motion. We can use the sets of positions and velocities to construct other mechanical quantities such as the total kinetic energy and the total potential energy. In Fig. 1, we give the total kinetic energy as a function of time for two different MD calculations.

Both of the calculations shown in Fig. 1 consisted of about 1000 particles whose initial positions were those of a bcc lattice. The initial velocities were chosen randomly and with a different amount of total energy for each calculation. The curve labeled "fluid" is for an MD system whose final state was in the fluid phase, and the curve labeled "solid" had a final state in the solid phase. We start the calculation at time = 0 and wait for the system to come to *mechanical* equilibrium. The solid system reaches equilibrium around time = 500 (arbitrary units), while the fluid calculation does not reach equilibrium until time = 700. These two curves represent the type of basic information that one gets directly from MD. The first step is to construct a suitable ensemble to do statistical mechanics; we shall call this ensemble the MD ensemble. In order to construct statistical ensembles we must wait until the system is in mechanical equilibrium and then form the appropriate averages. The procedure for choosing the start time for averaging will be discussed in more detail below.

In statistical mechanics, an ensemble is a very large number of systems in contact with the appropriate reservoir and with constraints for each system. We will be concerned with two different ensembles: (1) the MD ensemble that we will construct from time averages of dynamical variables such as that shown in Fig. 1; and (2) the canonical ensemble that provides the direct link to defining thermodynamics from statistical mechanics. The canonical ensemble constrains each system to a constant number of particles N at fixed

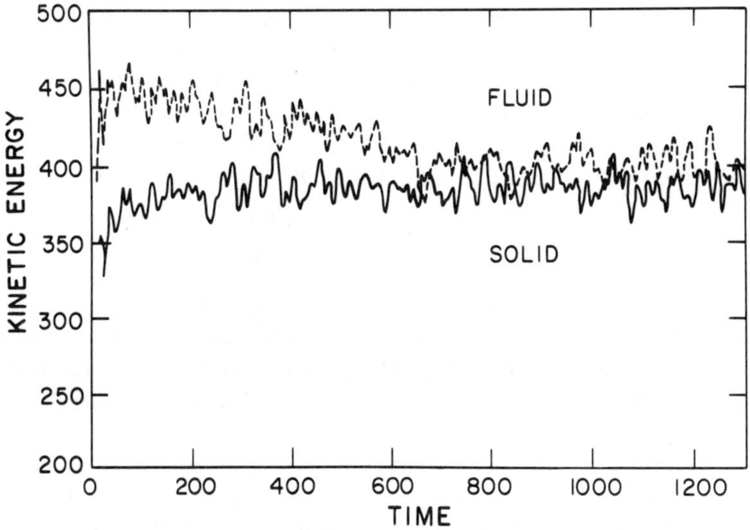

Fig. 1. Kinetic energy vs. time for an MD calculation. The curve labeled "fluid" is for an MD system whose final state was in the fluid phase, and the curve labeled "solid" had a final state in the solid phase. Both calculations were started in a bcc lattice with about 1000 particles, but with different initial kinetic energies. The solid system reaches equilibrium around time = 500 (arbitrary units), while the fluid calculation does not reach equilibrium until time = 700. Averaging for thermodynamic quantities must wait until the system is in equilibrium.

volume V in contact with a reservoir at constant temperature T. The MD ensemble consists of a large number of systems having constant N and V with the total energy held constant. The reservoir for each MD system is an angular momentum reservoir. The MD ensemble is *not* a microcanonical ensemble, since the microcanonical ensemble has no restriction on the total angular momentum.

We want to define thermodynamic functions in the MD ensemble in terms of averages of the dynamic variables as one does for the canonical ensemble, but here we run into a problem. The averages of the same dynamical variable will be different in different ensembles. For quantities that only involve averages of dynamical variables, the resulting difference between ensembles is of relative order 1/N. Thus, for instance, we can calculate the temperature T from an MD ensemble by averaging the kinetic energy, but this average will differ from averages of a canonical ensemble by

$$\langle \text{K.E.} \rangle_{\text{MD}} = \langle \text{K.E.} \rangle_{\text{canonical}} + O(1/N)$$

where O(1/N) signifies terms of relative order 1/N. This poses no problem at all since we can easily make $N \sim 10^3$. On the other hand, if we are interested in calculating averages of the fluctuation of a dynamical variable, then the corrections are of O(1). The calculation of the elastic constants involves averages over the fluctuations of the interparticle potential energy, and correcting the MD averages to the canonical ensemble averages is essential. These last statements may be summarized by the following: let $A \equiv$ dynamical variable, $\langle A \rangle \equiv$ average, and the fluctuation of A, $\delta A \equiv A - \langle A \rangle$, then

$$\langle A \rangle_{MD} = \langle A \rangle_{canonical} + O(1/N)$$
$$\langle \delta A \rangle_{MD} = \langle \delta A \rangle_{canonical} + O(1)$$

For the present discussion, we will not have to deal with averages of fluctuations. Formulas for the corrections of fluctuations between MD and canonical ensembles have been derived in detail in Ref 1 and 2. The calculation of the elastic constants from the MD ensemble and the ensemble corrections are given in Ref 3 and 4.

In metals, the net force on any particular ion is the result of screened long-range coulomb interactions. A metal consists of a coupled system of conduction electrons and charged ions. The conduction electrons may be considered to remain in their ground state and are responsible for the screening of the ionic coulomb potential. The Born-Oppenheimer adiabatic approximation may be used to determine the ion-ion interaction.[(5)] The adiabatic approximation allows us to solve the electronic structure problem for the ions at some set of fixed positions. The positions are arbitrary, but not moving. The adiabatic potential is the sum of the ion-ion interaction plus the electronic energy, where both are functions of the ionic positions. The vibrational energy of the ions may then be solved using the adiabatic potential. The conduction electrons contribute in two ways. First, as an electron gas, their kinetic, exchange, and correlation energy may be written as a function of volume only. Second, they act to screen the ions from one another and reduce the range of the coulomb forces. A useful way of looking at the forces between ions is that the screened interaction is a direct function of the electron density. As a result, the forces between a system of ions are an *explicit* function of the volume.

The problem to be addressed for the simulation of a metal by MD is to accurately calculate the vibrational energies of the ions, which means that we must properly represent the net many-body forces between the ions for integration of the equations of motion. The adiabatic approximation allows us to write the potential energy as the sum of two terms, a volume-dependent term resulting mainly from the conduction electrons, and a many-body ion term that is an explicit function of volume. The result of using the many-body ion term alone would be an unstable system, since in general neighboring ions do not lie at the minimum of a potential well. In MD, we can use the periodic boundary conditions to confine the system to a fixed volume.

Both terms are essential for the description of a metallic material. Because of the importance of the volume dependence in the potential energy, Section 2 is devoted to a more detailed discussion of this point.

The procedure for applying statistical mechanics to an MD simulation giving the route from a mechanical theory to a thermodynamic one is given in Section 3. This involves some formal definitions of thermodynamic quantities in terms of statistical averages as mentioned above. In Section 4, the calculation of the total energy and pressure from MD will be given. Then, the analysis of these quantities will be given whereby one may construct the Helmholtz free energy throughout phase space. In the solid we make use of the theory of lattice dynamics.[(6)] Lattice dynamics gives the correct answers for thermodynamics in the low-temperature quantum-mechanical regime where MD, because of its restriction to classical equations of motion, fails. Below the Debye temperature Θ_0 (Al = 428 K, Cu = 345 K, Na = 153 K) quantum-mechanical effects may not be neglected as illustrated by experimental data for the heat capacity. Above Θ_0, vibrational anharmonicity becomes important and the decomposition of the energy and pressure into harmonic and anharmonic contributions will be given. For the fluid phase, the classical equations of MD are adequate, although the electron gas energy requires a high-temperature correction. Unfortunately, one does not have a reference theory, such as harmonic lattice dynamics for the solid, and the fluid-phase thermodynamics must be treated using numerical fits of the MD results.

Another difficulty arises in the treatment of thermodynamics of the fluid phase, namely, the determination of the constant in the entropy. In a classical description the value of the entropy is not unique, but only defined to within an arbitrary constant. This problem is handled in Section 4 by making a high-temperature cluster expansion of the free energy (which contains the entropy constant) and evaluating the free energy at one point in phase space with the aid of MD. The free energy can then be integrated down in temperature to the region of interest, in our case so that we may construct the solid-fluid phase boundary. A problem with the constant of entropy for the solid does not arise since we treat the solid phase as a quantum-mechanical system at low temperatures with MD as an approximate description appropriate only above the Debye temperature. In quantum mechanics, the constant of the entropy may be uniquely determined.

Throughout the text, the calculational methods will be illustrated for the example of solid and fluid metallic sodium. The same methods may be applied to transition metals, but the effective ion-ion interaction needed for MD is more complicated. In Appendix 1 at the end of this chapter, a simple treatment of metals containing d-electrons and formulas for the resulting interactions are given.

2. Total Adiabatic Potential for a Metal

The Born-Oppenheimer adiabatic approximation for a coupled system of electrons and ions allows us to write the potential energy of a metal in terms of the ion coordinates, but with the electron coordinates as explicit parameters.[5] This decomposition is possible because the vibrational motion of the ions has a characteristic time scale of about 10^{-14} s, while electronic time scales range from about 10^{-17} to 10^{-18} s. In terms of a physical picture, this means that for any configuration or change in configuration of the ions, the electrons respond so quickly that they remain in an energy ground state. The term "adiabatic" refers to this situation.

We take as our model a system of N ions and NZ electrons in a volume V. In the adiabatic approximation, the total energy is the sum of the kinetic energies of all the ions, plus the total adiabatic potential Φ. We pick the zero of Φ to correspond to having neutral atoms infinitely separated, and I_z is the ionization energy. Φ consists of an electron gas term E_G, the ion-ion coulomb interaction $\Omega_{i\text{-}i}$, the electron-ion interaction $\Omega_{e\text{-}i}$, and the electron-electron interaction $\Omega_{e\text{-}e}$. The electron-ion and electron-electron terms are combined in the framework of pseudopotential theory with the result[7,8]

$$\Phi = NI_z + \Omega(V) + \sum \phi(r;V) \qquad (1)$$

$\Omega(V)$ is a volume-dependent term and $\sum \phi(r;V)$ is the summation over the many-body *effective* ion-ion interaction. To this point, there is no restriction to two-body interactions with the only approximation being the use of perturbation theory to solve for the electron energies. In practice, we truncate the perturbation expansion at second order, rearrange terms, and write ϕ as an effective two-body interaction between ions that includes the effects of the electron-ion interactions and electron-electron screening interactions. (Third order in the perturbation expansion gives three-body interactions, etc.) In any case, $\phi(r;V)$ is an explicit function of the volume as well as the interparticle separation r.

Figure 2 is a plot of the total effective pair potential for sodium calculated for the point-ion pseudopotential. The volume dependence of the pair interaction is clearly apparent. At the zero-pressure volume in a bcc lattice, the first neighbors lie on the repulsive side of the minimum and the second neighbors on the attractive side. ϕ is also very long range and oscillates about zero (they are not visible on the scale of Fig. 2; see Ref 9 and 10). These long-range oscillations are the result of the electron screening and are called Friedel oscillations. We need not be concerned with them in the present discussion, but in practice, a total of ten shells of neighbors (~200) was routinely included in $\sum \phi$.

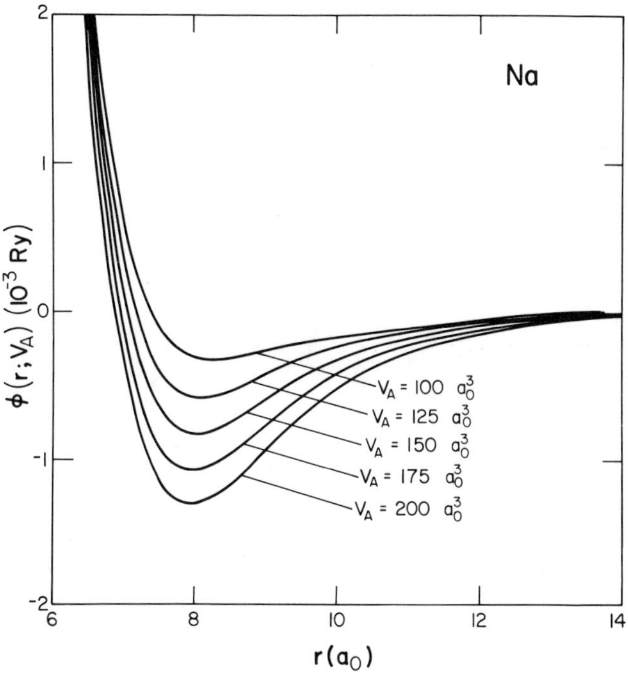

Fig. 2. Total effective pair potential for sodium calculated from the modified point-ion pseudopotential. Our system consists of N ions and NZ electrons in a volume V, and is described by an adiabatic potential Φ evaluated in the pseudopotential perturbation formulation. Φ consists of two parts: a volume-dependent term $\Omega(V)$, and a sum over all distinct pairs of ions of the effective ion-ion potential $\phi(r;V)$. ϕ includes the effects of the electron-ion interactions and the many-body electron-electron screening interactions. The potential energy is in units of rydbergs (Ry) and the separation is in units of Bohr radii (a_0).

The contributions to the total adiabatic potential Φ as a function of volume are plotted in Fig. 3. Nearly 90% of the total potential energy comes from the volume-dependent terms in Ω, while only 10% comes from the $\sum \phi(r;V)$. The energy is a minimum at the experimental value of the volume at zero pressure. Thus, to first order in the pseudopotential perturbation expansion, there is no need for a pair potential at all (we still need a pseudopotential) and quite reasonable estimates and trends in simple and transition metals may be made.[8,11] However, for thermodynamic properties as a function of temperature we require the vibrational properties of the ions, and this information is contained in $\sum \phi(r;V)$ where the summation is over the sets of positions of the ion $\{r(t)\}$.

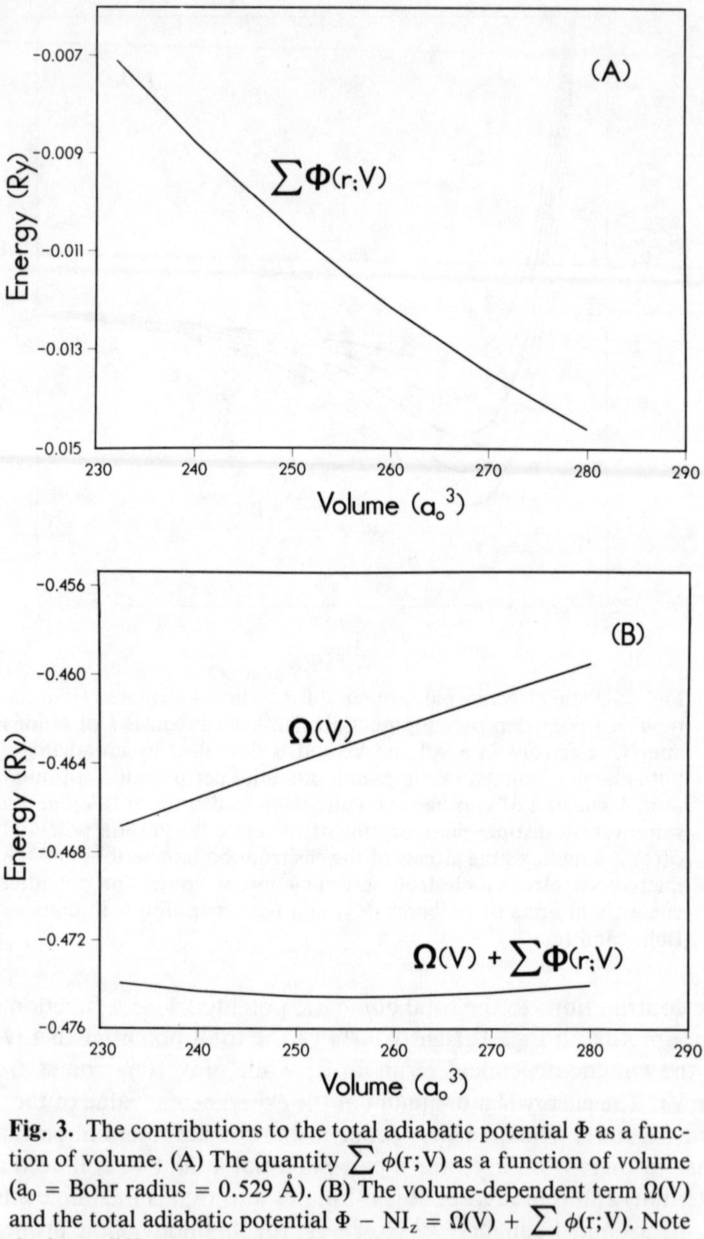

Fig. 3. The contributions to the total adiabatic potential Φ as a function of volume. (A) The quantity $\sum \phi(r;V)$ as a function of volume (a_0 = Bohr radius = 0.529 Å). (B) The volume-dependent term $\Omega(V)$ and the total adiabatic potential $\Phi - NI_z = \Omega(V) + \sum \phi(r;V)$. Note the large difference in the scales of panels (A) and (B). Nearly 90% of the total energy comes from the volume-dependent terms alone. The energy is a minimum at the experimental value of the volume at zero pressure.

For transition metals, Eq 1 holds as it is written, except now ϕ contains terms that describe the d-d and d-conduction electron interactions. At first thought, one might expect that the addition of the d-electrons might make such an approach much too complicated. In fact, the treatment of the transition metals is as straightforward as for the simple metals. This is possible because the cohesive energy in transition metals is primarily due to the presence of the d-electrons. One of the most complex aspects of the treatment of the simple metals is the electron screening among the conduction electrons. Because of its reduced relative importance in the transition metals, we may further approximate this screening to simplify the theory, but without a great overall loss in accuracy.[(12)]

The presence of the d-electrons in the transition metals causes the formation of d-bands that are coupled to the conduction electron bands. The bonding results from the partially filled d-bands where the average electron energy is lower than that of the free atom. There is also a repulsive interaction among the d-orbitals on neighboring ion sites that causes a shift upward of the entire band. The banding and repulsive potentials have been calculated from tight-binding theory,[(12)] with parameters that may be determined entirely from a first-principles theory of the electronic structure.[(13)] Further details of the theory and the resulting potentials for nickel are given in Appendix 1.

3. Statistical Mechanics of an MD Simulation

We shall pick as our starting point for discussion the end of an MD simulation where we have some set of dynamical variables A, B, . . . , written as functions of the sets of positions and velocities, {r(t)} and {v(t)}, respectively. We shall only concern ourselves with systems that are in equilibrium. The initial portion of an MD run has not reached equilibrium and must be allowed to do so. The choice of the starting point for equilibrium is discussed in Ref 14 and we will assume that this has been done properly. Briefly, equilibrium for an MD system is achieved when the probability distributions for all the dynamical variables of interest are independent of time. A more useful definition is that an MD system is in equilibrium when the mean and variance of a dynamical variable are independent of time. We may also make a guess of the equilibrium starting time by a graph such as that in Fig. 1, but this is a subjective choice.

The total kinetic energy shown in Fig. 1 is one example of a dynamical quantity that is a function of {v(t)}. The potential energy would be an example of a dynamical quantity that is a function of only {r(t)}. In the constant-energy MD simulation we are considering, the potential energy could be obtained by subtracting the kinetic energy from the total energy and would have exactly the same shape, but with a change in sign.

An ensemble is a very large number of systems in contact with the appropriate reservoir and with constraints for each system. Strictly speaking, in order to create an MD ensemble we would have to run a great many MD simulations with each calculation being one system. We may do this from a single MD run by invoking the quasiergodic hypothesis.[(2)] The quasiergodic hypothesis states that the time average of a dynamical quantity averaged over a sufficiently long period of time, such that the average is independent of the averaging time, is the same as the ensemble average.

Thermodynamic functions are defined in terms of the canonical ensemble.[(15)] The canonical ensemble consists of N particles in a volume V where V is so large that boundary conditions and surface effects can be neglected. All systems in the ensemble are in contact with a reservoir at constant temperature T. For the MD ensemble we have constant N and V, but neither of these quantities will be large enough to neglect the surfaces or boundary conditions. The reservoir for each MD system is an angular momentum reservoir.[(1)] The path to equilibrium thermodynamics from MD consists of defining thermodynamic functions as averages over a canonical ensemble, relating the canonical ensemble averages to MD ensemble averages, and creating the MD ensemble by taking the time average of a single MD system.

3.1. Thermodynamic Functions in the Classical Canonical Ensemble

We will follow the approach of Wallace[(2)] for the discussion of the canonical ensemble and the definition of thermodynamic functions. For the canonical ensemble, we have constant N, V, and a reservoir that keeps each system at constant T. The canonical weight function is

$$W(\beta) \propto e^{-\beta H} \tag{2}$$

where $\beta = 1/kT$ and k is Boltzmann's constant. H is the system Hamiltonian such that $H = K.E. + \Phi$ and K.E. is the total system kinetic energy. The classical canonical partition function is

$$Z(\beta) = 1/N!h^{3N} \int \ldots \int e^{-\beta H} \Pi_k \, dx_k \, dp_k \tag{3}$$

where p_k is the momentum coordinate of particle k. For a fluid, each positional coordinate x_k is integrated over the entire volume. In a crystal each particle coordinate may be restricted to the volume surrounding a lattice site, and then Z must be multiplied by N! to count the particle permutations. The partition function may be written in the form

$$Z(\beta) = (V^N/N!\Lambda^{3N})Q(\beta) \tag{4}$$

where $\Lambda^2 = (2\pi\hbar^2/mkT)^{1/2}$ is the thermal de Broglie wavelength of the particles. The configuration integral $Q(\beta)$ is

$$Q(\beta) = 1/V^N \int \ldots \int e^{-\beta\Phi} \Pi_k \, dx_k \tag{5}$$

Thermodynamic quantities can be connected to the statistical mechanics of the ensemble through a single equation involving the Helmholtz free energy F:

$$F = -kT \ln Z(\beta) \tag{6}$$

and all remaining thermodynamic functions are obtained from F. The entropy S is

$$S = -[\partial F/\partial T]_V \tag{7}$$

the internal energy U is

$$U = F - TS \tag{8}$$

and the pressure P is

$$P = -[\partial F/\partial V]_T \tag{9}$$

The functions S, U, and P can all be shown to be given by ensemble averages. It is important to note that β is a parameter that identifies a particular canonical ensemble (along with N,V). If we note the canonical ensemble average of a dynamical quantity A by $\langle A|\beta\rangle$, and define $\vec{b} = -\beta\vec{u}$ where $\vec{u}$ is the mean center-of-mass velocity, then Ref 1 gives the following results:

$$\langle H|\beta\rangle = U + Nm\vec{b}^2/2\beta^2 \tag{10}$$

$$\langle K.E.|\beta\rangle = \frac{3}{2} kT + Nm\vec{u}^2/2 \tag{11}$$

For a metal with a Hamiltonian H = K.E. + Φ we rewrite the adiabatic potential assuming pair interactions only. Then,

$$\Phi = NI_z + \Omega(V) + \frac{1}{2} \sum_{i,j} \phi(r_{ij};V) \ , \quad i \neq j \tag{12}$$

The ensemble average of the pressure is given by

$$P = -d\Omega/dV + NkT/V - \langle Y|\beta\rangle/V \tag{13}$$

where Y is the generalized virial function given by

$$Y = \frac{1}{2} \sum_{i,j} \left(V\partial\phi/\partial V + \frac{1}{3} r_{ij}\partial\phi/\partial r_{ij} \right) \tag{14}$$

For the dynamical variable A, the fluctuation of A is given by $\delta A = A - \langle A\rangle$ where the average is over a particular ensemble. Canonical ensemble averages of the fluctuations of the potential and kinetic energies are

$$\langle \delta\Phi\delta K.E. | \beta \rangle = 0 \quad (15)$$

$$\langle (\delta K.E.)^2 | \beta \rangle = \frac{3}{2} N(kT)^2 + Nm\vec{b}^2/\beta^3 \quad (16)$$

$$\langle (\delta\Phi)^2 | \beta \rangle = N(kT)^2 \left(c - \frac{3}{2} \right) \quad (17)$$

$c = 1/Nk[\partial U/\partial T]_V$ is the heat capacity per particle at constant volume.

3.2. Ensemble Corrections

We are able to calculate dynamical quantities using MD and form an MD ensemble from the time averages using the quasiergodic hypothesis. We have defined the thermodynamic functions in terms of a different ensemble, namely, the canonical ensemble. These ensembles are different and averages of some dynamical quantity for the two ensembles will be different. Since we want to obtain highly accurate thermodynamic functions (in the canonical ensemble), but we have dynamical quantities calculated in the MD ensemble, we will need to make corrections between the two. These corrections are given in detail in Ref 1. Here we will merely state the results.

For any dynamical quantity involving only averages of the ensemble, then

$$\langle A \rangle_{MD} = \langle A | \beta \rangle + O(1/N) \quad (18)$$

The term O(1/N) refers to corrections of order 1/N where for an N = 100 MD calculation, the correction would be 1%, for N = 1000, 0.1%, etc. The kinetic energy averaged over the canonical ensemble is 3/2 NkT (for a mean center-of-mass velocity equal to zero). Then,

$$\langle K.E. \rangle_{MD} = \frac{3}{2} NkT + O(1/N) \quad (19)$$

$$U_{MD} = \langle K.E. \rangle_{MD} + \langle \Phi \rangle_{MD} + O(1/N) \quad (20)$$

$$P_{MD} = -d\Omega/dV + \langle 2/3V\ K.E. - Y \rangle_{MD} + O(1/N) \quad (21)$$

The formula relating the pressure in the MD and canonical ensembles is

$$P_{MD} = P - kT/V \left(1 - \frac{3}{2} \gamma + T/2\ \partial\gamma/\partial T \right) \quad (22)$$

$\gamma = V[\partial P/\partial U]_V$ is the Grüneisen parameter. The last term in Eq. 22 is of relative order 1/N with respect to P.

For fluctuations of dynamical quantities, the ensemble corrections are of O(1) and must be included. For instance,

$$\langle (\delta K.E.)^2 \rangle_{MD} = 3N/2(kT)^2(1 - 3/2c) \quad (23)$$

$$\langle \delta Y \delta\Phi \rangle_{MD} = 3N/2(kT)^2(1/c - \gamma) \quad (24)$$

The calculation of the elastic constants from MD requires the evaluation of fluctuations such as in Eq 23–24. The ensemble corrections have been worked out in Ref 3.

4. Calculation of Thermodynamic Quantities

MD simulations can give "exact"[16] solutions to the problem of solving for the vibrational motion of a system of point particles interacting with some potential that depends on the relative positions of all the particles. In the last section, the construction of the MD statistical ensemble was discussed, as well as the definition of thermodynamic functions in the classical canonical ensemble. Relationships between the MD and canonical ensemble were given along with formulas for the calculation of ensemble corrections. We are now ready to proceed with a more detailed analysis of the MD results.

The strength of MD simulations lies in their ability to produce "exact" answers that can augment or extend other theoretical approaches. This situation is well exemplified for the study of thermodynamics in solids where the theory of lattice dynamics in the quasiharmonic approximation gives amazingly accurate results at low to moderate temperatures, generally up to the Debye temperature. In fact, we shall see that since MD is solving classical equations of motion, it will give *wrong* answers below the Debye temperature. Above the Debye temperature, the vibrational behavior of the ions can be adequately described as classical point particles and then MD can give extremely reliable and useful information. In the solid phase, we will use the theory of lattice dynamics to analyze the MD results. The necessary information needed from lattice dynamics will be presented in Appendix 2 at the end of this chapter.

4.1. The Internal Energy and Pressure

From an equilibrium MD simulation at constant volume, we are able to calculate the total energy and pressure by making time averages (indicated by $\langle\,\rangle_{MD}$) of the kinetic energy, the potential energy, and the generalized virial. The total internal energy is given by

$$U_{MD} = \langle K.E.\rangle_{MD} + \langle \Phi \rangle_{MD} \tag{25}$$

where from Eq 1 we may write out the last term for ϕ as a two-body central potential ($r_{ij} = r_i - r_j$):

$$\langle \Phi \rangle_{MD} = NI_z + \Omega(V) + \left\langle \frac{1}{2} \sum_{i,j} \phi(r_{ij};V) \right\rangle_{MD}, \quad i \neq j \tag{26}$$

The total pressure is

$$P_{MD} = -d\Omega/dV + 2/3V\langle K.E.\rangle_{MD} - \langle Y\rangle_{MD}/V \tag{27}$$

Y is the generalized virial for a volume-dependent potential and is

$$Y = \frac{1}{2}\sum_{i,j}\left(V\partial\phi/\partial V + \frac{1}{3}r_{ij}\partial\phi/\partial r_{ij}\right) \tag{28}$$

From Eq 19 we can calculate the temperature and we have

$$kT = 2/3N\langle K.E.\rangle_{MD} = M/3N\left\langle\sum_i v_i^2\right\rangle_{MD} \tag{29}$$

In Fig. 4, the results for the total internal energy and the total pressure at several different values of the initial energy are given as the solid points for three different volumes. The temperature was determined at the end of each run using Eq 29. The break in each curve separates runs whose final equilibrium state was in the solid phase from those in the fluid phase. All calculations were started at time zero with all particles on bcc lattice sites and a random distribution of velocities. The time evolution of the kinetic energy for two such calculations is shown in Fig. 1.

4.2. Free Energy

In Section 3 the Helmholtz free energy F was used to define all the thermodynamic functions. We shall now write F in terms of a harmonic part F_H and an anharmonic part F_A. F_H can be calculated in the high-temperature limit of quasiharmonic lattice dynamics theory as given in Appendix B. The result is

$$F = F_H + F_A \tag{30}$$

$$F = \Phi_0 - 3NkT\ln(T/\Theta) + F_A \tag{31}$$

where

$$\ln(k\Theta) = \langle\ln\hbar\omega\rangle_{BZ} \tag{32}$$

$\langle - \rangle_{BZ}$ is an average of the Brillouin zone of the phonon frequencies calculated for the central potential $\phi(r;V)$. We may calculate the other thermodynamic functions from Eq 30–32 and for the internal energy U, we have

$$U = F - TS = F - T[\partial F/\partial T]_V \tag{33}$$

$$U = \Phi_0 + 3NkT + U_A \tag{34}$$

where the harmonic part is $U_H = \Phi_0 + 3NkT$ and the anharmonic part is

$$U_A = F_A - T[\partial F_A/\partial T]_V \tag{35}$$

The pressure may be similarly calculated from F and written as harmonic and anharmonic contributions.

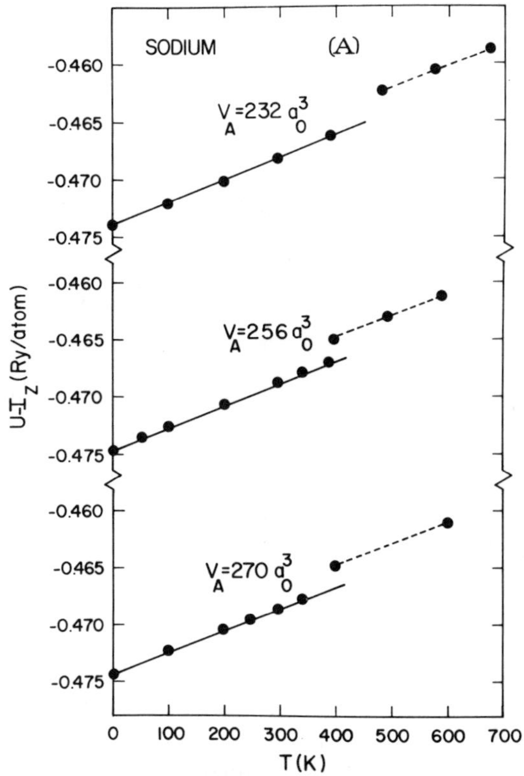

Fig. 4. MD results for the total internal energy and the total pressure. The points are the results of MD calculations at three different volumes. (A) The total internal energy per atom U minus the ionization energy I_z vs. temperature from the MD calculations. The solid lines are the quasiharmonic part of $U = \Phi_0 + 3NkT + U_A$; that is, they express the quantity $\Phi_0 - NI_z + 3NkT$. Φ_0 is Φ evaluated at $T = 0$. (B) The total pressure P vs. temperature from the MD calculations. The solid lines are the quasiharmonic part of $P = -d\Phi_0/dV - 3NkTd\ln\Theta/dV + P_A$. The dashed lines through the fluid phase points are not theoretically derived. The right-hand scale is in units of kilobars; T is in units of kelvins.

$$P = -[\partial F/\partial V]_T = -[\partial F_H/\partial V]_T - [\partial F_A/\partial V]_T \tag{36}$$

$$P_H = -d\Phi_0/dV - 3NkTd\ln\Theta/dV \tag{37}$$

$$P_A = -[\partial F_A/\partial V]_T \tag{38}$$

The equations for the harmonic and anharmonic portions of the entropy are

$$S = -[\partial F/\partial T]_V = S_H + S_A \tag{39}$$

$$S_H = 3Nk + 3Nk\ln(T/\Theta)\ ;\quad S_A = -[\partial F_A/\partial T]_T \tag{40}$$

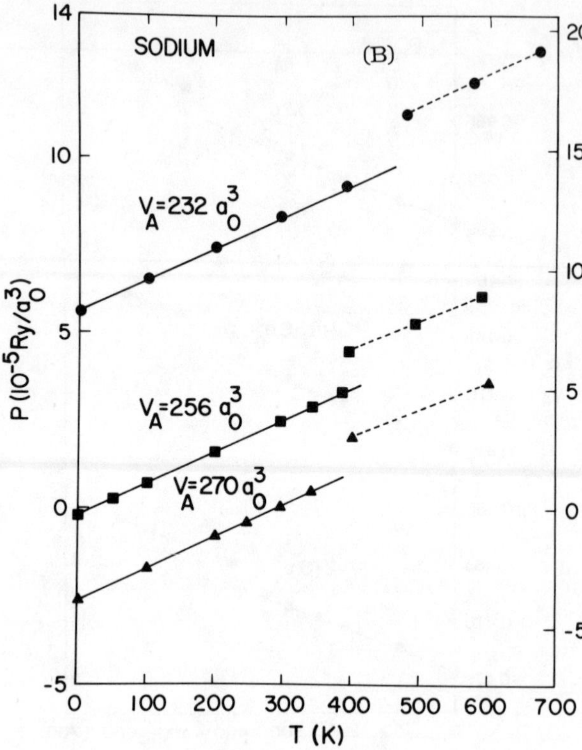

Fig. 4. continued

At high temperatures, there are also contributions from the thermal excitation of the electrons from the ground state. These are particularly important in the fluid and we give the appropriate formulas here.[17] The electronic internal energy for the conduction electron gas is given by

$$U_e = \frac{1}{2}\Gamma T^2 \tag{41}$$

and the pressure is given by

$$P_e = \frac{2}{3}U_e = \frac{1}{3}\Gamma T^2 \tag{42}$$

Γ was evaluated in the free-electron approximation from the expression

$$\Gamma = N\pi^2 k^2 mZ/\hbar^2 k_f^2 \tag{43}$$

where m is the electron mass and k_f is the free-electron Fermi wave number. Because of the fairly strong temperature dependence of the electronic terms, they cannot be neglected in the fluid phase and are necessary for good agreement with experiment. In metallic fluid sodium, P_e becomes about 3% of the ionic contribution to the pressure at T = 1800 K.

4.3. Fitting the Molecular Dynamics Data

The solid lines in Fig. 4 are from U_H and P_H and lie very close to the MD points. We are interested in the deviations of the MD points from the straight line. The MD results must be accurate enough to determine such deviations. Otherwise, we can obtain all thermodynamic quantities from quasiharmonic lattice dynamics theory and will not need to do an MD simulation at all. More importantly, Eq 33–38 suggest a simple technique to extract the maximum information from MD and extremely accurate thermodynamics. Since we believe that the total pressure and energy are correct from the MD calculations (denoted as P_{MD} and U_{MD}), then the anharmonic quantities are given by

$$U_A = U_{MD} - U_H \tag{44}$$

$$P_A = P_{MD} - P_H \tag{45}$$

This has been done for bcc sodium, and the results are given in Fig. 5.

The solid curves are the results of numerical fits of U_A and P_A using all the MD results. The functional form of this fit is determined by studying the behavior of the high-temperature expansion of the anharmonic free energy, which suggests that F_A should be a power-series expansion of leading order T^2. Accordingly, we let

$$F_A = A_2(V)T^2 + A_3(V)T^3 + A_4(V)T^4 \tag{46}$$

where each of the coefficients have explicit volume dependence, but no further temperature dependence. The $A_i(V)$ may be expanded as a power series in the volume V which we took to be

$$-A_i(V) = d_{i0} + d_{i1}V + d_{i2}V^2 \tag{47}$$

for i = 2, 3, 4. Then the anharmonic internal energy is

$$U_A = -A_2T^2 - 2A_3T^3 - 3A_4T^4 \tag{48}$$

The anharmonic pressure is

$$P_A = -(dA_2/dV)T^2 - (dA_3/dV)T^3 - (dA_4/dV)T^4 \tag{49}$$

and the anharmonic entropy is

$$S_A = -2A_2T - 3A_3T^2 - 4A_4T^3 \tag{50}$$

The alternative to the analysis given here is to take the MD points directly and numerically fit them to some functional representation of the Helmholtz free energy that allows one to easily take derivatives with respect to V and T. Although certainly possible and even required for the fluid phase, one

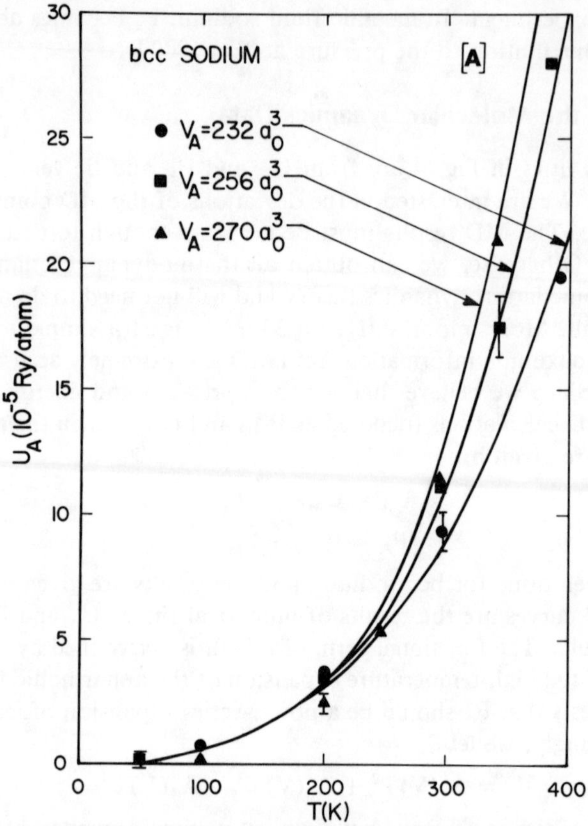

Fig. 5. Anharmonic internal energy and pressure vs. temperature. The anharmonic contributions to (A) the internal energy U_A and (B) the pressure P_A at three different volumes. The solid curves are from the fits to all the MD results for the internal energy and pressure by fitting the coefficients in Eq 46.

loses a great deal of information and understanding of the relative contributions in the vibrational behavior for the thermodynamic functions.

4.4. Thermodynamic Functions

We can now use the formulas of Eq 44–50 to evaluate any thermodynamic function of interest. Ref 9, 10, and 18 contain the results for the evaluation and comparison with experiment of the thermodynamic functions for bcc sodium. Here we will give the results for the isothermal bulk modulus B_T, the heat capacity at constant volume C_V and pressure C_P, and the thermodynamic Grüneisen parameter γ. The isothermal bulk modulus is given by

$$B_T = -V[\partial P/\partial V]_T = -V\{-d^2\Phi_0/dV^2 - 3NkTd^2 \ln \Theta/dV^2 + [\partial P_A/\partial V]_T\} \quad (51)$$

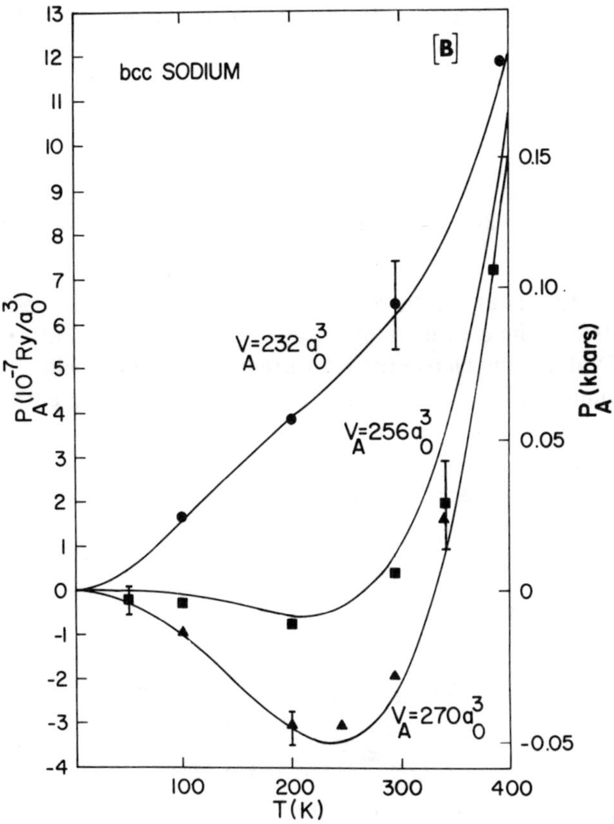

Fig. 5. continued

where the anharmonic contribution is

$$-V[\partial P_A/\partial V]_T = -(d^2A_2/dV^2)T^2 - (d^2A_3/dV^2)T^3 - (d^2A_4/dV^2)T^4 \tag{52}$$

The anharmonic contribution is not entirely negligible, but is only a few percent from T = 200 K to melting for sodium.

The constant volume heat capacity is

$$\begin{aligned} C_V &= [\partial U/\partial T]_V = 3Nk + [\partial U_A/\partial T]_T \\ &= 3Nk - 2A_2T - 6A_3T^2 - 12A_4T^3 \end{aligned} \tag{53}$$

The heat capacity at constant pressure is

$$C_P = T[\partial S/\partial T]_P \tag{54}$$

When $P = 0$, then $dU = TdS$ and $C_P(P = 0) = [\partial U/\partial T]_{P=0}$. This will not produce the same formula as Eq 53 since the $P = 0$ constraint is not equivalent to the V = constant constraint; the correct treatment of the temperature dependence of the volume is required.

The thermodynamic Grüneisen parameter is

$$\gamma = [\partial P/\partial U]_V = [\partial P/\partial T]_V/[\partial U/\partial T]_V \tag{55}$$

$$[\partial P/\partial T]_V = -3Nkd \ln \Theta/dV + [\partial P_A/\partial T]_V \tag{56}$$

$$[\partial P_A/\partial T]_V = -2(dA_2/dV)T - 3(dA_3/dV)T^2 - 4(dA_4/dV)T^3 \tag{57}$$

and $[\partial U/\partial T]_V = C_V$ from above.

The heat capacity at constant pressure is plotted in Fig. 6 for solid sodium from T = 0 K to the melting temperature T_M. This figure illustrates the success of MD above the Debye temperature and its failure below the point at which a quantum-mechanical treatment is required. The curve in Fig. 6 labeled "classical theory" is the MD-based result; the curve labeled "quantum theory" comes from quasiharmonic lattice dynamics. The experimental data is plotted as measured and with a correction that is appropriate for the

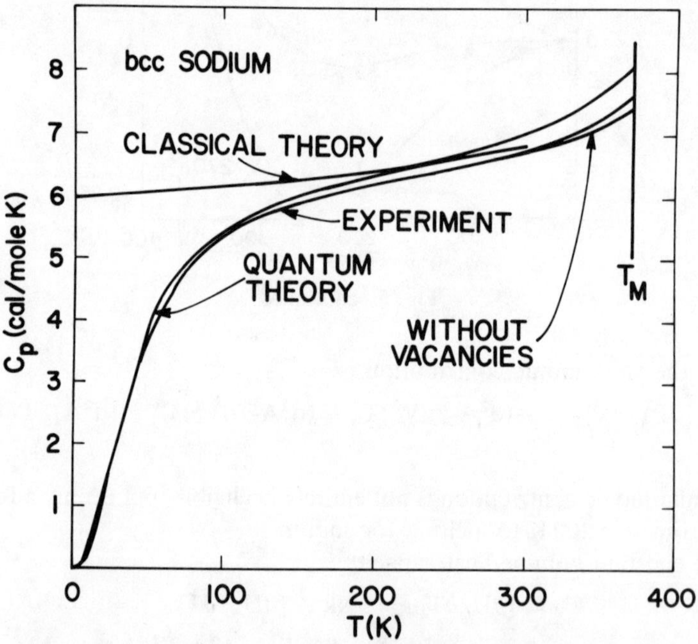

Fig. 6. Heat capacity at constant pressure C_P in solid sodium. The heat capacity vs. temperature is plotted from classical theory using MD, from quantum theory using quasiharmonic lattice dynamics, and compared with experiment. The experimental results have been corrected to the case without vacancies as indicated near T_M.

situation where there are no vacancies. The MD simulations do not contain vacancies near T_M. At low temperatures, MD will always give the classical value of 3Nk. In this region one must use lattice dynamics theory. Above the Debye temperature, classical theory becomes a good approximation and the MD calculations are meaningful. MD is able to correctly include the many-body anharmonicity with considerable accuracy. Even though the anharmonic contributions are small, as may be seen in Fig. 5, they are primarily responsible for the excellent agreement with experiment above the Debye temperature. One should note that the same potential is used for the quantum and MD calculations.

The thermodynamic Grüneisen parameter γ as a function of temperature is plotted in Fig. 7 using Eq 55–57. γ may be thought of as a highly anharmonicity-dependent quantity since it involves high-order derivatives of the potential; in fact, the anharmonicity contributes directly in the differentials of the pressure and the internal energy. Near T_M, the slopes of U_A and P_A increase rapidly. The agreement with experiment, both the qualitative shape and the numerical values, is excellent. The points are determined from

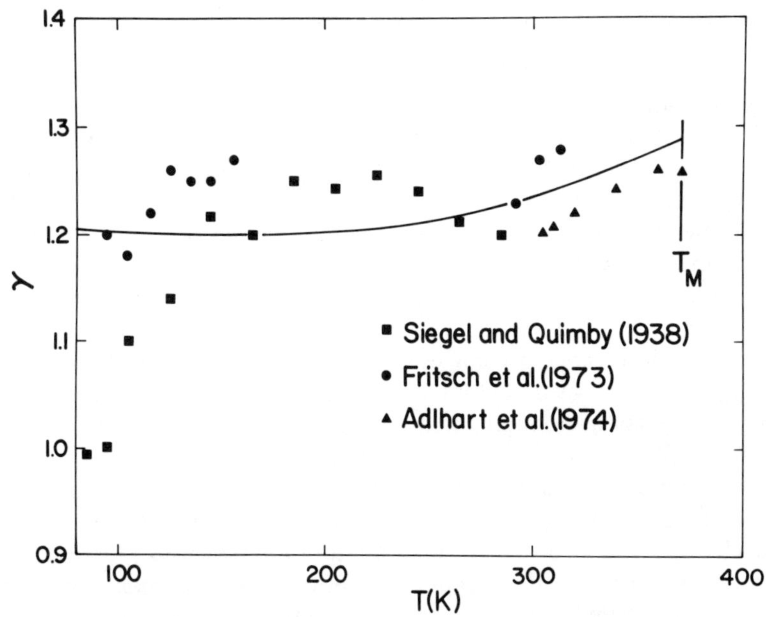

Fig. 7. Grüneisen parameter γ vs. temperature in the solid. The points are determined from the experimental values of the thermal expansion β, the isothermal bulk modulus B_T, and the heat capacity at constant volume C_V using Eq 58. The solid line was calculated from Eq 55–57.

the experimental values of the thermal expansion β, the isothermal bulk modulus B_T, and the heat capacity at constant volume C_V using the expression

$$\gamma_{exp} = V\beta B_T/C_V \tag{58}$$

4.5. The Metallic Fluid Phase

If one increases the energy in an MD simulation to a sufficiently high level, the final equilibrium state for the system will be in the fluid phase, as may be seen on the right-hand sides of Fig. 4(A) and (B). MD will give the total internal energy and pressure as a function of temperature at constant volume. This has been done for a large number of volumes and temperatures in the fluid, and the results are given in Ref 17. The electronic contributions to the internal energy and pressure, U_e and P_e as given in Eq 41–42, need to be included.

Unfortunately, we do not have a theory comparable to quasiharmonic lattice dynamics to use as a guide in the fluid phase. In the present case, U and P were fitted to cubic polynomials as a function of temperature at each volume. From these fits C_V, γ, B_T, and

$$C_P = C_V + T\gamma^2 C_V^2/VB_T \tag{59}$$

were evaluated. These quantities may be evaluated without constructing the Helmholtz free energy. Below we shall discuss how F and the fluid-phase entropy constant are obtained from the high-temperature cluster expansion of the pressure and energy.

For the fluid phase, we only present the results for C_V, C_P, and γ and the comparison with experiment in Fig. 8 and 9. Agreement with experiment is quite good. If one were to plot C_P for the solid and fluid phases on the same graph, the characteristic lambda shape would be seen at T_M. The slope of the heat capacity changes sign in passing from the solid to the fluid. The agreement between the MD result and the experiment for the Grüneisen parameter is not as good as for the solid. However, the change in slope of γ upon melting is correctly given in the fluid.

The potential used in the fluid calculations was the same as for all the solid calculations. It is quite remarkable that a single potential is capable of giving reliable results from T = 0 K to about three times T_M. At some point, the approximations of the pseudopotential will begin to break down, but an analysis of detailed electronic band structure calculations shows that the approximations should hold to above 2000 K. In comparing the P = 0 vs. T curve with experiment, at T = 2100 K the entire discrepancy was equivalent to a 2.1-kbar difference in the total pressure. Thus we conclude that the pseudopotential method is valid for volumes investigated.

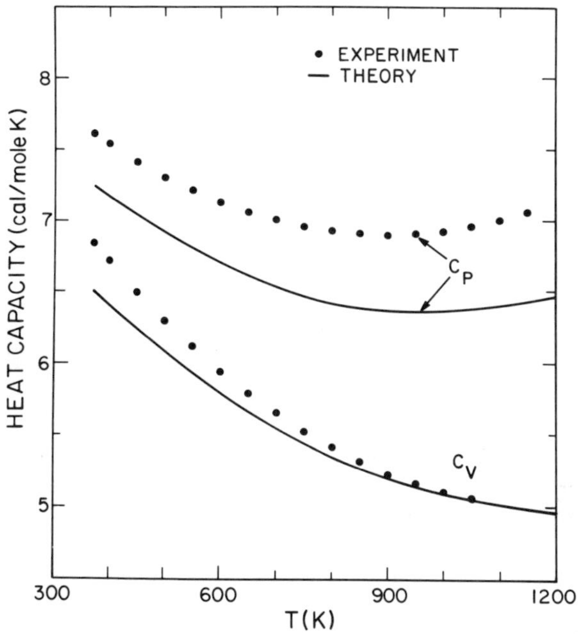

Fig. 8. Heat capacity at constant pressure C_P and constant volume C_V in metallic fluid sodium. The solid lines are calculated from MD simulations in the fluid state.

4.6. Entropy Constant of a Metallic Fluid

In the theory of classical statistical mechanics, the value of the entropy is not unique, but is only defined to within an arbitrary constant. In the solid, we start with a quantum-mechanical theory and make a high-temperature expansion of F (see Eq 12 in Appendix 2). This expansion, which contains the entropy constant, is the basis of the connection to the classical equations of motion in MD. In quantum mechanics, the constant of the entropy is uniquely determined.[15] We handle the problem in the fluid by making a high-temperature cluster expansion of the free energy. If we are able to evaluate $F = F(V,T)$ at any particular point in phase space, then we have implicitly determined the entropy constant contained in $F = U - TS$. The constant in the entropy is universal throughout the fluid phase.

The high-temperature cluster expansion for a metal contains terms that result from the direct dependence of the potential upon the volume. From the classical partition function for a fluid based on the potential given in Eq 1, the cluster expansion of the Helmholtz free energy written in reduced form F^* is[19]

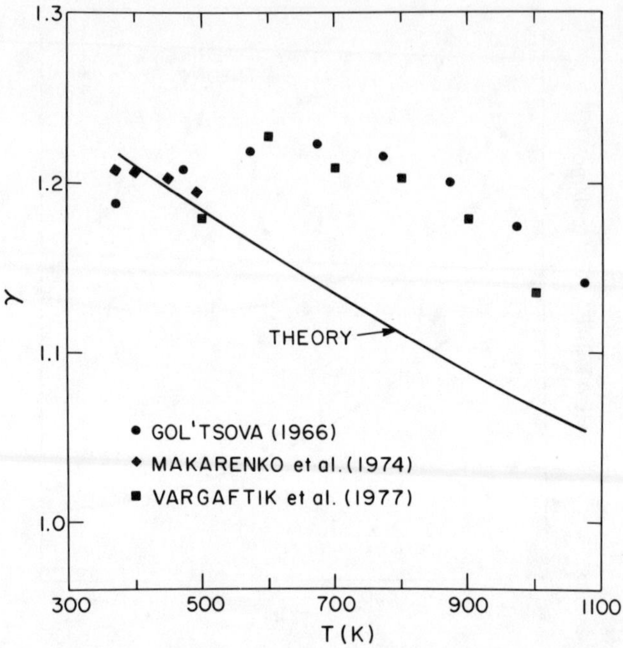

Fig. 9. Thermodynamic Grüneisen parameter γ vs. temperature in metallic fluid sodium. The solid lines are calculated from MD simulations in the fluid state, and the experimental points were obtained using Eq 58.

$$F^* = (F - I_z - \Omega)/kT = \ln[\Lambda^3/V] - 1 + \sum_{n=2}^{\infty} B_n\rho^n/(n-1)$$

$$= \ln[\Lambda^3/V] - 1 + B_2/V + \ldots \qquad (60)$$

where $\rho = N/V$ is the atomic density and $\Lambda^2 = (2\pi\hbar^2/mkT)^{1/2}$ is the thermal de Broglie wavelength. The first coefficient of the summation is

$$B_2 = 2\pi \int_0^{\infty} (1 - e^{-\beta\phi(r;V)})r^2\,dr \qquad (61)$$

In general, $B_n = B_n(T,V)$. The reduced internal energy is

$$U^* = (U - I_z - \Omega)/kT = \frac{3}{2} - (T/V)\partial B_2/\partial T + \ldots \qquad (62)$$

The reduced pressure is

$$P^* = (PV - d\Omega/d\ln V)/kT = 1 + (B_2/V - \partial B_2/\partial V) + \ldots \qquad (63)$$

We may evaluate B_2 by a direct integration of Eq 61. If we go to a sufficiently high temperature, then the expansions truncated beyond terms containing B_2 will be converged and will give an accurate description.

MD again plays a vital role by determining the convergence of the series. MD calculations of U and P as a function of T were carried out for a fixed volume. The MD results are then compared with the second-order cluster expansions for U^* and P^*. It is seen that the approximate series becomes more accurate as the temperature increases, as would be expected. The same behavior is assumed for F^*. Thus, at a very high temperature ($kT = 1.16$ Ry for sodium) we may use Eq 60 to evaluate F^*, which gives us the result that we want, namely, the implicit determination of the constant of the entropy contained in F^*. The temperature at which the second-order expansions converge is not and need not be a physically meaningful one. The procedure is to evaluate a constant of the *entire* fluid phase, and the trick we have used to accomplish this is theoretically justifiable. (See the discussion of this point in Ref 20.)

The next step is to integrate down in temperature at constant volume to regions of interest. This process is illustrated in Fig. 10. From the cluster expansion, we have evaluated the free energy at the point marked $F^*_{\text{High T}}$.

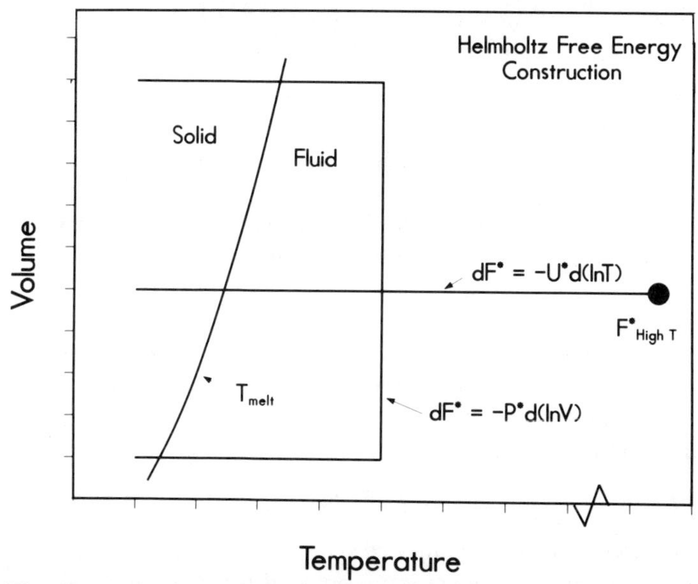

Fig. 10. Determination of the entropy constant for the fluid phase and construction of the equilibrium melt line from the Gibbs free energy. The free energy is evaluated at a temperature T_{high} using the cluster expansion of Eq 60 to second order and using MD to determine the convergence of the expansion. This implicitly determines the constant of entropy contained in F for the entire fluid phase. The free energy can then be constructed throughout the phase using the MD results for U^* to integrate along lines of constant volume and MD results for P^* for lines of constant temperature.

Using the same MD results in the form of numerical fits of U_{MD}, we calculate U^* and integrate down in temperature at constant volume with

$$dF^* = -U^* d(\ln T) \quad (64)$$

From Eq 64, the free energy along any line of constant volume may be determined. To obtain the free energy at some other volume, we may fix the temperature and integrate with respect to volume from the equation

$$dF^* = -P^* d(\ln V) \quad (65)$$

P^* is also determined from fits to P_{MD}. Continuing in this manner F may be calculated for all volumes where MD runs have been made.

4.7. Solid-Fluid Phase Boundary

Having constructed the free energies in both the solid and fluid phases, we may establish the phase boundary for equilibrium melting. With the Gibbs function $G = F + PV$, and the subscripts f or s for fluid or solid, respectively, the fluid-solid coexistence equations are

$$P_f(V_f, T_m) = P_s(V_s, T_m) = P_m(T_m) \quad (66)$$

$$G_f(P_m, T_m) = G_s(P_m, T_m) \quad (67)$$

These equations were solved at each of the volumes of the MD runs for T_m and P_m. The dashed line marked "Theory" in Fig. 11 is the result. The solid line is the experimental melting curve. The solid squares are the MD results at only one of the volumes, $V = 256a_0^3$; dynamic melting takes place between the highest point on the solid-phase branch (385 K) and the lowest point on the fluid-phase branch (396 K). The agreement with experiment is excellent, and further comparisons are made in Ref 20.

As a final point, the question of the relative importance of the volume terms $\Omega(V)$ to $\sum \phi$ can be examined. When sodium melts at zero pressure, there is a 2.4% change in volume with a principal contribution coming from $\Omega(V)$. Denoting the change in a quantity upon melting by Δ, we see from our calculations:

$$\Delta\Omega = 296 \text{ cal/mole}, \ \Delta\sum\phi = 252 \text{ cal/mole}, \ \Delta U_{total} = 548 \text{ cal/mole}$$

The last number may be compared with $\Delta U_{exp} = 621$ cal/mole. Thus, we see that the volume-dependent terms account for more than half the change in the internal energy upon melting and are therefore essential.

Acknowledgments

It is a pleasure to acknowledge useful conversations with Dr. D. C. Wallace and Professor J. R. Beeler. The figures were prepared with the capable

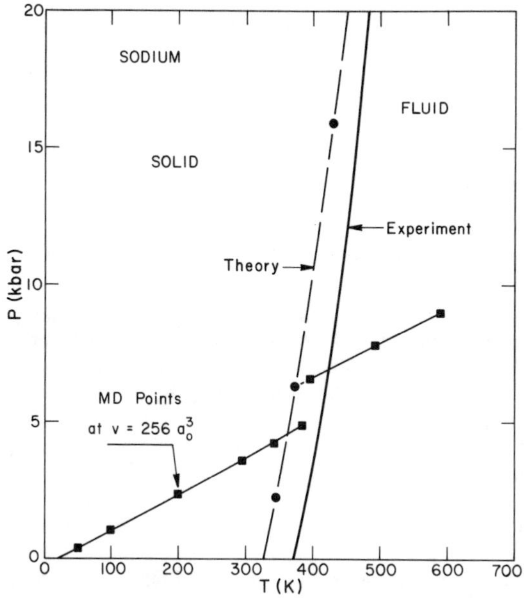

Fig. 11. P-T phase diagram for sodium. The equilibrium melting line is the result of solving Eq 66–67 for P_m and T_m. The solid line is the experimental melting curve, and the solid circles show the calculated equilibrium melting points. The solid squares are MD results for the fixed volume of 256 a_0^3/atom; dynamic melting takes place between the highest point on the solid-phase branch (385 K) and the lowest point on the fluid-phase branch (396 K).

assistance of Dr. M. C. Brady. This work was supported by the U.S. Department of Energy.

References

1. D. C. Wallace and G. K. Straub, Phys. Rev. A*27*, 2201 (1983).
2. D. C. Wallace, in *Electronic Structure, Dynamics, and Quantum Structural Properties of Condensed Matter*, edited by J. T. Devreese and Piet Van Camp (Plenum, New York, 1985), p. 521.
3. D. C. Wallace, S. K. Schiferl, and G. K. Straub, Phys. Rev. A*30*, 616 (1984).
4. S. K. Schiferl and D. C. Wallace, Phys. Rev. B*31*, 7662 (1985).
5. D. C. Wallace, *Thermodynamics of Crystals* (Wiley, New York, 1972).
6. See Chapter 3 and Appendix B in Ref. 5.
7. W. A. Harrison, *Pseudopotentials in the Theory of Metals* (Benjamin, New York, 1966).
8. W. A. Harrison, *Electronic Structure and the Properties of Solids* (Freeman, San Francisco, 1980).

9. R. E. Swanson, G. K. Straub, B. L. Holian, and D. C. Wallace, Phys. Rev. B*27*, 2873 (1983).
10. G. K. Straub, R. E. Swanson, B. L. Holian, and D. C. Wallace, in *Ab Initio Calculation of Phonon Spectra*, edited by J. T. Devreese, V. E. Van Doren, and P. E. Van Camp (Plenum, New York, 1983), p. 137.
11. J. M. Wills and W. A. Harrison, Phys. Rev. B*28*, 4363 (1983).
12. W. A. Harrison and J. M. Wills, Phys. Rev. B*25*, 5007 (1982).
13. G. K. Straub and W. A. Harrison, Phys. Rev. B*31*, 7668 (1985).
14. S. K. Schiferl and D. C. Wallace, J. Chem. Phys. *83*, 5203 (1985).
15. F. Reif, *Fundamentals of Statistical and Thermal Physics* (McGraw-Hill, New York, 1965).
16. The term "exact" refers to a full nonlinear solution to the coupled many-body equations of motion limited only by the size of the computer. For a constant-volume calculation, the total energy may be conserved to the accuracy of the computer by making the time step small enough; this is a practical limit but is not a significant problem. We must worry also about the accuracy of Φ; this must ultimately be resolved by comparison with experiment.
17. G. K. Straub, S. K. Schiferl, and D. C. Wallace, Phys. Rev. B*28*, 312 (1983).
18. G. K. Straub and D. C. Wallace, Phys. Rev. B*30*, 3929 (1984).
19. D. C. Wallace, B. L. Holian, J. D. Johnson, and G. K. Straub, Phys. Rev. A*26*, 2882 (1982).
20. B. L. Holian, G. K. Straub, R. E. Swanson, and D. C. Wallace, Phys. Rev. B*25*, 7807 (1982).
21. See Eq 1 and the discussion there.
22. See p. 350 of Ref 8.

APPENDIX 1

Effective Pair Potentials for d-States in Transition Metals

One of the fundamental problems of condensed-matter physics is the calculation of a real-space many-body potential starting at the level of the electronic structure. At present, we are concerned with determining potentials for metals where the effective interaction between ions is the result of the electron screening and a function of the ion positions. We can write the potential energy of a system of ions (nucleus + deep core electrons), outer valence electrons (that is, d-electrons), and nearly free electrons (conduction electrons) as[21]

$$\Phi = \Omega(V) + \sum \phi(r;V) \tag{1.1}$$

The summation is over $\{r_i\}$, the set of all ion positions, and is the many-body effective ion-ion interaction. The last term may be written as a sum of two-, three-, four-body, etc., terms, but is an explicit function of the volume by virtue of the electrons.

In the transition metals, we treat the s- and p-like conduction electrons in a nearly-free-electron model pseudopotential. The d-d and the d-conduction electron interactions are treated within the framework of linear-combination of atomic-orbitals theory (LCAO).[13] A real-space-effective ion-ion potential of the form of the last term in Eq 1.1 is determined.[11] For the conduction electrons, $\phi(r;V)$ is a second-order term in the energy and may be cast in the form[12]

$$V_F = Z_s^2 e^2 \cosh^2(\kappa r_c) e^{-\kappa r}/r \qquad (1.2)$$

where $\kappa^2 = 4k_f me^2/\hbar^2\pi$, r_c is the Ashcroft empty-core pseudopotential radius,[22] and Z_s is the number of conduction electrons. This potential is calculated using the Thomas-Fermi linear-screening approximation, which leads to the purely repulsive nature of the interaction. Treating the electron screening in the Lindhard approximation for the Hartree dielectric function gives a much longer-range potential that oscillates about zero as a function of r.[7]

An effective d-d interaction is determined by coupling d-orbitals on different sites with the conduction electrons. This is evaluated using LCAO theory and Hartree-Fock free-atom wave functions.[13] The tight-binding matrix elements are represented in the theory by the quantity r_d, and there are no adjustable parameters in the theory. Tables of r_d are given in Ref 13. The potential is composed of an attractive term $V_B(r)$ describing the formation of bands from the atomic d-orbitals, and a repulsive term $V_C(r)$ describing the overlap repulsion of the d-orbitals on neighboring sites. The resulting formulas are:[11]

$$V_B(r) = -Z_d(1 - Z_d/10)(28.1/\pi)\hbar^2 r_d^3/mr^5 \qquad (1.3)$$

$$V_C(r) = Z_d^2(225/\pi^2)\hbar^2 r_d^6/mr^8 \qquad (1.4)$$

Z_d is the number of d-electrons; for nickel $Z_s = 1.5$ and $Z_d = 8.5$. In Fig. 12 we have plotted the effective interaction and the d-state contributions for nickel. The net d-electron potential is $V_D = V_B + V_C$ and the total ion-ion potential is $V_{Total} = V_F + V_D$. As is always the case for metals, the total adiabatic potential energy includes volume-dependent terms in the term $\Omega(V)$, just as in the simple metals. The largest contribution to the binding energy in the transition metals comes as a result of the d-electrons.

APPENDIX 2
Lattice Dynamics

The theory of quasiharmonic lattice dynamics has been extremely successful for the calculation and interpretation of the thermodynamic properties of insulators, semiconductors, metals, and compounds. In this appendix, we will sketch the results from lattice dynamics necessary to provide a comprehensive analysis of MD simulations. We will follow closely the presentation

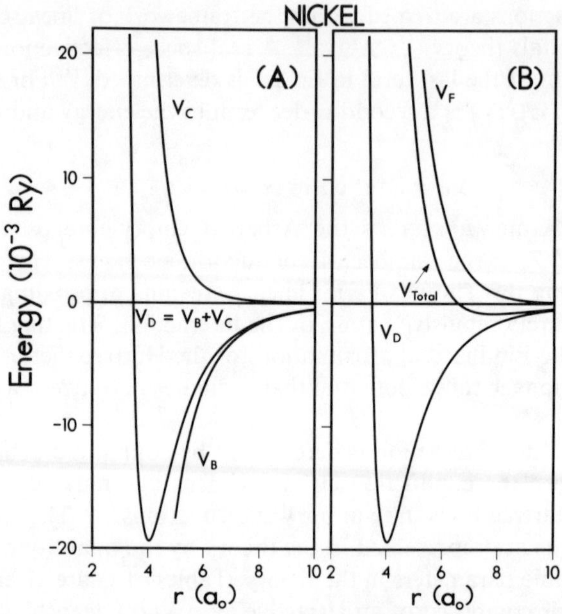

Fig. 12. Effective ion-ion interaction and d-state contribution for nickel. (A) The effective ion-ion interaction resulting from the d-electrons, which is responsible for most of the bonding in the transition metals, is composed of an attractive term $V_B(r)$ describing the formation of bands from the atomic d-orbitals, and a repulsive term $V_C(r)$ describing the overlap repulsion of d-orbitals on neighboring sites. For nickel, the parameters for Eq 1.3–1.4 are $Z_d = 8.5$ and $r_d = 1.34a_0$. The total d-electron potential is $V_d = V_B + V_C$. (B) The conduction electrons result in an effective ion-ion interaction given by the curve labeled V_F. The parameters for Eq 1.2 are $Z_s = 1.5$ and $r_c = 1.4a_0$ for nickel. V_{Total} is the sum of the d- and conduction electron terms. The total potential energy also includes a volume-dependent term $\Omega(V)$ as in the simple metals.

of Wallace in *Thermodynamics of Crystals*,[(5)] where the theory is developed in detail and applied to insulators and metals. We will present the formulas necessary for the evaluation of the phonon frequencies using the effective two-body potential of Appendix 1.

The theory of lattice dynamics is a fully quantum-mechanical theory treating a system of electrons and ions. One can write the total crystalline Hamiltonian in the form $H = \text{K.E.} + \Phi$ where K.E. is the kinetic energy of the ions and Φ is the total adiabatic potential in the Born-Oppenheimer approximation and contains the kinetic energy of the electrons and the potential energy of the interacting ion-electron system. One then expands Φ about the equilibrium positions (crystalline lattice sites); for the harmonic theory one includes terms to second order in the displacements where higher-order terms may be included in perturbation theory. The quasiharmonic Hamiltonian is $H = \Phi_0 + H_2$. Now Φ_0 is the adiabatic potential evaluated at the equilibrium positions of the ions and includes the entire volume dependence of the metallic potential as discussed in Section 2.

We solve H_2 for its normal vibrational modes (phonons). In matrix notation the eigenvalue-eigenvector equation is

$$\vec{v}^*(\vec{k}) \cdot D(\vec{k}) \cdot \vec{v}(\vec{k}) = \omega^2(\vec{k}) \tag{2.1}$$

This set of equations is solved for the vibrational frequencies at each wavenumber $\vec{k}$. The $\vec{v}(\vec{k})$'s are the polarization vectors that diagonalize $D(\vec{k})$ and give the set of ionic displacements for a particular mode $\vec{k}$. D is the dynamical matrix

$$D_{ij}(\vec{k}) = 1/M \sum_n{}' \Phi_{ij}(r_n)(\cos \vec{k}\cdot\vec{r}_n - 1) \tag{2.2}$$

where M is the ion mass, the summation is over all lattice sites (the prime means to exclude the term for $r_n = 0$), and i,j are indices for the cartesian coordinates x,y,z. The potential energy coefficients are the second-order derivatives of the adiabatic potential with respect to the displacements and are given by

$$\Phi_{ij} = \partial^2\Phi/\partial u_i \partial u_j \tag{2.3}$$

with the cartesian components of the displacements $u_i = r_i - r_{i0}$, r_{i0} = cartesian component of the lattice positions. In terms of the $\phi(r;V)$ of Appendix 1,

$$D_{ij}(\vec{k}) = 1/M \sum_n{}' \{2\phi'(r_n^2)\delta_{ij} + 4\phi''(r_n^2) r_{ni} r_{nj}\}(1 - \cos \vec{k}\cdot\vec{r}_n) \tag{2.4}$$

where $\delta_{ij} = 1$ for $i = j$ and 0 for $i \neq j$. The primed quantities are defined to be derivatives of the two-body potential with respect to r^2:

$$\phi = \phi(r^2) \;;\quad \phi' = \partial\phi/\partial(r^2) \;;\quad \phi'' = \partial^2\phi/\partial(r^2)^2 \tag{2.5}$$

For the Wills-Harrison potential in the transition metal series

$$\phi = V_F + V_B + V_C \tag{2.6}$$

as given by Eq 1.2–1.4, and

$$\phi' = -V_F(1/r + \kappa)/2r + V_B' + V_C' \tag{2.7}$$

$$\phi'' = V_F(\kappa^2 r^2 + 3\kappa r + 3)/4r^4 + V_B'' + V_C'' \tag{2.8}$$

$$V_B' = -5V_B/2r^2 \;,\quad V_B'' = 35V_B/4r^4 \tag{2.9}$$

$$V_C' = -4V_C/r^2 \;,\quad V_C'' = 20V_C/r^2 \tag{2.10}$$

The contributions from the d-electron repulsion V_C and the formation of bands V_B are to be included with the conduction electron potential V_F.

Eq 2.1 is solved by making Fourier expansions of the displacements u, leading to the phonon coordinates q. The q's constitute a complete set of 3N coordinates (one atom per unit cell) capable of a complete description of the vibrational motion. The set of 3N allowable wavenumbers $\vec{k}$ is determined by constructing the Brillouin zone in the reciprocal lattice and contains the full symmetry properties of the crystal. The reader unfamiliar with the Brillouin zone terminology is referred to Ref 5, p. 106ff.

Phonon Thermodynamics

The Helmholtz free energy for the phonons in the harmonic approximation is

$$F_H = \sum_k \left\{ \frac{1}{2}\hbar\omega(\vec{k}) + kT\ln(1 - e^{-\hbar\omega(\vec{k})/kT}) \right\} \tag{2.11}$$

and the first term is the zero-point vibrational energy. The internal energy, pressure, entropy, heat capacity, etc., are defined by the appropriate derivatives of F as discussed in Section 2, although here the description is quantum mechanical and contains the correct behavior at low temperatures, such as that of the heat capacity as shown in Fig. 6. The connection between harmonic phonon theory and the classical MD simulations is made by taking the high-temperature limit of the harmonic free energy given by

$$F_H = -kT\sum_k \left\{ \ln(kT/\hbar\omega(\vec{k})) - \frac{1}{24}(\hbar\omega(\vec{k})/kT)^2 + \ldots \right\} \tag{2.12}$$

In this limit, we need to evaluate

$$\langle \ln k\Theta \rangle = \langle \ln \hbar\omega \rangle_{B.Z.} = 1/3N \sum_k \ln \hbar\omega(\vec{k}) \tag{2.13}$$

The summation is over the 3N values of $\vec{k}$, and this average may be evaluated for a given Brillouin zone using the technique in Section 37 of Ref 5. This is the quantity given in Eq 32 and is necessary to analyze the MD simulations. Since we also need volume derivatives of $\langle \ln k\Theta \rangle$, a linear numerical fit as a function of V of Eq 2.13 was made and determined to be adequate.

5

Grain-Boundary Modeling*

D. Wolf

Materials Science Division
Argonne National Laboratory
Argonne, IL 60439

Abstract

A rather comprehensive set of computer codes has been developed in recent years enabling one to investigate the physical properties of grain boundaries in metals and predominantly ionic ceramic materials. In this brief overview the modeling of grain boundaries in metals at absolute zero temperature by means of lattice-statics (iterative energy minimization methods) is described. In contrast to experiments, in computer simulations it is rather easy to systematically vary the five macroscopic geometrical parameters describing a planar, bicrystalline grain boundary. In this manner, computer modeling can be used to extract rather fundamental insight into the physics of solid interfaces. By comparing the results of such computer "experiments" with experimental results whenever possible, the validity range of these simulation methods can be tested. As reviewed in this chapter, the combination of computer-simulation studies with related experiments is capable of providing insight which, from experiments alone, would be very difficult to obtain.

1. Introduction

Since grain boundaries (GBs) play an important role in the cohesion and deformation of all polycrystalline materials, a workshop on the application of computer-simulation methods to forging is not complete without an overview of the simulation of GBs.

The GB studies we will discuss all involve planar bicrystal models, in contrast to the often curved interfaces in the polycrystalline materials used in the "real world." A bicrystal is formed by joining together two identical perfect lattices whose relative orientation is characterized by eight geometrical parameters (see Section 2).

*Work supported by the U.S. Department of Energy, BES-Materials Sciences, under Contract W-31-109-Eng-38.

Both lattice-statics and, more recently, molecular-dynamics and Monte Carlo simulation methods have been applied to investigate the physical properties of GBs, and many reviews have been presented. A good summary of the earlier work, all involving static methods, is found in the article of Harrison et al.[1] An overview of the dynamical methods applied to GBs was presented recently by Yip.[2] For that reason, in this brief overview we will focus less on the computer simulation methods as such rather than on what insight can be gained from such simulations into the basic physical properties of GBs.

One of the fundamental questions in the GB area concerns the relationship between the very well understood low-angle and the poorly understood high-angle boundaries. Whereas the description of low-angle boundaries in terms of networks of so-called GB dislocations due to Read and Shockley[3] has been extremely successful, no such single simple model has evolved which permits the prediction of properties of high-angle boundaries that are consistent with experimental observations. Most of these models are strictly geometrical in nature and based essentially on Bollmann's coincident-site-lattice (CSL) model and its extensions (such as the O-lattice model).[4]

Within the framework of the CSL model it is usually assumed that the energy and structure of high-angle GBs are governed by the parameter Σ, which characterizes the inverse density of CSL superlattice sites in common to the two crystal lattices forming the bicrystal. As illustrated schematically in Fig. 1, low values of Σ are thus thought to give rise to cusps in the energy vs. misfit angle curve; the corresponding boundaries are usually referred to as "special," that is, low-Σ boundaries (see also Table 1). Since small-angle boundaries (typically for $\theta \lesssim 15°$ for which the dislocation cores do not overlap) generally correspond to very large values of Σ, the predicted and experimentally-observed smoothness of the energy curve for $\theta \lesssim 15°$ (see Fig. 1) is not in conflict with the CSL model. As another consequence of the CSL

Table 1. Angle–Σ combinations for rotations about $\langle 100 \rangle$ in cubic crystal lattices

θ	Σ
36.87	5
22.62	13
28.07	17
16.26	25
43.61	29
18.92	37
12.68	41
31.89	53

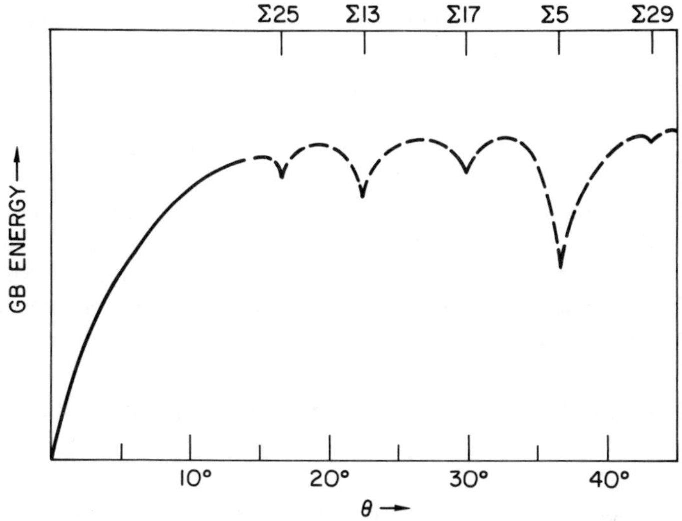

Fig. 1. Transition from low-angle to high-angle boundaries (schematical). The smooth energy vs. misfit angle curve for low-angle GBs (for $\theta \lesssim 15°$) is governed by the Read-Shockley expression[3] (solid line). The CSL model for high-angle GBs assumes energy cusps for "special" angles for which Σ is particularly small (dashed line; see also Table 1).

model, the lowest-energy GB plane for a given misorientation (characterized by Σ) is postulated to be the *symmetrical* configuration, that is, the densest plane in the CSL superlattice.

Much of the experimental evidence is in conflict with these predictions. Thus, in their thermal-grooving experiments on ⟨110⟩ tilt GBs in aluminum, Hasson et al.[5] observed energy cusps for the Σ3 (111) orientation but not for the Σ3 (211) GB. Also, whereas the Σ11 (113) boundary gave rise to an energy cusp, the Σ9 (221) and Σ9 (114) GBs did not. Hasson et al.[5] also report that in many cases their bicrystals preferred asymmetrical plane configurations, in contrast to what is expected within the framework of the CSL model. An extensive overview of the earlier experimental work in comparison with the different geometrical models was presented by Goodhew.[6] Very recent high-resolution TEM results by Merkle et al.[7-9] and Krakow et al.[10] on NiO and Au, respectively, show asymmetrical GB-plane orientations in many instances, even for low-Σ boundaries—in contrast to what the CSL model postulates.

It is the purpose of this brief overview to offer an answer to the central question in the GB area—"What are 'special' high-angle GBs?"—from a physical rather than geometrical point of view. We do this in an attempt to illustrate how computer modeling can be used to gain rather simple yet fundamental insight into the basic physics of solid interfaces. One tremendous advantage of simulations over experiments is the relative ease with which, by means of the computer, the different degrees of freedom of a GB can be varied in a completely controlled manner, an extremely difficult, if not virtually impossible, experimental task. By comparing the results of such "computer experiments" with "real" experiments wherever possible, the validity range of the simulation methods can be treated. As illustrated below, the combination of computer-simulation studies with corresponding experimental work is capable of providing new insight which, from experiments alone, could not be obtained.

2. Geometrical Characterization of Grain Boundaries

As is well known, eight geometrical parameters are necessary—in addition to the crystal structure and lattice parameter—to fully characterize a planar grain boundary. These eight degrees of freedom (dof) may be subdivided into macroscopic and microscopic ones.[11] The latter are represented by the three translational dof of the two crystals with respect to one another parallel and perpendicular to the GB plane. They are usually characterized by the translation vector $\vec{T}$. By the very nature of $\vec{T}$, only experiments which can measure the translation on an atomic scale are capable of determining the three components of $\vec{T}$.

The five macroscopic dof characterize the misorientation of the two crystal lattices relative to each other and the GB plane. In the conventional GB characterization scheme,[11] the misorientation is characterized by the rotation axis, $\vec{n}_R$ (a unit vector, hence constituting only two dof), and the rotation angle, θ_R, whereas the GB plane is defined by the plane normal, $\vec{n}$, with respect to either of the two crystal coordinate systems. The eight dof of a GB may hence be summarized as follows:

$$\vec{n}_R, \theta_R, \vec{n}, \vec{T} \tag{1}$$

This choice of parameters emphasizes the fact that a bicrystal may be thought of as having been generated by a single rotation, $R(\vec{n}_R, \theta_R)$, of two identical interpenetrating crystal lattices relative to each other with the subsequent removal of corresponding material to accommodate the particular GB plane chosen.[4,11] It suffers a certain disadvantage in that symmetry properties of GBs are not readily apparent from this choice of the macroscopic parameters. For example, to the unexperienced reader it may not be obvious that the GB characterized by $\vec{n}_R, \theta_R, \vec{n} = \langle 100 \rangle$, 36.87°, (013) is a *symmetrical* tilt boundary which, in fact, is characterized by only *three* independent macroscopic parameters.

A simple alternative choice of the eight dof which avoids this difficulty is the following[12] (see Fig. 2):

$$\vec{n}_1, \vec{n}_2, \theta, \vec{T} \tag{2}$$

where $\vec{T}$ is the same translation vector as in scheme (1). The unit vectors $\vec{n}_1$ and $\vec{n}_2$ are now chosen to represent the GB-plane normals relative to the principle coordinate systems of crystals 1 and 2, respectively, whereas θ is the twist angle about this plane normal.

In this scheme, a symmetrical GB is readily recognized by the condition $\vec{n}_1 = \pm\vec{n}_2$. For example, the above (013) twin GB in the fcc lattice may be characterized as the 180° twist GB[12] (013), (013), 180°, from which it is apparent that only three variables are needed for its macroscopic characterization, namely the GB plane (two dof) and the twist angle. Also, the separation of tilt and twist components for a general GB is readily accomplished: Whereas θ represents the angle of a twist rotation about the GB-plane normal $\vec{n}$ (= $\vec{n}_1$ or $\vec{n}_2$, depending on which coordinate system is chosen), the tilt component is readily extracted from $\vec{n}_1$ and $\vec{n}_2$, considering that the tilt axis is perpendicular to both normals. Finally, the determination of the total misorientation, $R(\vec{n}_R, \theta_R)$, requires merely a multiplication of the rotation matrices for the twist and the tilt rotation. For further details see Ref 13.

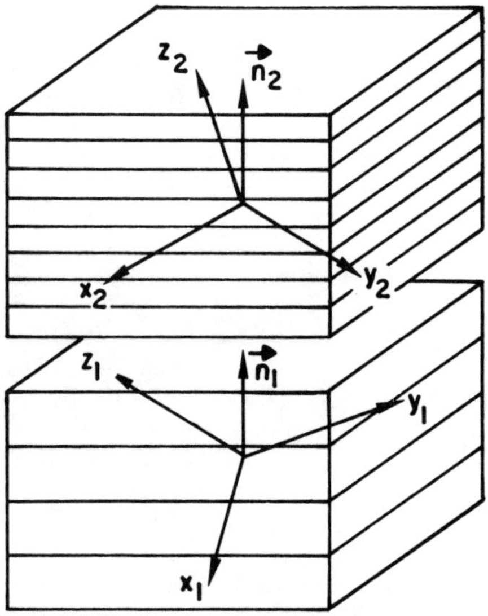

Fig. 2. GB-plane normals $\vec{n}_1$ and $\vec{n}_2$ in the principal coordinate systems of crystals 1 and 2, respectively.

3. Geometrical Relationship Between Tilt and Twist GBs

One of the most characteristic differences between tilt and twist GBs arises from the fact that a tilt rotation results in a change of the boundary-plane normal, $\vec{n} = (hk\ell)$, as function of the tilt angle, θ, whereas $(hk\ell)$ remain unchanged for a twist rotation. For simplicity, we consider here only the relationship between *symmetrical* tilt boundaries (STGBs) and twist boundaries.

The main effect of a symmetrical tilt rotation is the inversion of the stacking sequence of lattice planes parallel to the GB plane on one side of the interface with respect to the other (see Fig. 3). As recognized earlier[14,15] such an inversion may also be accomplished by choosing as starting configuration a single crystal ($\Sigma = 1$) already oriented along the $(hk\ell)$ direction (with stacking sequence A/J-A/J-A/J in Fig. 3) and rotating, say, the bottom half by 180° with respect to the top half (which results in the symmet-

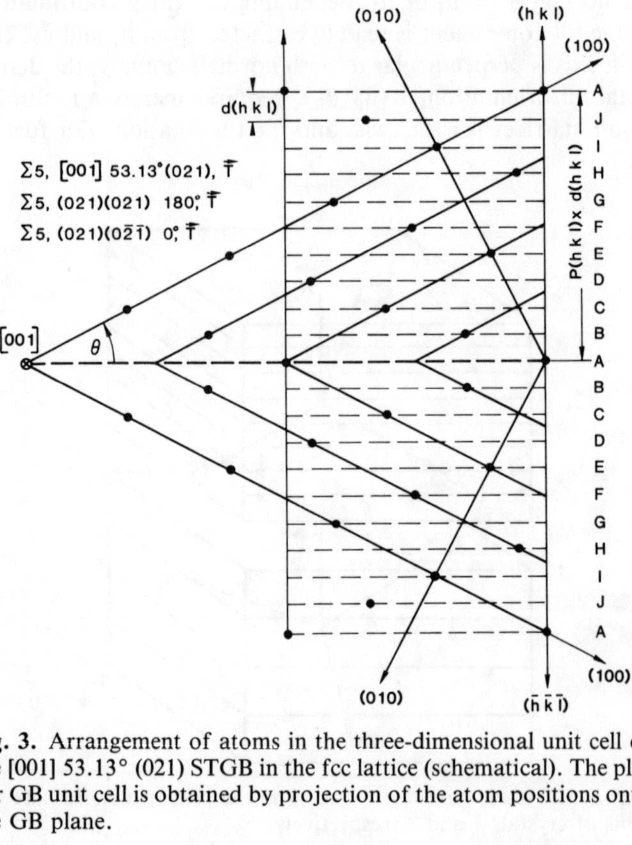

Fig. 3. Arrangement of atoms in the three-dimensional unit cell of the [001] 53.13° (021) STGB in the fcc lattice (schematical). The planar GB unit cell is obtained by projection of the atom positions onto the GB plane.

rical planar stacking sequence A/J-A/A-J/A). The particular configuration shown in Fig. 3 is then readily obtained by a translation of the lower crystal parallel to the boundary plane such that A → B, B → C, etc.

We define the period, P(hkℓ), as the number of Bravais-lattice planes in the repeat unit of the ideal crystal in the (hkℓ) direction, whereas d(hkℓ) denotes the spacing between these planes. Except for directions for which P(hkℓ) = 2 (such as the (100) and (110) directions in cubic crystals), such translations parallel to the GB plane cannot result in an inversion of the planar stacking sequence. The twin-like translational configuration in Fig. 3 (with P(hkℓ) = 10) avoids the energetically extremely unfavorable situation in which identical A planes face each other directly across the interface. In reality, all possible (even noninteger) translation vectors parallel to the GB plane have to be considered in the search for the one with the lowest energy. In principle, however, all STGBs may be looked upon as twins such as the one in Fig. 3.

For rotation angles other than 180°, a twist GB on the (hkℓ) plane is generally created. If the unit cell of the single crystal in the (hkℓ) direction contains an m-fold rotation axis, all rotations about (hkℓ) by angles 180°/m + k 360°/m (k = 1,2,...m − 1) yield the same inverted planar stacking sequence. All STGBs may, therefore, be considered as special 180°/m twist boundaries on the (hkℓ) plane.

A well-known example is the (111) twin in the fcc lattice which may be considered as either a Σ3, ⟨110⟩ 70.53° (111) STGB or a Σ3, 60° (111) twist boundary. To generate this boundary as a special twist boundary (see Fig. 4) the ideal-crystal (111) stacking sequence, C/ABC/ABC/A, may be inverted by twist rotations of 60° + k 120° (with m = 3) which yield the energetically unfavorable stacking C/ABC/CBA/C. A translation of the second crystal

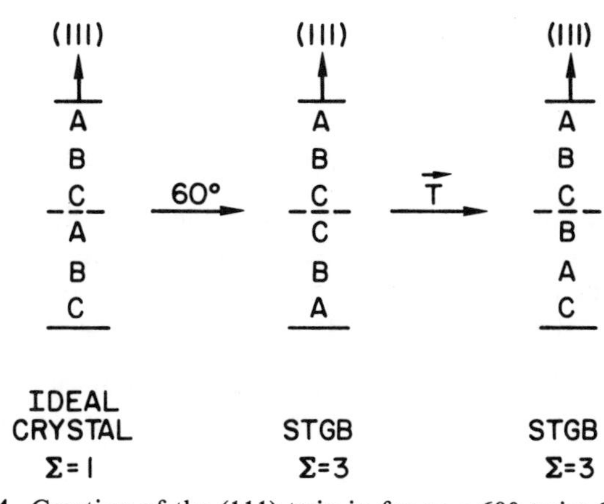

Fig. 4. Creation of the (111) twin in fcc as a 60° twist GB.

parallel to the GB plane may be chosen such that C → B, etc., and we obtain the familiar geometry of the (111) twin: C/ABC/BAC/B.

The fact that a 180°/m rotation leads to an inversion of the stacking sequence originates in the inversion symmetry of Bravais lattices. Since a crystal lattice consists of a Bravais lattice plus a set of basis atoms, this property of STGBs in a crystal lattice holds only if the basis also has inversion symmetry. This is always the case, however, if the basis contains only one atom.

From the above discussion it follows that among all possible grain boundaries on a given crystallographic plane (hkℓ), we may have one STGB (for twist angle $\theta = 180°/m$) and infinitely many twist boundaries (for $\theta \neq 180°/m$). The fact that this STGB represents a very special twist boundary has some fascinating consequences[12] which are summarized below:

1. Neither the inversion of the stacking sequence on one side of the GB plane nor the translations parallel to the GB plane alter the size or shape of the three-dimensional GB unit cell (see Fig. 3). The ideal crystal ($\Sigma = 1$) and the STGB on the (hkℓ) plane, therefore, have identical unit-cell dimensions. In particular, the planar GB unit cell on the (hkℓ) plane is the same in all three cases. This unit cell is the smallest possible for any GB on this plane. Its area, $A_1(hk\ell)$, is given by

$$\Omega = A_1(hk\ell)d(hk\ell) \tag{3}$$

   where $d(hk\ell)$ denotes the spacing between (hkℓ) Bravais-lattice planes, whereas Ω represents the volume of the unit cell of the Bravais lattice. Hence, for a Bravais lattice with only one atom in the basis, Ω is the atomic volume.

2. The number of (hkℓ) lattice planes in the repeat unit in the (hkℓ) direction, referred to as the *period* P(hkℓ), represents an important geometrical parameter of the ideal crystal or any interface on the (hkℓ) plane (see Fig. 3–6). For cubic crystal lattices

$$P(hk\ell) = \alpha(h^2 + k^2 + \ell^2) \qquad (\alpha = 1 \text{ or } 2) \tag{4}$$

   where the value of α has to be determined in each case by inspection of the particular crystal lattice. When considering a stacking fault or GB on the (hkℓ) plane, a stack containing at least P(hkℓ) planes on each side of the interfacial plane has to be considered.

3. Since the STGB on the (hkℓ) plane differs from the ideal crystal ($\Sigma 1$) only by the inversion of the stacking sequence, its corresponding Σ value is determined by

$$\Sigma = \beta P(hk\ell) \qquad (\beta = 0.5 \text{ or } 1) \tag{5}$$

where $\beta = 0.5$ if $P(hk\ell)$ is even (see Fig. 5), whereas $\beta = 1$ if $P(hk\ell)$ is odd (see Fig. 6). The proper value for β has to ensure that Σ is odd in cubic crystals.

4. In cubic Bravais lattices the spacing between lattice planes is given by

$$d(hk\ell) = \gamma a(h^2 + k^2 + \ell^2)^{-1/2} \tag{6}$$

where a is the lattice parameter, and the proportionality factor γ may assume the values 0.5 or 1.

5. Equations 5 and 6 may be combined and we obtain

$$d(hk\ell) = \epsilon a \Sigma^{-1/2} \tag{7}$$

where the proportionality constant ϵ may, in general, assume the values[12] $\sqrt{2}/4$, 1/2, $\sqrt{2}/2$, and 1. Which of these values applies has to be inspected in each case. For example, in the bcc lattice only the values $\epsilon = 1/2$ and $\epsilon = \sqrt{2}/2$ are found.

According to Eq 5, the value of Σ for a STGB is solely determined by the number of lattice planes in the repeat unit in the ideal crystal. Because $P(hk\ell)$ is a *volume* quantity associated with the *ideal* crystal, it is hard to imagine why Σ should have any significance for the physical properties of *planar* defects. Instead, the geometry of STGBs may be fully described by the Miller indices of the GB plane.

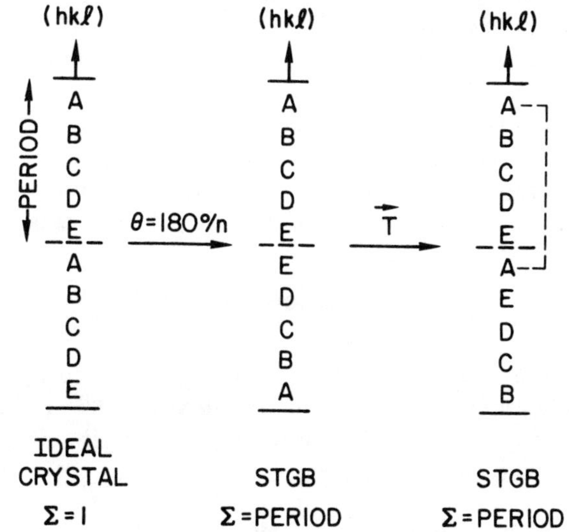

Fig. 5. Inversion of an ideal-crystal stacking sequence by a 180°/m twist rotation about (hkℓ) for a direction with *odd* period P(hkℓ).

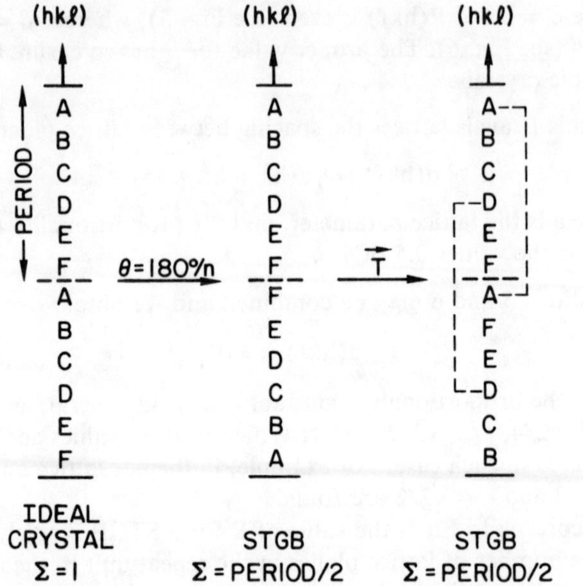

Fig. 6. Inversion of an ideal-crystal stacking sequence containing an *even* number of planes in the repeat unit.

As illustrated above, the STGB on a given crystallographic plane has the smallest planar unit-cell area of any GB on that plane, namely that of the ideal crystal and the stacking fault. For any twist angle $\theta \neq 180°/m$, a twist boundary with generally much larger unit-cell area is obtained. The STGB is therefore the most "special" of all GBs on that lattice plane.

The above discussion demonstrates the importance of the GB plane and of the planar unit-cell area in the geometrical description of grain boundaries. In the computer-simulation results discussed below, both have been varied systematically to analyze their effect on the GB energy.

4. Computational Procedure

All of the results presented below have been obtained by means of the HELL codes developed at Argonne in recent years.[16-18] The lattice-statics computational procedure applied in these codes consists essentially of the four steps described in the following sections.

4.1. GB Geometry: The CSL Unit Cell

For a given choice of the five macroscopic degrees of freedom characterizing the GB to be investigated (see Section 2), the atom positions in the periodic unit cell of the interface are determined in two steps. First, the CSL unit cell is determined for the underlying *Bravais* lattice by a twist rotation of two semi-infinite single crystals about the desired GB-plane normal. (Notice that even tilt GBs are generated by such a twist rotation; see Section 3.) An example of the site arrangement in the planar CSL unit cell is shown in Fig. 7. Next, the basis atoms are attached to every Bravais-lattice site thus generating the CSL unit cell of the *crystal* lattice. Obviously, for a crystal lattice with only one atom in the basis, the Bravais and crystal CSL unit cells are identical.

4.2 Region I–Region II Strategy

As in virtually all defect calculations of this nature, a region I–region II strategy is applied. Far away from the interface plane the atoms are in their respective perfect-crystal positions. A typical region I–region II set up is shown in Fig. 8. In region I the atoms are allowed to relax explicitly to find the equilibrium configuration. The two-dimensional periodicity parallel to the GB plane is exploited to minimize the number of atoms that must be directly included. This periodicity is determined by the size of the CSL unit cell (see Fig. 7 and 8). Although in the outer regions (so-called region II) the atoms remain in their respective ideal-crystal positions, in general these regions must be allowed to move rigidly to allow dilation of the bicrystal due to the introduction of the grain boundary. This dilation is represented by the three microscopic degrees of freedom discussed in Section 2. They are included in the minimization, to represent the relative position of the two halves of region II.

The GB energy, E_Σ, can be defined as

$$E_\Sigma = [U(\text{faulted block}) - U(\text{perfect block})]/A_\Sigma, \tag{8}$$

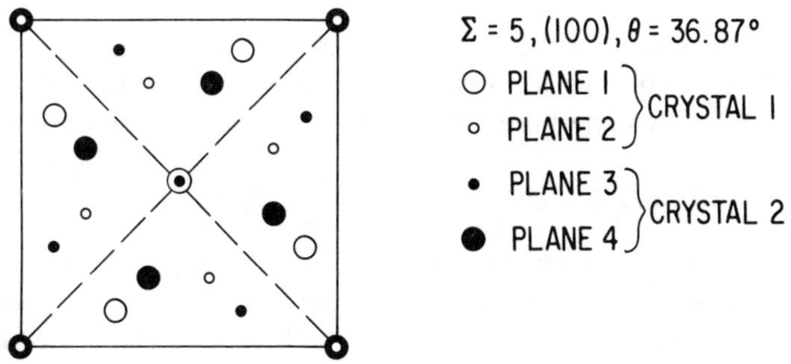

Fig. 7. Bravais-lattice sites in the CSL unit cell for the [100] 36.87° (100) (Σ5) twist GB in the fcc lattice. Shown are the sites in the two lattice planes nearest to the GB plane on either side of the interface (which is situated between planes 2 and 3).

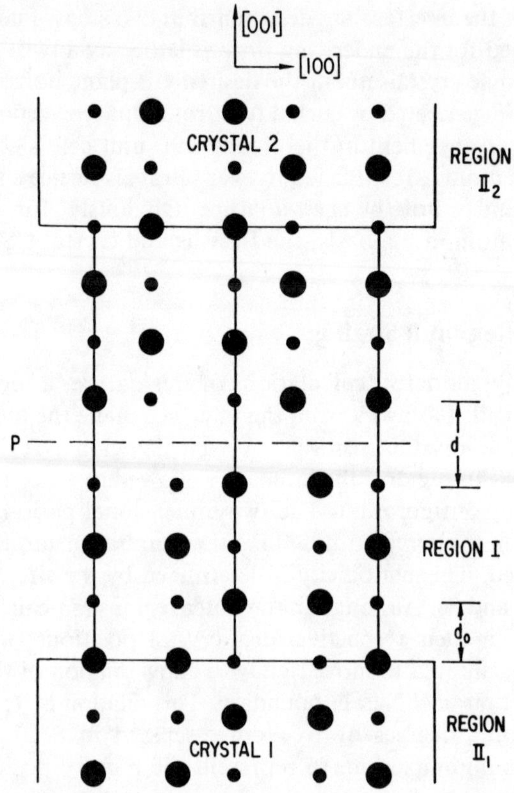

Fig. 8. Subdivision of the bicrystal containing a grain boundary P into region I (in which the atoms may relax freely) and region II (in which all atoms occupy their respective ideal-crystal positions). Columns of coincidence atoms are shown at the edges and in the center of the CSL unit cell (compare also Fig. 7).

where U is the energy of the blocks with and without the interface, and A_Σ is the area of the planar CSL unit cell. Assuming only central-force potentials, $V_{ij}(r)$, where i and j are ion indices, U may be calculated as follows:

$$U = \frac{1}{2} \sum_{i \in I} \sum_{j \in I} \sum_{\ell}{}' V_{ij}(|\vec{r}_{ij} + \vec{r}_\ell|) + \sum_{i \in I} \sum_{j \in II} \sum_{\ell}{}' V_{ij}(|\vec{r}_{ij} + \vec{r}_\ell|) \qquad (9)$$

where $\vec{r}_\ell$ is a two-dimensional lattice vector in the GB plane, and the prime indicates omission of the term for $\vec{r}_\ell = 0$ whenever j = i. The important feature in Eq 9, arising from the region I–region II subdivision, is the omission of the factor 1/2 in the second term. Whereas in the first term on the right-hand side the factor of 1/2 arises from the double counting of interactions when both i and j are in region I, no such double counting occurs for ions j in region II. As a consequence, U(perfect block) in Eq 8 differs from the cohesive energy, U_{coh}, times the number of ions, N, in region I and depends on where the region I–region II boundary is situated relative to the GB plane. However, with increasing distance of this boundary from the interface, U(perfect block) asymptotically approaches NU_{coh}. In reality, however, the evaluation of Eq 8 requires *two* calculations, one for the perfect crystal and one for the bicrystal, owing to this subdivision into regions I and II.

4.3. Interatomic Potentials

As pointed out earlier in this volume,[19,20] a central problem in all computer simulations in metals lies in the choice of the pair potential used to predict materials properties. Whereas empirical (usually spline-fitted) potentials as well as simple analytical (such as Morse or Lennard-Jones) potentials are usually of rather short range, the longer-range charge-density oscillations are included in potentials derived by means of pseudopotential theory. To avoid some of the limitations associated with the choice of a single potential, in the investigations reviewed below a broad variety of interatomic potentials has been applied. These potentials are shown in Fig. 9.

One of the most characteristic structural features of GBs is associated with the local expansion at the GB plane as the result of the destruction of the perfect-crystal stacking sequence. This expansion gives rise to a local dilution of the electron gas near the GB plane. Atoms interacting across the interface, therefore, experience an electron density different from that of atoms interacting with each other on the same side of the interface. This intrinsically anisotropic character of the interaction of atoms near the interface is obviously not taken into account by *any* pair potential. Another limitation of most potentials for metals is that they are not of an equilibrium type; that is, the lattice is often not in equilibrium at the lattice parameter to which the potential has been fitted. Examples of such potentials are the Born-Mayer potential (which would yield an infinitely large lattice parameter in equilibrium), potentials derived by means of pseudopotential theory, and many spline-fitted potentials. By contrast, most of the simple analytical potentials (Morse, Lennard-Jones, etc.) are of an equilibrium type. This leaves us with the dilemma that this local expansion at the GB may only be considered by using the least sophisticated of all potentials, whereas the more complex potentials based on the physical properties of the electron gas (such as the pseudopotentials) cannot be used for a matter density other than the ideal-crystal density for which they were derived. Faced with this dilemma, in the calculations discussed below the bicrystals have been kept under axial compression to prevent this volume expansion altogether. Unrealistic as this constant-volume restriction may appear, to this author it seems to be the "cleanest" and best-defined way to use central-force non-equilibrium poten-

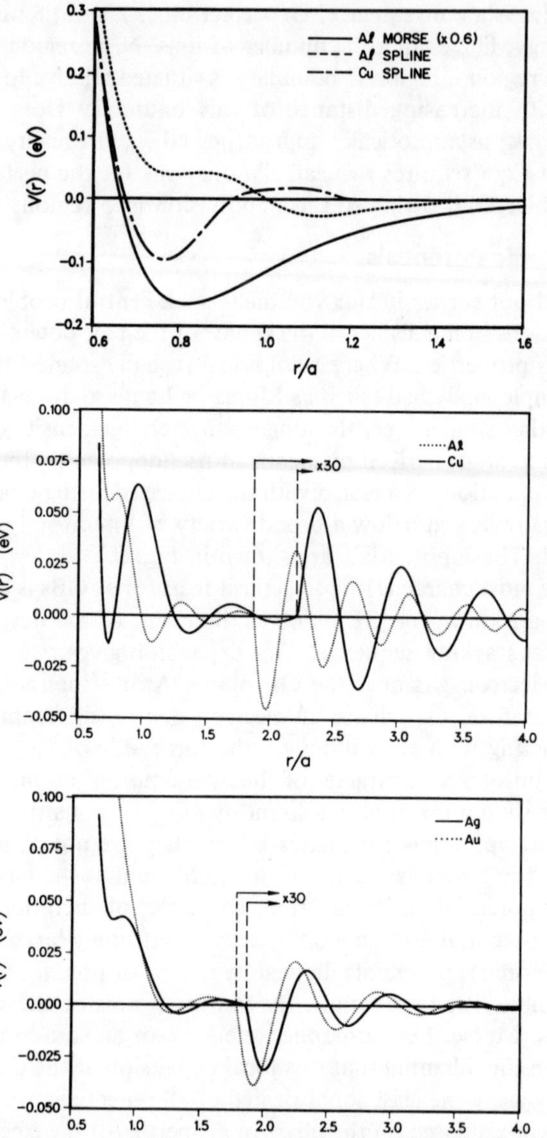

Fig. 9. Interatomic potentials for fcc metals applied in the GB studies discussed in Section 5. Top: Morse potential for Al[(21)] and spline-fitted potentials for Al[(22)] and Cu.[(23)] Center: pseudopotentials for Al[(24)] and Cu.[(25)] Bottom: pseudopotentials for Ag and Au.[(25)] For larger distances, the charge-density oscillations have been enhanced by a factor of 30 for the purpose of clarity.

tials in GB calculations. As discussed in the following chapter in these proceedings, however, the embedded atom method provides hope that these somewhat unrealistic restrictions can be overcome in the near future.

4.4. Lattice Summation and Relaxation Procedure

Applying central-force potentials, the purpose of the lattice summation is the calculation of the energy and forces for all atoms needed to relax the structure towards an energy minimum. In an earlier investigation for fcc metals,[17] the efficiency of different relaxation procedures was tested in some detail. It was concluded that procedures using the force constants (that is, the second derivatives of the potential) in addition to the forces are far superior to gradient methods which use only the forces to find the energy minimum. For further details, see Ref 23.

An interesting feature is the form of the pair correlation function for atom interactions across the GB plane. A typical example is shown in Fig. 10 for three unrelaxed (100) twist boundaries and for an ideal crystal. Whereas in a bulk ideal crystal the ions are situated in well-defined shells (see Fig. 10a), the atoms in the GB region show strong deviations from such an arrange-

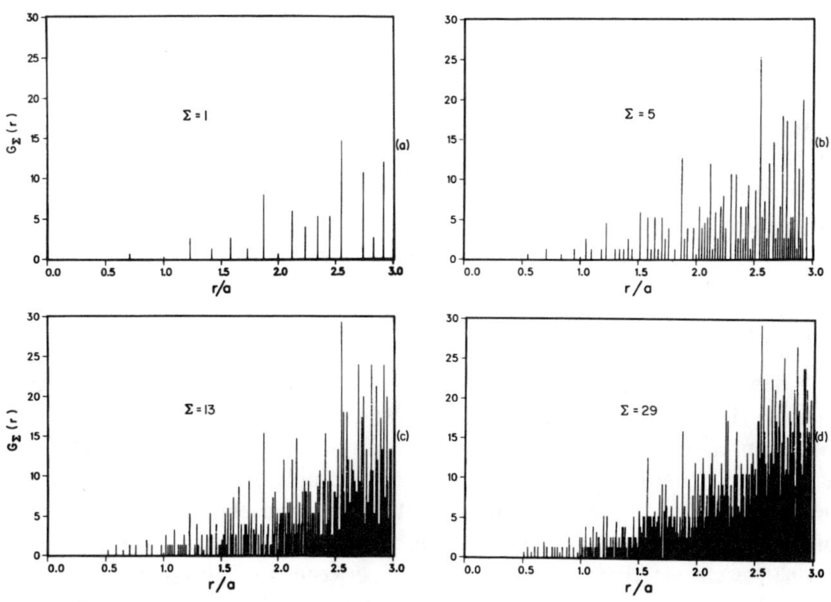

Fig. 10. Sets of interatomic distances between atoms on opposite sides of the $\Sigma = 5$, 13, and 29 (100) CSL twist boundaries in the fcc lattice. For comparison, the limiting case for $\Sigma = 1$ (ideal crystal) is also shown (in arbitrary units).

ment. Their complete pair correlation function may be decomposed into contributions due to interactions with atoms on the same side of the GB plane (similar to Fig. 10a) and interactions with atoms on the opposite side of the interface plane (see Fig. 10b–d). The latter are almost gas-like. It is this part of the pair correlation function which controls the GB properties.

5. Two Criteria for the Selection of Low-Energy Symmetrical GBs

The above computational procedures have been applied in a systematic investigation of grain boundaries in fcc metals.[12,17,27] The purpose of these calculations was twofold: first, to elucidate which of the five macroscopic degrees of freedom of the GB most strongly influences the GB energy and, hence, to find geometrical criteria on the basis of which "special" GBs (with particularly low energy) can be identified; and second, to investigate the role played by the interatomic potential chosen for the calculations on the predicted GB energy and structure.

As pointed out in Section 3, the GB plane and the size of the planar unit-cell area play an important role in the geometrical description of GBs, in particular, in the comparison between tilt and twist GBs. Given that symmetrical tilt GBs are very special twist GBs (see Section 3), for the purpose of investigating the conditions for which symmetrical GBs with particularly low energy are obtained, only twist GBs thus need to be considered. According to scheme 2 in Section 2, this requires the investigation of the role of the GB-plane normal $\vec{n}$, and the twist angle, θ, on the GB energy. The two simple geometrical criteria thus obtained for the identification of "special" GBs will be applied to asymmetrical GBs in Section 6. The role of the interatomic potential on the predicted GB energy and structure will be discussed in Section 7.

5.1. Role of the GB Plane

When creating a planar defect on a given crystallographic plane (with normal $\vec{n}$), the perfect stacking of lattice planes in the $\vec{n}$ direction is destroyed. As a consequence, the perfect coordination of atoms near the defect plane is destroyed; that is, atoms which were, for example, nearest neighbors in the perfect crystal may have been forced much more closely together as a result of the defect creation (see Fig. 10). One might ask about the role played by $\vec{n}$ on the closest distance of approach between any two such perturbed atoms.

The $\vec{n}$ direction in the perfect crystal is characterized by a certain number of staggered lattice planes in the repeat unit, defined in Section 3 as the period $P(\vec{n}) = P(hk\ell)$ (the latter for cubic crystals). As seen from Fig. 10, in the unrelaxed GB the closest distance between any two atoms facing each other across the interface plane is the interplanar spacing, $d(\vec{n})$. In fact, in the energetically worst case, two identical lattice planes which were staggered relative to one another prior to the creation of the interface may end up being right on top of each other at the distance $d(\vec{n})$. Since, in general, $d(\vec{n})$ may be substantially less than the nearest-neighbor distance, d_{NN}, in the perfect crystal, all atoms facing each other across such an interface repel each other substantially.

As a consequence of Pauli's principle, all interatomic interaction potentials are strongly repulsive (with an approximately exponential dependence on the interatomic separation) for short distances of approach between two atoms (see Fig. 9). The GB energy is therefore expected to increase substantially with decreasing separation, $d(\vec{n})$ of the lattice planes facing each other across the GB plane. This effect was confirmed in recent calculations for (100)[17,27] and (111) twist GBs[12,18] (see Fig. 11–13).

The smoothness of most of the energy curves in Fig. 11–13 is remarkable and in strong contrast to the CSL model assumption sketched in Fig. 1. The general shape of the $E_\Sigma(\theta)$ curve was investigated experimentally for (100) twist boundaries in Au by Chan and Balluffi.[29] These rotating-crystallite experiments demonstrated that in contrast to Fig. 1, there are no major cusps on the energy curve for (100) twist GBs with the exception of a minor cusp near $\theta = 36.87°$ (the $\Sigma = 5$ orientation). As seen from Fig. 11–13, a minor cusp near that orientation is predicted for some potentials, particularly the Al pseudopotential, whereas the Au pseudopotential does not yield such a deviation from a smooth curve. This illustrates the limitations in interatomic potentials: one can obviously not expect a specific pair potential to represent a specific material in too much detail.

Figures 11–13 demonstrate, however, that for all seven potentials considered, the (100) GBs always have higher energies (by about a factor of 2–3) than the (111) GBs. This difference is due to the difference in interplanar lattice spacings parallel to the GB plane: $d(100) = 0.5a$ whereas $d(111) = 0.578a$. The less than 20% difference in d-spacing thus gives rise to energy differences exceeding a factor of two, in agreement with the qualitative prediction discussed above that GB energies should depend approximately exponentially on the interplanar spacing.

As discussed in Section 3, STGBs are fully characterized by solely specifying their respective GB plane. The comparison of the energies of symmetrical tilt boundaries, therefore, provides an excellent opportunity to further test this dependence of E_Σ on $d(hk\ell)$. The energies of ⟨110⟩ STGBs calculated by means of the spline potential for Cu in Fig. 9(a) are shown in Fig. 14 (top half), together with the d spacings for the corresponding GB planes (bottom half). Similar to earlier calculations of Hasson et al.[5] in which the Al

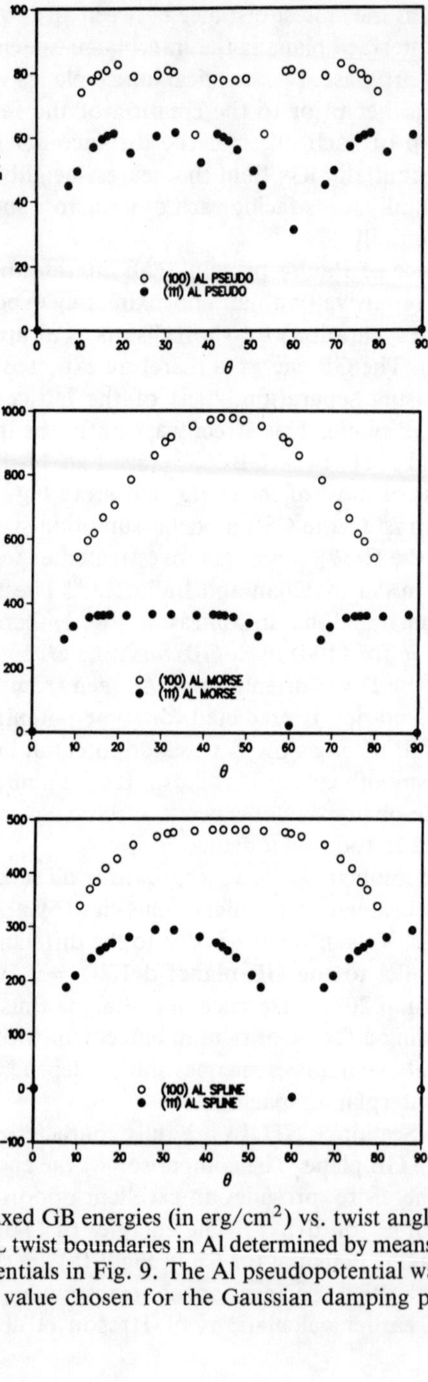

Fig. 11. Relaxed GB energies (in erg/cm^2) vs. twist angle θ for (100) and (111) CSL twist boundaries in Al determined by means of the three different potentials in Fig. 9. The Al pseudopotential was truncated at 3.64a; the value chosen for the Gaussian damping parameter[28] was 0.25/a.

Morse potential in Fig. 9 was employed, the STGB on the (111) plane (with d(111) = 0.578a) shows the lowest energy, followed by the STGB on the (113) plane (with d(113) = 0.302a). Again, although d(111) and d(113) differ by less than a factor of two, the corresponding GB energies differ by about a factor of fifty. As already pointed out by Hasson et al.,[(5)] these results are in excellent qualitative agreement with their thermal-grooving experiments which yielded energy cusps for the same tilt boundaries.

In conclusion, the above calculations demonstrate that larger d-spacing gives rise to lower GB energy. The fact that the planar atom density is pro-

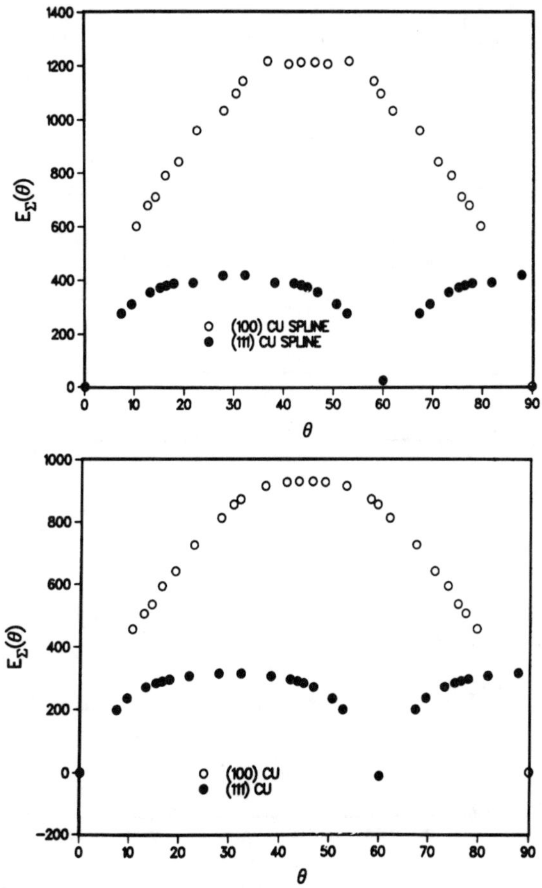

Fig. 12. $E_\Sigma(\theta)$ curves (in erg/cm^2) for (100) and (111) CSL twist boundaries in Cu for the two Cu potentials shown in Fig. 9. For the pseudopotential, a truncation radius $r_c = 3.87a$ and a Gaussian damping parameter of 0.3/a were chosen.

portional to the interplanar spacing leads to the conclusion[12] that in a given crystal structure, the densest planes are the most favored lattice planes for the accommodation of low-energy symmetrical GBs.

5.2 Role of the Unit-Cell Area

As already mentioned, one of the most pronounced structural features of GBs is the translation, $\vec{T}$, of the two grains for which a state of minimum energy is obtained. As illustrated in Ref 12, the GB-energy variation for

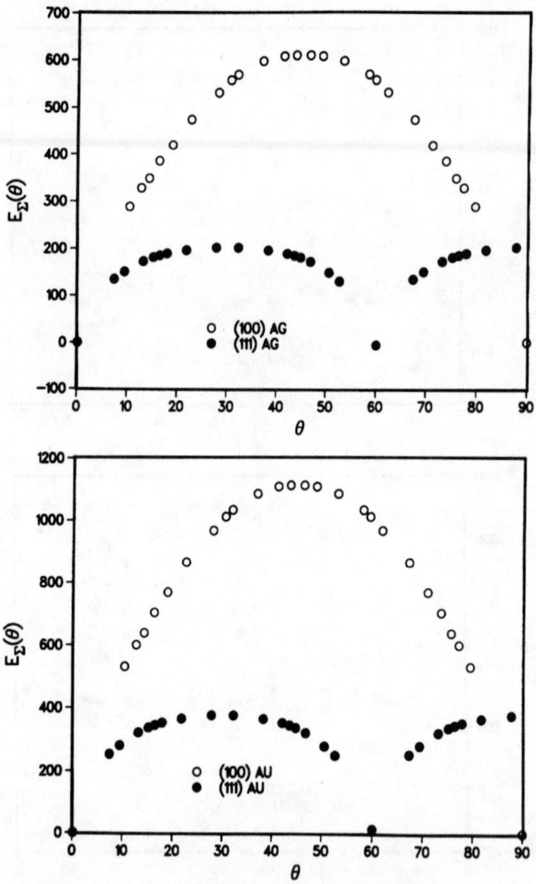

Fig. 13. $E_\Sigma(\theta)$ curves (in erg/cm^2) for (100) and (111) CSL twist boundaries in Ag and Au for the pseudopotentials in Fig. 9. The truncation radii selected were 3.81a and 3.72a, respectively, whereas the same value of 0.25/a was applied for the Gaussian damping parameter.

translations parallel to the GB plane is particularly sensitive to the size of the planar unit cell (for coherent interfaces). More precisely, it is the number of atoms per lattice plane in the unit cell which controls how much energy can be gained by translating from the least to the most favorable configuration

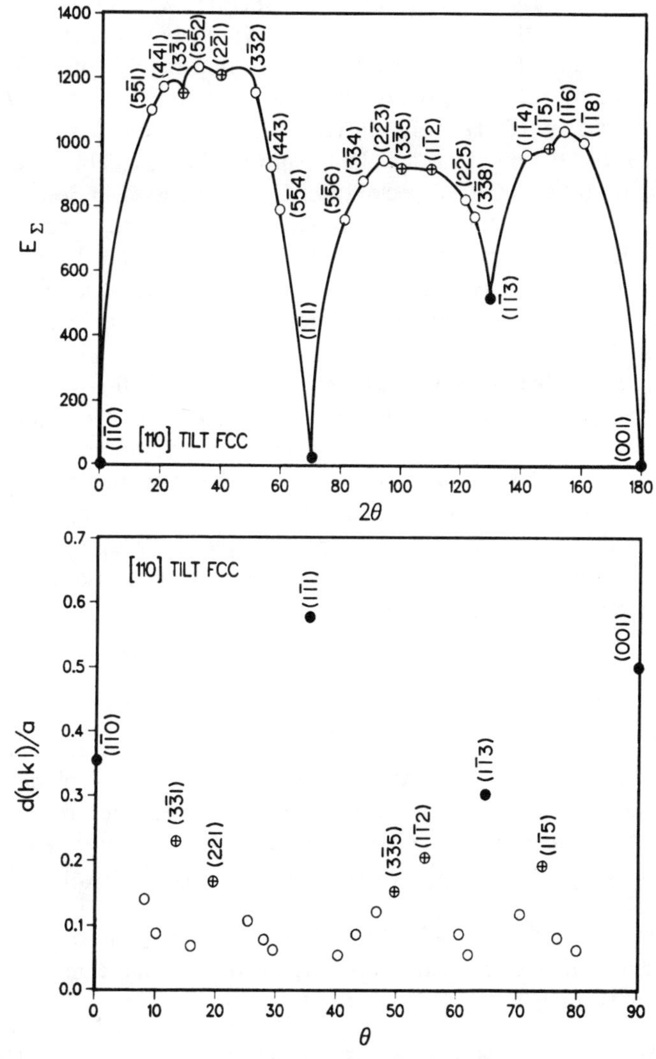

Fig. 14. Relaxed energies (in erg/cm^2) of ⟨110⟩ symmetrical tilt GBs vs. tilt angle for the Cu spline potential in Fig. 9. The bottom half shows the corresponding interplanar lattice spacings parallel to the GB plane (in units of the lattice parameter a).

(see, for example, Fig. 4 in Ref 12). Since the symmetrical tilt GB on a given lattice plane contains only one atom per lattice plane in the unit cell (see Section 3), the translational gain of energy is greater for this "special" GB than for any other (twist) boundary on this plane, thus giving rise to an energy cusp when the twist angle $\theta = 180°/m$ is approached. This effect is apparent in Fig. 11–13 for the (111) twist GB energies which show a cusp near $\theta = 60°$ for which the STGB configuration (namely the (111) twin boundary) is approached (see also Fig. 4). (Notice that m = 3 for (111) planes.) The twist GBs on the (100) plane (with m = 2) also show a cusp. However, the corresponding STGB configuration is identical to the ideal crystal, as is the case for STGBs on all lattice planes with $P(hk\ell) \leq 2$. This is readily seen by considering the STGB as special twist GB as discussed in Section 3: Starting with the ideal-crystal stacking, B/AB/AB/A, and rotating one half by 180°/m, the planar arrangement B/AB/BA/B is obtained. However, translation parallel to the GB plane (to eliminate the unfavorable matching of B planes) such that B → A (and therefore A → B) results in the ideal-crystal stacking.

Our main conclusion may hence be formulated as follows:[12] The translational gain of energy favors GBs with the smallest number of atoms in the planar unit cell. In particular, on a given lattice plane an energy cusp is expected for the special rotation angle for which the symmetrical tilt configuration (with only one atom per plane in the unit cell) is approached. This conclusion demonstrates that the "special" geometry of symmetrical tilt boundaries with respect to all other GBs on the same plane does, indeed, give rise to especially low GB energy.

6. Asymmetrical GBs

As mentioned in the Introduction, for a given misorientation of two crystal lattices the CSL model postulates the lowest-energy GB plane to be the symmetrical configuration; that is, the plane with the highest density of CSL sites. The conditions under which this assumption holds were elucidated in a recent article[30] in which the energies of some asymmetrical tilt GBs were determined.

As discussed in more detail in Ref 30, for most lattice misorientations an asymmetrical GB-plane orientation may be found such that the effective interplanar spacing at the GB (see Fig. 15),

$$d_{eff} = [d(\vec{n}_1) + d(\vec{n}_2)]/2 \tag{10}$$

is greater than for the corresponding symmetrical configuration $d(\vec{n}_s)$. Here $d(\vec{n}_1)$ and $d(\vec{n}_2)$ denote the spacing of lattice planes parallel to the GB

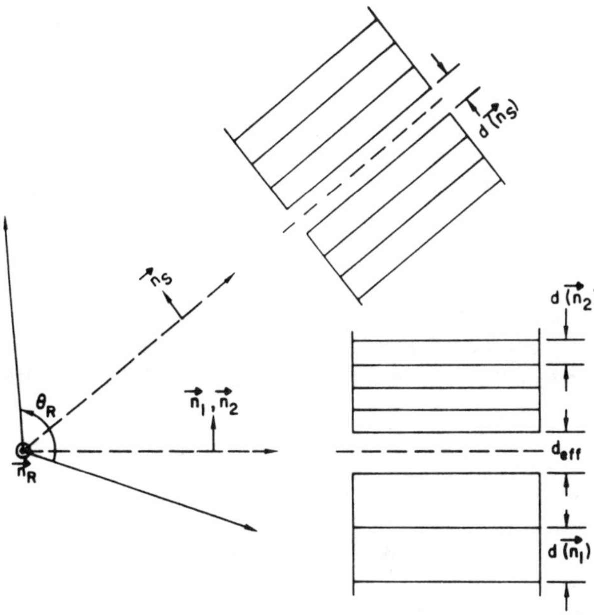

Fig. 15. Symmetrical vs. asymmetrical GB-plane orientations (schematical).

plane in the ideal semi-infinite crystals 1 and 2, respectively, which form the bicrystal. According to Section 5.1, this increase in the interplanar spacing is expected to give rise to a decrease in the GB energy. However, the simultaneous increase in the planar unit-cell area as one goes from a symmetrical to an asymmetrical GB-plane orientation—and the corresponding increase in GB energy (see Section 5.2)—often compensates for this energy gain. The net result is a delicate tradeoff between the increased d-spacing and the increased unit-cell area associated with asymmetrical plane orientations. Hence, the relative stability of symmetrical and asymmetrical configurations has to be investigated in each case, and it may to some degree depend on the detailed nature of the interatomic interaction potentials.[30]

7. Role of the Interatomic Potential in Metals

The effect of the shape and cutoff radius of the interatomic potential on the predicted GB energies and structures was investigated in some detail.[17,18,27] The main conclusions are based on the observation that whereas the GB energy is governed by the very short-range part of the interatomic potential, V(r) (that is, for distances less than the nearest-neighbor distance), the detailed GB structure is influenced by the detailed shape of the potential out to distances of about 1.5a to 2a. These conclusions may be summarized as follows:

(a) The absolute value of the GB energy for a given GB depends strongly on the potential since the magnitude of the short-range (practically exponential) increase in V(r) varies from one potential to another.
(b) The relative magnitudes of the energies of different GBs (i.e., energy vs. misfit-angle curves such as the ones shown in Fig. 11–14) are rather insensitive to a change in potential.
(c) The detailed atomic structure depends strongly on the detailed shape of V(r); in particular, whether or not charge-density oscillations are included in the potential may have a major effect on the computed atomic structure of a given GB.
(d) The lowest-energy translational state of a given GB depends strongly on the shape of V(r).

Acknowledgment

This work was supported by the U.S. Department of Energy, BES-Materials Sciences, under Contract W-31-109-Eng-38.

References

1. R. J. Harrison, G. A. Bruggeman, and G. A. Bishop, in *Grain Boundary Structure and Properties*, edited by G. A. Chadwick and D. A. Smith, Academic Press, New York, 1976, p. 45.
2. S. Yip, Comments Solid State Phys. *11*, 125 (1984).
3. W. T. Read and W. Shockley, Phys. Rev. *78*, 275 (1950).
4. W. Bollman, *Crystal Defects and Crystalline Interfaces* (Springer-Verlag, New York, 1970).
5. G. Hasson, J. Y. Boos, I. Herbeuval, M. Biscondi, and C. Goux, Surf. Sci. *31*, 115 (1972).
6. P. J. Goodhew, in *Grain Boundary Structure and Kinetics* (ASM, Metals Park OH, 1980), p. 155ff.
7. K. L. Merkle, J. F. Reddy, and C. L. Wiley, Ultramicroscopy *18*, 281 (1985).
8. K. L. Merkle and D. J. Smith, Ultramicroscopy *22*, 57 (1987).
9. K. L. Merkle, J. F. Reddy, C. L. Wiley, and D. J. Smith, Proc. Intl. Conf., "Ceramic Microstructures 1986: Role of Interfaces," Berkeley CA, 1986 (to be published).
10. W. Krakow, J. T. Wetzel, and D. A. Smith, Phil. Mag. A *53*, 739 (1986).
11. C. Goux, Canadian Metall. Quarterly *13*, 9 (1974).
12. D. Wolf, J. de Physique Colloque C4, *46*, C4-192 (1985).
13. D. Wolf, Phil. Mag. A (to be published).
14. H. Grimmer, W. Bollmann, and D. H. Warrington, Acta Cryst. A*30*, 197 (1974).

15. See, for example, H. Mykura, in *Grain Boundary Structure and Kinetics* (ASM, Metals Park OH, 1980), p. 445ff.
16. D. Wolf, J. Am. Ceram. Soc. *67*, 1 (1984).
17. D. Wolf, Acta Metall. *32*, 245 (1984).
18. D. Wolf, Physica *131B*, 53 (1985).
19. See R. A. Johnson, Chapter 2 in these proceedings.
20. See J. R. Beeler, Chapter 3 in these proceedings.
21. R. M. J. Cotteril and M. Doyama, in *Lattice Defects and Their Interactions*, edited by R. R. Hasiguti (Gordon and Breach, New York, 1967), p. 1.
22. D. M. Esterling and A. Swaroop, Phys. Stat. Sol (b) *96*, 401 (1979).
23. A. G. Crocker, M. Donegan, and K. W. Ingle, Phil. Mag. A *41*, 21 (1980).
24. L. Dagens, M. Rasolt, and R. Taylor, Phys. Rev. B *11*, 2726 (1976).
25. N. Q. Lam, L. Dagens, and N. V. Doan, J. Phys. F: Metal Physics *13*, 2503 (1983).
26. See M. I. Baskes, Chapter 6 in these proceedings.
27. D. Wolf, Acta Metall. *32*, 735 (1984).
28. M. S. Duesbury, G. Jacucci, and R. Taylor, J. Phys. F: Metal Phys. *9*, 413 (1979).
29. Siu-Wai Chan and R. W. Balluffi, Acta Metall. *26*, 1113 (1985).
30. D. Wolf, Proc. Intl. Conf., "Ceramic Microstructures 1986: Role of Interfaces," Berkeley CA, 1986 (to be published).

6

Fracture From an Atomistic Point of View*

M. I. Baskes and M. S. Daw

Sandia National Laboratories
Livermore, CA 94550

Abstract

The requirements of atomistic fracture calculations are reviewed. Based on these requirements, it is concluded that the Embedded Atom Method (EAM) is ideally formulated to be applied to such atomistic calculations. The EAM is briefly reviewed with an emphasis placed on user application. Specific examples of brittle fracture, dislocation motion, and ductile fracture are presented.

Introduction

Fracture in metals must to some extent be governed by processes which are most easily viewed on the atomistic level,[1] since the processes occurring near a crack tip, such as brittle fracture and dislocation generation, quite obviously involve events on the atomic scale. Away from the crack tip, dislocation motion and pinning of dislocations affect the crack propagation; yet little is known about the core of a moving dislocation and the influence of alloying additions or impurities. Atomistic calculations will eventually evolve to the point where they can give meaningful insight into the relative importance of the different processes. This chapter will describe a significant step that has been taken in advancing this evolution.

Historically, the energies and forces of an atomistic region have been calculated primarily by one of two methods. Both approaches have their inherent limitations. The more fundamental approach has been to solve the best possible approximation to the full Schrödinger equation.[2] Using state-of-the-art high-speed computers (such as the Cray-1), a highly accurate calcu-

*Work supported by the U.S. Department of Energy, Office of Basic Energy Sciences, Division of Materials Sciences.

lation of the energy of one configuration of 10 atoms takes approximately 10 hours of CPU time. The effects of the approximations and the level of accuracy are probably known; however, because of the complexity of these calculations, the fundamental approach leaves one attempting to describe grain boundaries and crack tips by small clusters of atoms. The large surface-to-volume ratio of these clusters makes interpretations very tenuous, especially with regard to fracture processes. On the opposite end of the spectrum is the use of interatomic pair potentials.[(3)] In this approximation the total energy is given by a sum over pairs of atoms. The use of pair potential allows one to carry out calculations on relatively large systems (say 10,000 atoms), providing some hope of describing large-scale effects. For such a system a single energy calculation would take about 10 seconds on a high-speed computer. Clearly, computer time is not as critical a factor as in the first principles calculations using the Schrödinger equation discussed above. However, pair potentials do not have a strong theoretical basis.

One can show the plausibility of the pair-potential scheme in the following way.[(4)] For electrons in a metal, the scattering by the atomic centers can be viewed as relatively weak; therefore a perturbation series can be set up which expresses the energy of the solid in successively higher powers of this scattering. If this expansion is terminated at second order, a pair approximation is obtained. This systematic derivation shows that pair potentials as they are generally applied neglect two important effects. The derivation of pair potentials (or two-body interactions) shows the existence of lower-order effects (one-body interactions) and higher-order effects (N-body interactions, where N is greater than 2). The one-body interaction is known as the *volume-dependent energy*: some view it as the effects of the ambient electron gas created in the solid. This one-body interaction is *required* in addition to the two-body interactions in order to describe the elastic properties of the solid.[(5)] Use of this interaction requires a knowledge of the *volume* of an arrangement of atoms. How does one define volume when the solid is plastically deforming, cracking, or forming internal voids? In general, this volume-dependent term causes research to take one of two tacks: either to ignore it (this changes the elastic properties), or to require the atoms to fit inside a fixed volume (thus no change occurs in the volume-dependent energy). Pair-potential calculations of fracture processes almost without exception neglect the volume dependence of the energy, thus leading to quite serious questions as to the elastic properties of the solid; in a sense, the pair potentials are describing an unrealistic solid with unphysical elastic constants.

Daw and Baskes[(6)] have recently proposed a new framework for calculating the energetics of transition metals which they call the Embedded Atom Method (EAM). In this approach the dominant energy of the metal is viewed as the energy to embed an atom into the local electron density as provided by the remaining atoms of the system. This energy is supplemented by a short-range electrostatic repulsion. The resulting approach combines the computational simplicity needed for defects and amorphous systems with a physical picture that includes many-body effects and avoids the ambiguities of the pair-potential schemes. This method has been applied to such problems as phonons,(7) liquid metals,(8) defects,(6,9) alloys,(9) impurities,(6) fracture,(10,11) surface structure,(12) surface adsorbate ordering,(13) and surface segregation.(9,14) The computer time is not significantly greater than that required for pair-potential calculations.

This chapter will give a qualitative review of the EAM with the emphasis placed on how a user would specifically apply the method. We will present a number of examples of calculations applicable to the atomistics of fracture. Examples of brittle fracture, dislocation motion, and ductile fracture will be given.

Basic Theory: The Embedded Atom Method

Any questions of interest to the fracture problem can be addressed in terms of the energy of the system. From the energy, the forces can be calculated, as well as any thermodynamic, kinetic, or dynamical quantities. Therefore, the basic quantity in atomistic calculations is the energy. In the simplest schemes, approximations are made which enable the energy to be calculated in a simple manner explicitly in terms of the positions of the atoms. The Embedded Atom Method(6) is an example of such a scheme.

The energy of a system of atoms is specified uniquely by the electron density (the distribution of electrons in the solid). One can argue that the major contribution to the energetics is the energy to embed each atom into the electron density of the neighboring atoms. This energy is supplemented by a short-range, doubly screened pair interaction that accounts for the electrostatic repulsions. These ideas have been developed formally,(15,16) where it is shown that the total energy of the system can be written as

$$E_{tot} = \sum_{i} F_i(\rho_{h,i}) + \frac{1}{2} \sum_{i,j} \phi_{ij}(R_{ij}) \tag{1}$$

In this expression, $\rho_{h,i}$ is the total electron density seen by atom i due to the rest of the atoms in the system, and F_i is the embedding energy for placing an atom into that electron density. The ϕ_{ij} is a short-range pair interaction representing the electrostatic repulsion between atoms i and j separated by R_{ij}. The subscripts on F and ϕ indicate that these quantities depend on the *type* of atom (for example, Ni or H); otherwise the functions are universal in that they are used for that element regardless of the environment.

In order to apply this equation, a semi-empirical scheme has been developed by Daw and Baskes.[6] In this scheme, the host electron density is calculated by the simple, yet reasonably accurate, approximation of the superposition of atomic densities:

$$\phi_{h,i} = \sum_{j \neq i} \rho_j^a(R_{ij}) \tag{2}$$

In this expression, $\rho_j^a(R)$ is the atomic electron density contributed by atom j at the distance R from its nucleus. This choice for the electron density makes the actual calculations with this method not significantly more computer intensive than the use of pair-potential models. In addition, simple parameterized forms are assumed to describe the functions $F(\rho)$ and $\phi(R)$. The parameters are then adjusted so that the expression for the total energy (Eq 1) yields an accurate description of the bulk solid. In particular, the functions are required to give the correct equilibrium lattice constant, heat of sublimation, and elastic constants, and to conform to a universal binding function. The details of the fitting procedure are described in Ref 9 and 10.

At this point it is useful to sit back and compare the complexity of the EAM to that of pair-potential calculation. Since the application of both methods is quite similar, only the differences will be pointed out here. An atom is chosen for which the energy (and forces) is desired. To save computer time, a finite interaction distance is always used; the atoms within this distance are obtained. Depending on the problem of interest these neighbors may be obtained from a stored list of atoms or recalculated at each time step. The pair interaction of each neighbor with the initial atom is computed either by table-lookup and interpolation or with an analytic function. In the EAM the electron density is computed using the same neighbors; in complexity this is similar to the pair interaction. The embedding energy, a function of this density, is added to the pair term. Thus we may see that the EAM differs only marginally in complexity from the simple pair-potential calculation. Because of the great similarity it is easy to modify current pair-potential codes to perform EAM calculations.

Calculations

This section describes how the atomistic calculations are performed. In particular, calculations are made of (a) the effects of hydrogen and helium on brittle fracture, (b) dislocation motion in a perfect lattice, and (c) dislocation emission from a crack tip. All calculations have been done on fcc nickel.

The calculations proceed by setting up a configuration of atoms representing a crack or dislocation, with or without some impurity atoms present. The motion of the atoms is obtained by solving Newton's equation of motion with the internal forces obtained from Eq 1 (molecular dynamics). (The velocities are started off with a random distribution particular to a specified temperature.) External conditions are applied to achieve a specified state of stress or strain. The resulting evolution of the system is then observed and compared to runs at other stresses, temperatures, and impurity distributions. Solving Newton's equation conserves the total energy of the system, so that the temperature (which is computed from the velocities) can and does vary. Typically the runs are carried out for tens of picoseconds and in the range of temperatures between 0 and 300 K.

Brittle Fracture Calculations

For the calculations involving a crack tip, we start with the basic geometry shown in Fig. 1(a) and introduce a thin crack whose length runs parallel to the slip planes. Because of the periodicity of the calculation, this crack is reproduced in other unit cells, which gives an array of parallel cracks, separated by the periodicity in the [110] direction. Forces are applied to the surfaces along the surface normals. One important aspect of the enforced periodicity should be pointed out here. In constraining the periodic area to be fixed and applying tension to the sample, the calculation does not allow the Poisson effect.

The brittle fracture calculations are done using either energy minimization (appropriate only for calculations at 0 K) or molecular dynamics. In either case, the stress is increased gradually until the sample fractures, since once the crack tip has begun to propagate, the fracture has, in every case, progressed to completion. We define the stress at which propagation begins as the fracture stress. Hydrogen atoms can be introduced in the slab and the fracture stress recomputed. (Note that the periodicity means that each hydrogen atom is actually one of an infinite string of atoms, and each string is one of an infinite array of strings.) The change in fracture stress due to the introduction of the hydrogen is then considered. The fracture stress was obtained for (a) pure nickel, (b) pure nickel with a crack, (c) nickel hydride (a hydrogen atom in every octahedral interstitial site) with a crack, and (d) pure nickel with nickel hydride on the crack plane in advance of the tip. The results were tested for different slab thicknesses (in the [111] direction) and periodic lengths (in the $[1\bar{1}0]$ and $[11\bar{2}]$ directions) and are independent of those conditions. Different stress rates were applied, and a rate of $10^{10}C_{44}$/s was found to be slow enough to yield convergent results. Constant strain-rate calculations were also performed, but the results were no different from those of the stress-rate calculations. Two different temperatures were used in the starting configurations (0 and 300 K).

The results of these molecular dynamics calculations can be studied by

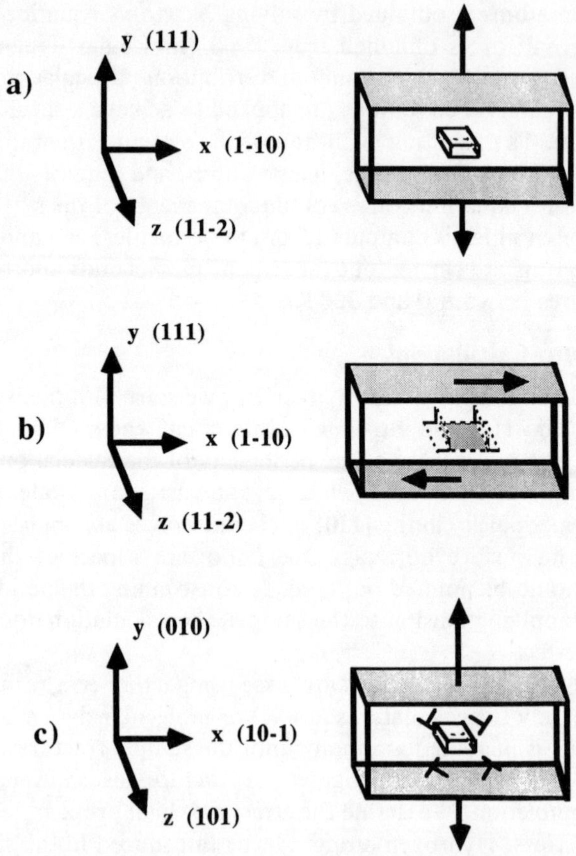

Fig. 1. Basic geometry used in the calculations. The slabs shown are periodic in the x and z directions. The y direction is finite (bounded by free surfaces). The slabs are oriented so that (a) brittle fracture is enhanced, (b) dislocation motion is unimpeded by free surfaces, and (c) dislocation emission from the crack tip is enhanced (ductile fracture).

plotting the strain (computed from the average separation of the two surfaces) versus the stress (a linear function of the time). This is done for the four cases discussed above, as shown in Fig. 2. The stress-strain curve for the perfect crystal is shown, as well as the cases for a slab with a crack, a slab with a monolayer of hydrogen in the plane of the crack, and a hydrided material with the same crack. The fluctuations in the stress-strain curves are artifacts of the molecular dynamics simulation on a *finite* system; the strain oscillations are actually lattice vibrations, where the slab is vibrating in a breathing mode, plotted versus time (because the stress is a linear function

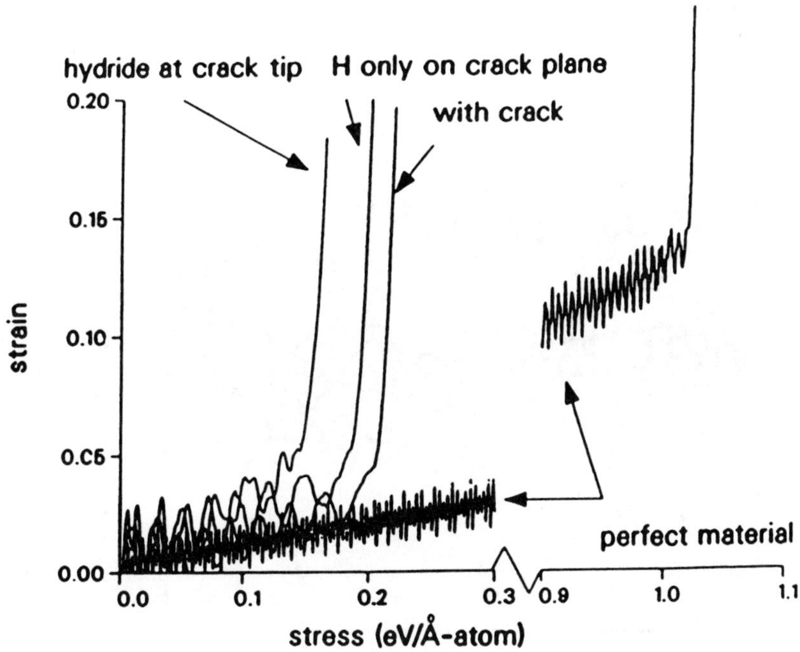

Fig. 2. Calculated stress-strain curves for four different slabs, all related to the one shown in Fig. 1(a). The perfect material is like the slab in Fig. 1(a), but without the crack. The curve labeled "with crack" is that for the slab in Fig. 1(a). The effects of adding hydrogen only to the crack plane and then hydriding the crack tip region can be seen. The oscillations, due to the molecular dynamics simulations, are discussed in the text.

of time). The periods of the fluctuations are related to the periods of the slab's normal modes. The strength of the oscillations are related to the size of the system; that is, for an infinite system, the relative strength of the oscillations goes to zero. Thus, in the limit of an infinite system, the oscillating curve reduces to the smooth stress-strain curve that one is accustomed to viewing. One can see from the curve for the perfect material that the slab has been strained into the anharmonic region and fractures at a high stress (in this case, the theoretical strength of the material). Introducing the crack greatly reduces the fracture stress, putting hydrogen along the plane of the crack continues this reduction, and hydriding the material reduces the stress even further.

A dramatic picture (Fig. 3) of the effect of hydrogen on fracture may be seen by looking at a snapshot of another slab just after fracture has been initiated. In this case we have created hydride on only the left half of the

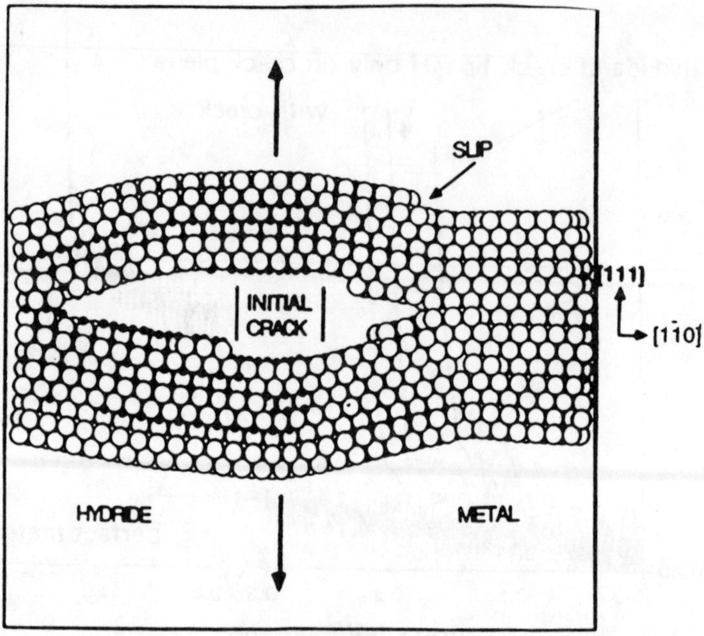

Fig. 3. Atom positions just after fracture. The open circles represent nickel atoms; the filled, hydrogen atoms. A hydride was initially created on the left half of the slab. As the stress was increased, the hydride fractured brittlely while the crack tip in the nickel blunted.

slab. We see a brittle fracture initiated in the hydride, while in the more ductile matrix, the crack has blunted. Slip has occurred in the pure nickel at this very high stress in a high-energy direction.

A second example is the effect of helium on the brittle fracture stress. The results presented in Fig. 4 were obtained by the energy minimization technique using the geometry of Fig. 1(a) with no initial crack. The stress was increased until an unstable configuration was found. This stress (the brittle fracture stress) is shown in Fig. 4 as a function of the defect concentration, that is, periodicity in the z direction. Note that at high density (short periodicity) the stress tends to a constant value. The small cluster of helium atoms (He_6V) has a much greater effect than a substitutional helium (HeV) or a vacancy (V).

Dislocation Calculations

In order to verify the validity of the EAM, we have performed some simple calculations of dislocation motion. Calculations of dislocation motion are carried out with the same basic slab, except that a dislocation is introduced

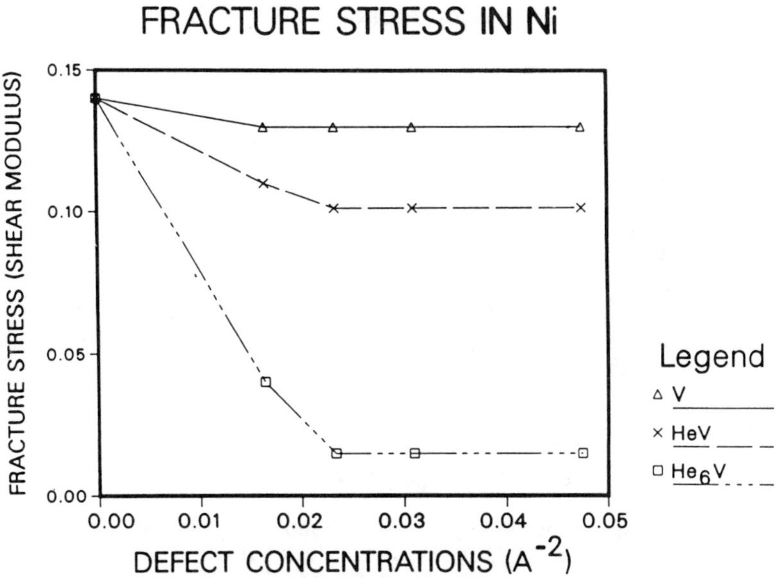

Fig. 4. Brittle fracture stress (relative to the shear modulus) as a function of defect concentration. The fracture stress is reduced only slightly by the introduction of vacancies (V). Putting helium in the vacancies (HeV) reduces the stress further, while a small helium cluster (He_6V) decreases the stress dramatically.

initially by inserting two planes in the top of the slab, as in Fig. 1(b). Shear stresses are achieved by applying external forces on the surface atoms in the slip direction. In the dislocation calculations, the stresses required to move the dislocation are so small that we have found it unnecessary to apply the stress in a ramped fashion as was required in the brittle fracture calculations. In the dislocation dynamics calculations, the stresses quoted are applied at the beginning of the runs and held constant. When subjected to a shear stress, the dislocation accelerates to a terminal velocity, as shown in Fig. 5 for several stresses at 0 K. This motion may be considered as acceleration with a drag. Note that the terminal velocity is reached for only the case of the highest stress. This terminal velocity is close to the speed of sound in a finite slab of the dimensions used here. A complete analysis of the dislocation motion including comparison to continuum models may be found in Ref 10.

It is important to note that observations of dislocation motion in a real material reflect the average speed of the dislocation as it jumps from one pinning site to another. This average speed can be considerably less than the terminal speed attainable in a perfect solid.

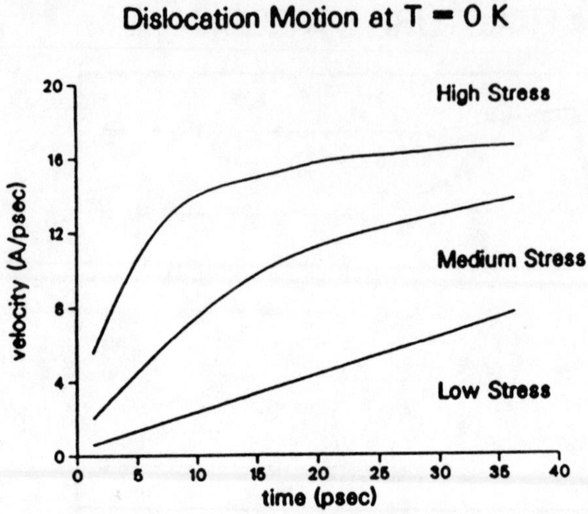

Fig. 5. Edge dislocation velocity, in the slip direction $[1\bar{1}0]$, for the dislocation in Fig. 1(b), under various shear stresses. The low, medium, and higher stresses are 0.015%, 0.06%, and 0.24% of C_{44}. The dislocation accelerates to a terminal velocity, but in the lower-stress cases the time illustrated is insufficient for the velocity to reach its terminal value.

Ductile Fracture Calculations

Using the geometry in Fig. 1(c), we are able to investigate the propensity for dislocation emission from a crack tip. This geometry was chosen so that the slip planes are at a favorable angle (with respect to the applied stress) to promote dislocation emission. For a given initial temperature a constant tensile stress is applied and molecular dynamics is used to follow the atom positions. Using a previously published algorithm[10] we are able to detect the presence of a dislocation when it is about two nearest-neighbor distances from the crack tip. Since the probability for dislocation emission is a random process dependent upon the specific thermal atomic positions, we repeat these calculations many times with different initial thermal velocities in order to obtain the probability of dislocation emission as a function of time.

It is likely that the probability of the emission of the first dislocation, p, at time, t, is given by

$$p = 1 - e^{-\alpha(t-t_0)} \tag{3}$$

where α is a function of applied stress and temperature, and t_0 represents a "dead time" due to nucleation processes and the transport of the dislocation to a point where our algorithm detects it. If Eq 3 is correct, a plot of $-\ln(1 - p)$ vs. t should yield a straight line. In Fig. 6 we have plotted the

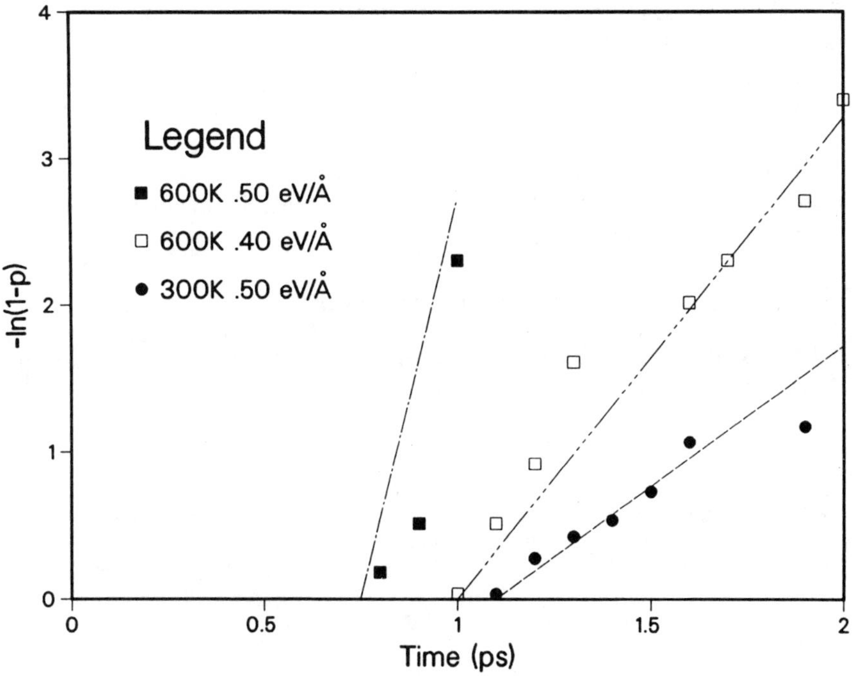

Fig. 6. The probability of the emission of the first dislocation, p (plotted as $-\ln(1 - p)$), vs. the time after the stress is applied. The points shown represent the results of 30 molecular dynamics runs for each temperature and force per surface atom.

results from our atomistic calculations in the above form. The data points represent results of 30 individual runs at each temperature and stress. Note first of all that the points lie approximately on straight lines. In addition, the steepest slope (highest probability) occurs at the highest stress and temperature. The slope decreases, as expected, for lower temperature or stress. A more complete analysis is in progress.

Summary

The various atomistic methods have been briefly reviewed with the conclusion that the EAM provides an ideal formulation for fracture calculations. Details of the EAM have been presented emphasizing the needs of the user. The EAM has been applied to brittle fracture in the presence of impurities, namely hydrogen and helium, which are found to reduce the stress required for brittle fracture. Dislocation motion as a function of stress and time has been presented. Finally, an application to the emission of dislocations from a crack tip at various temperatures and stresses has been given as an example of the ductile fracture process.

Results of these calculations are in qualitative agreement with experimental deformation in hydrogen and helium embrittlement observations in "real" materials. Of course, we are a long way off from modeling polycrystalline, multiphase, structural alloys, but hopefully these atomistic results are the start of a series of calculations that will systematically evaluate proposed deformation and failure mechanisms.

Acknowledgment

We would like to thank K. Hadley for his help in performing the dislocation emission calculations.

References

1. See papers in *Advances in the Mechanics and Physics of Surfaces*, Vol. 2, edited by R. M. Latanision and T. E. Fischer (Harwood, N.Y., 1983).
2. See, for example, C. L. Briant and R. P. Messmer, Phil. Mag. B*42*, 569 (1980); A. C. Switendick, Sol. State Comm. *8*, 1463 (1970); and C. D. Gelatt, Jr., H. Ehrenreich, and J. A. Weiss, Phys. Rev. B*17*, 1940 (1978).
3. See papers contained in *Interatomic Potentials and Crystalline Defects*, edited by J. K. Lee (Metallurgical Society of AIME, New York, 1981).
4. W. A. Harrison, *Pseudopotentials in the Theory of Metals*, (W. A. Benjamin, Inc., New York, 1966); for a specific example, see also L. Dagens, M. Rasolt, and R. Taylor, Phys. Rev. B*11*, 2726 (1975).
5. R. A. Johnson, Phys. Rev. B*6*, 2094 (1972).
6. M. S. Daw and M. I. Baskes, Phys. Rev. B*29*, 6443 (1984).
7. M. S. Daw and R. L. Hatcher, Sol. State Comm. *56*, 697 (1985).
8. S. M. Foiles, Phys. Rev. B*32*, 3409 (1985).
9. S. M. Foiles, M. I. Baskes, and M. S. Daw, Phys. Rev. B*33*, 7983 (1986).
10. M. S. Daw, M. I. Baskes, C. L. Bisson, and W. G. Wolfer in *Modeling Environmental Effects on Crack Growth Processes*, edited by R. H. Jones and W. W. Gerberich (Metallurgical Society, Inc., Warrendale PA, 1986), p. 99.
11. M. I. Baskes, MRS Bulletin XI, 14 (1986).
12. M. S. Daw, Surf. Sci. 166, L161 (1986).
13. S. M. Foiles and M. S. Daw, J. Vac. Sci. Technol. A*3*, 1565 (1985); T. E. Felter, S. M. Foiles, M. S. Daw, and R. H. Stulen, Surf. Sci. 171, L379 (1986); M. S. Daw and S. M. Foiles, Phys. Rev. B *35*, 2128 (1987).
14. S. M. Foiles, Phys. Rev. B*32*, 7685 (1985).
15. M. S. Daw (in preparation).
16. K. W. Jacobsen, J. K. Nørskov, and M. J. Puska, Phys. Rev. B (to be published).

7

Computer Simulation of Dislocations on an Atomistic Level

DONALD M. ESTERLING
School of Engineering and Applied Science
George Washington University
Washington, DC 20052

Abstract

The atomic structure near the core of a dislocation can influence the macroscopic yield stress, including orientation and temperature dependence. Certain techniques for the computer simulation of the atomic geometry, for both an unstressed and a stressed system, are outlined. A crucial aspect of any simulation is the interatomic interaction. Most studies have assumed a central force model. A model based on the electronic tight-binding method is proposed as a viable alternative that accommodates noncentral forces. A survey of atomistic model results is provided along with some suggestions for future research.

1. Introduction

Dislocations control plastic flow, and an understanding of their behavior is central to the development of constitutive equations, such as might be required in modeling a high-temperature, large-deformation process like forging. The stress-strain fields associated with dislocations in a homogeneous continuum are available in various texts.(1) To connect with the behavior of dislocations in materials of interest, we require the interaction of dislocations with large numbers of solute/interstitial atoms arranged in random or only partially ordered configurations. In addition, the motion of dislocations will be affected by non-point defects such as other dislocations, grain boundaries, voids, precipitates, and external surfaces. Textbook(1) results are available only for the interaction of dislocations in a continuum model with isolated point and line defects and simple interfaces. Further, even for a single, isolated dislocation, the continuum representation develops a well-known (l/r) singularity in the stress-strain field at the dislocation core. In sum, realistic input to constitutive equations requires: (1) dealing with complex defect arrays and (2) to the extent that core effects are important, correct (nonsingular) behavior in the near-field region.

This chapter will address the latter of these issues. Atomistic modeling of the dislocation core region will be summarized, along with suggestions as to where core effects may be significant. The interaction of a dislocation with a large number of randomly placed solute atoms will be analyzed within the context of a semicontinuum model in the succeeding chapter.[2] The atomistic model will assist the semicontinuum model by providing certain model parameters (such as Peierls stress, point-defect–dislocation interaction), whereas inadequacies in current semicontinuum models will suggest areas where more detailed atomistic calculations are required.

Atomistic modeling of dislocations has proceeded for over two decades.[3,4,5] At issue are the following: WHY use atomistic rather than continuum models; HOW to carry out these calculations; WHAT results have been obtained over this period; and WHERE is the field proceeding.

As already pointed out, continuum models produce unphysically large strains near the dislocation core. A convenient remedy—setting the core region to a constant stress or energy contribution—is analogous to the "plastic" region just in advance of a crack tip, wherein the stress is often set to a constant value. The simplification is only valid when core effects are not significant. However, as we will outline shortly, the core geometry can dominate slip behavior in bcc metals and the dominant (and perhaps controlling) dislocation–solute atom interaction occurs at short (atomic) distances.

Techniques for carrying out atomistic modeling are presented elsewhere in this volume.[6] Essentially the process consists of (1) selecting an "appropriate" interatomic interaction;[7] (2) setting up the atomic geometry in some initial configuration (typically that predicted by elasticity theory for the given defect); (3) establishing an appropriate boundary condition (such as fixing certain atoms outside the central relaxed region at their initial positions); (4) then iteratively moving the atoms until they reach equilibrium positions (minimizing, for example, total system potential energy, or until the maximum net force on any atom is sufficiently small), for static solutions, or, for dynamic solutions, following the atomic position changes. The fourth step entails 3N coupled nonlinear equations (corresponding to the equation of motion for the N atoms being relaxed in 3D space) and may be solved by any of an assortment of numerical techniques (Newton-Raphson, conjugate gradient, etc.).

There is some concern regarding the choice of boundary condition (that is, "rigid" versus "flexible") but, particularly with modern computing capacity, the relaxed atomic region can be large enough (a few thousand atoms for a line defect, with periodic boundary conditions along the line direction) so that the boundary does not influence the final geometry for STATIC problems periodic along one direction (such as a straight dislocation). If, however, a yield stress is sought, then the back stress from a rigid boundary can overwhelm and mask the true stress for dislocation motion.

Various solutions to the back-stress effect exist, such as iteratively updating the boundary condition for a displaced dislocation center (that is, making the far-field displacement field consistent with the current dislocation center). Woo and Puls[8] have proposed a clever scheme for determining the yield stress which actually exploits the presence of the back stress.

Relaxation methodology and boundary conditions are, for the most part, straightforward issues with fairly well-defined techniques. In contrast, interatomic interactions are a thorny issue that was raised (vehemently) in the first major symposium on atomistic modeling,[5] and the discussion has continued, unabated, to the present day. While various authors have claimed to have "solved" the interaction problem (either by sweeping the problems aside by saying that the important effects are all controlled by atomic geometry, not by the details of the interaction, or by proposing yet another interatomic potential), nay-sayers (such as this author) remain who suggest that, while certain phenomena are indeed geometry-controlled, the detailed understanding needed for constitutive relations requires much more sophisticated interactions than the central force models usually employed. One exception is the very careful work on potassium[9] where the interaction—determined from pseudopotential theory—is well established. As a step beyond pure central force models, volume-dependent contributions to the system energy and force field have been proposed (see Ref 10). We will suggest in a latter section a variation on the electronic tight-binding model as an approach converging to an interaction based on truly sound first principles. In sum, the subject of interatomic interactions remains in flux, but, as we will suggest in the next section, definite progress is being made. Certainly, there is virtually unanimous agreement that resolution of the interatomic interaction problem is crucial before we can perform atomistic modeling as a *predictive* (as opposed to *expository*) science.

2. Interatomic Interactions: Electronic Tight-Binding Approximation

A definitive set of forces on each atom could be obtained by a careful evaluation of the change in the electronic energy as each atom is displaced. Typically this energy subsumes the simple coulombic interaction between the nuclear cores. Unfortunately, because of computational constraints, a complete solution of the many-electron Schrödinger equation, including exchange and correlation effects, can only be carried out either: (1) for small clusters (the domain of quantum chemistry), or (2) for materials where the many-electron Schrödinger equation may be approximated by a one-electron Schrödinger equation, typically employing a local-density approximation. The latter is usually used in conjunction with a plane-wave expansion which is poorly convergent for d and f orbitals found in metals of metallurgical interest and, further, becomes inconvenient for aperiodic systems (such as a lattice with an extended defect). Recently, an extension of the pseudopotential method, long recognized as valid for simple metals such as sodium, has been proposed.(11) A systematic extension of traditional band-structure techniques to transition metals would be of enormous assistance to atomistic modeling. While the interaction range obtained from pseudopotential calculations is of very long range and hence makes the direct application to simulations computationally difficult,(7) such calculations will provide guidance as to the importance of noncentral forces and, more important, tests of various schemes such as we will propose, using the tight-binding method, to approximate the interaction by short-range terms.

The tight-binding method, which is one of the oldest of the electron band-structure methods to solve the one-electron Schrödinger equation,(12) is enjoying a renaissance of interest. This method can be adapted to arbitrary geometries, and there is at least empirical evidence that the method is physically reasonable. (As a simple example, the method provides an explanation of the variation in the cohesion energy across the 3d and 4d series in the periodic table.) As the name implies, the method is most applicable to the more tightly bound d and f valence band electrons, rather than the more extended s or p electron states. We will return to the treatment of the s and p electron effects (and hybridization terms) at the end of this section. The resurgence of interest in the tight-binding method is attributable, in part, to the development of efficient computational algorithms—moment methods from the French school(13) and also, more recently, recursion methods from Cambridge.(12) However, as we will show, even these recursion methods are computationally prohibitive in treating the relaxation of extended defects. After outlining the method, we will propose a variation of this recursion method which reduces computational time by orders of magnitude with only a modest sacrifice in accuracy.

The genesis of the tight-binding method was the linearized combination of atomic orbitals band-structure method. In this latter method, the solution, ψ_k, to the one-electron Schrödinger equation,

$$H\psi_k = \epsilon_k \psi_k \tag{1}$$

was assumed to be a linear combination of atomic orbitals ϕ_α (the α index refers to a particular atomic site and a particular orbital),

$$\psi_k(r) = \sum_\alpha c_{\alpha,k}\phi_\alpha(r) \tag{2}$$

Solution entails diagonalization of the Hamiltonian matrix elements:

$$h_{\alpha\beta} \equiv \langle\alpha|H|\beta\rangle = \int d^3r\phi_\alpha(r)H\phi_\beta(r) \tag{3}$$

In more recent applications, the "hopping" parameters $h_{\alpha\beta}$ are viewed as disposable parameters obtained, for example, as the natural output from a KKR or APW band-structure calculation.

The Cambridge recursion method is a transformation of this model into a new basis set such that the matrix formed from the Hamiltonian (the transformed $h_{\alpha\beta}$) is tridiagonal. Tridiagonal matrices are much easier to diagonalize than general matrices. Further, this transformation is achieved with computationally straightforward recursion relations that link the original hopping parameters ($h_{\alpha\beta}$) to the certain recursion parameters which we will denote as a_i or b_i. In this new framework, the one-electron Green's function can be expressed as a continued fraction:

$$G(\epsilon) = \cfrac{1}{\epsilon - a_1 - \cfrac{b_1^2}{\epsilon - a_2 - \cfrac{b_2^2}{\epsilon - a_3 - \ldots}}} \tag{4}$$

$G(\epsilon)$ is useful since the one-electron density of states, $\rho(\epsilon)$, is just:

$$\rho(\epsilon) = \lim_{\delta\to 0}\left[-\frac{1}{\pi}\,\mathrm{Im}\,G(\epsilon + i\delta)\right] \tag{5}$$

More accurately, each G and ρ refers to a particular site (and orbital) α; so Eq 5 really defines $\rho_\alpha(\epsilon)$. The total density of states is then:

$$\rho(\epsilon) = \sum_\alpha \rho_\alpha(\epsilon) \tag{6}$$

As mentioned, the coefficients (a_i,b_i) can be expressed in terms of $h_{\alpha\beta}$. For example (for site and orbital α):

$$b_1^2 = \sum_\beta h_{\alpha\beta}h_{\beta\alpha} \tag{7}$$

Higher-order a's and b's involve sums and products of the form

$$\sum_{\beta,\gamma,\ldots} h_{\alpha\beta}h_{\beta\gamma}h_{\gamma\ldots} \tag{8}$$

This formulation reveals some of the attractions of the tight-binding recursion method (TB-RM). No particular geometry is required, just some path counting. (However, for each inequivalent site or orbital α, a new $G(\epsilon)$ must be calculated.) The model is in real space (rather than the more abstract momentum space) and hence can focus on a local region of interest. As a drawback, many high-order a's and b's must be calculated to obtain a reasonable electron density of states.

Not to fear—calculating 11 a's and b's to obtain $\rho(\epsilon)$ takes about 3 CPU seconds on, for example, a VAX 11/780. Unfortunately, when we wish to perform a defect simulation, this is multiplied by:

(a) A factor of 4 (to compute a force vector we need 3 energy differences plus a reference energy);
(b) Another factor of 15; to get proper energy differences, we must compute change in energy on all affected atoms, as one is slightly displaced;
(c) Another factor of 25 (we need to do d orbitals, not s orbitals);
(d) Another factor of 20 (say we only relax 20 atoms in our TB-RM regions); and
(e) Another factor of 50 (roughly the number of iterations required for relaxation) leading to a CPU time on the order of 10^6 seconds!

Our modified TB-RM is based on two simple concepts. First, the crystalline environment "heals" fairly quickly to almost perfect crystal away from a defect center. Second, the higher-order a's and b's have a sort of orthogonalization property relative to the low-order a's and b's; that is, they tend to depend most strongly on the geometry of regions "far" from the atom/orbital of interest. Hence, we simply replace all a's and b's above a certain order by their perfect crystal values. We have done this for a range of electron densities and crystal defects with b_2 through b_{11} and a_3 through a_{11} fixed to perfect crystal values and a_1, a_2, and a_2 environment-dependent. The resulting forces compared quite favorably with those computed with all $a_1 \ldots a_{11}$ and $b_1 \ldots b_{11}$ taken as environment-dependent.[14,15]

As a slightly more extreme approximation (but probably at no serious loss of qualitative correctness), even a_2 may be fixed. The primary role of a_2 is to shift the zero of energy, which would have no influence on forces that are energy differences. Finally, there is no need to explicitly take energy differences since the expressions are available[15] for computing the forces directly from the recursion parameters.

An issue to resolve is the treatment of the s and p electron and hybridization terms. Pettifor,[16,17] and Gelatt et al.[18] have shown that for d-band metals, the primary role of s/p electrons is to fix such overall crystal properties as equilibrium lattice spacing and bulk modulus, whereas d-electron terms dominate geometry-sensitive problems. Hence, the approximation of the non-d-electron terms by a simple repulsive potential, chosen in conjunction with the d-electron contribution—to reproduce the correct lattice spacing and bulk modulus—is probably reasonable.

In the next section, we will include some results pertinent to dislocation modeling which have been obtained using both central force models and a modified tight-binding recursion method.

3. Atomistic Model Results

In keeping with our vertical theme of forging, the results we report will emphasize metallic systems. Indeed most simulation studies have been on metals. A review of dislocations in ionic crystals is provided by Puls.(19) The work on dislocations in semiconductors tends to focus on the effect of an unrelaxed defect on the electronic states.(20,21,22) The simulation of dislocations in metals can be classified by geometry (fcc, hcp, and bcc). Reviews of each class may be found in Ref 23. In the close-packed fcc metals, the dislocation dissociates into a pair of Shockley partial dislocations with 1/6 [112] Burgers vectors, forming a stacking fault. The width of the stacking fault depends on the interatomic potential, since the separation of the partials depends both on the interface energy (the energy per unit area of the fault for infinite separation of the partials, which we will refer to as the "stacking-fault energy") and on the repulsion of the partial dislocation. The latter depends on the partial dislocation core structure, which is dependent on the interatomic potential. An interesting experiment in atomistic modeling is provided by a series of potentials suggested by Baskes and Melius.(24) These semiempirical potentials, shown in Fig. 1, are obtained by fitting to an impressive array of properties (such as lattice spacing, elastic constant, vacancy formation energy) which include what we would call the "unrelaxed stacking-fault energy." This energy is that obtained for an infinite fault with the atoms positioned as in a hard-sphere model (see Ref 1). The "unrelaxed stacking-fault energy" is then a simple lattice sum. With the Baskes-Melius potentials, this number varies by almost an order of magnitude from aluminum through copper to silver. Yet, when these potentials are used in an atomistic simulation,(25) the stacking-fault widths vary by at most 10 to 20 percent, as shown in Fig. 2–4. Here the arrows represent the local σ_{xz} and σ_{yz} stress components, as explained in Ref 26. The final results reflect the results of relaxation and core structure effects. The moral is that, in semiempirical potential modeling, "more" is not necessarily "better." The empirical model must show that it is reasonable in explaining phenomena or parameters not fitted by the model or, in the above case, even a result (stacking-fault *width*) which is closely affiliated with the fitted parameter.

Atomistic modeling can be particularly helpful in edifying anomalous mechanical behavior. For example, in AB_3 ordered alloys having the $L1_2$ structure, the yield stress is relatively insensitive to temperature at low temperature, rapidly *increases* with increasing temperature up to a peak, and then drops sharply.(27) These effects have been shown to be linked to ease of

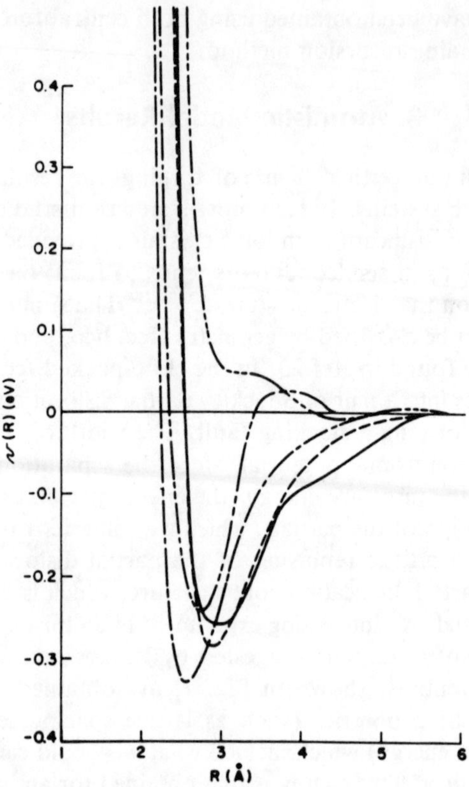

Fig. 1. Interatomic potentials for fcc stacking fault calculations: ——, the Baskes-Melius aluminum potential; ---, the Baskes-Melius silver potential; —–—, the Baskes-Melius copper potential; —-–—, the Englert-Tompa-Bullough copper potential; —---—, the Moriarty aluminum potential.

cross-slip which, in turn, depends critically on the atomistic dislocation core geometry.[28,29]

Reference 30 has reported finding dislocation segments in {111}- and {100}-type glide planes in aluminum in addition to the expected {110} glide plane. Esterling has carried out atomistic modeling of a screw dislocation using two rather similar potentials.[25] With the aluminum potential derived from pseudopotential theory,[31] the dislocation core geometry shows a very narrow splitting along a {100}, rather than {110}, plane, as shown in Fig. 5.[32] (Note correction on page 101 of Ref 25.) These phenomena may be related and aluminum is offered as a material which would benefit from further atomistic modeling, particularly at high temperature.

Little work has been carried out on hcp metals, except for the work of

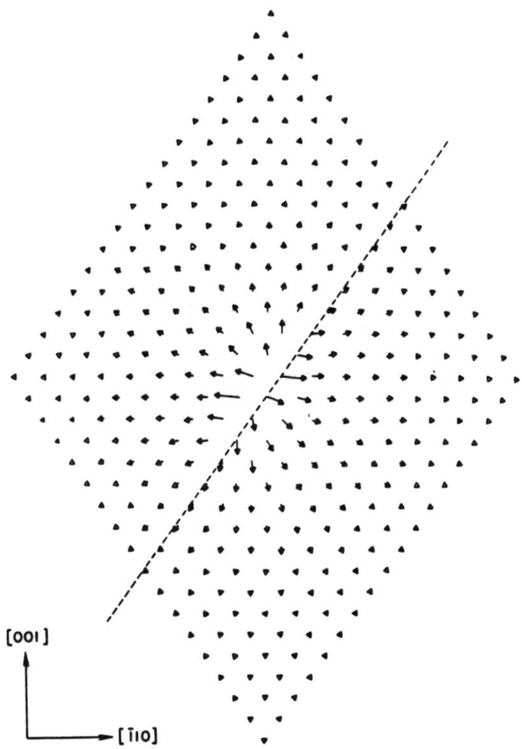

Fig. 2. Stress tensor field with the Baskes-Melius aluminum potential.

Bacon.[33] These metals show substantial variation from an ideal lattice-parameter ratio ($c/a = \sqrt{8/3}$) for close-packed spheres and display different preferred glide planes and loop habit planes.[34] Some understanding of this phenomena has been obtained by atomistic modeling,[33] but Bacon concludes that (to resolve some fundamental questions) "it will probably be necessary to develop many-body potentials which match particular metals."

The dominant slip system in fcc metals is that predicted by a simple hard-sphere model (close-packed directions and close-packed planes). As discussed earlier, the dominant slip system in hcp metals varies, with some clarification available from atomistic models. There is no well-defined slip system in bcc metals; rather, a "wavy" slip is observed. There is a strong temperature dependence to the yield stress as well as a variance from the Schmid orientation dependence of the yield stress. A number of atomistic simulations have been carried out which have been major contributors to our understanding of this phenomenon.[23]

The yield stress in bcc metals and alloys has a large component as the

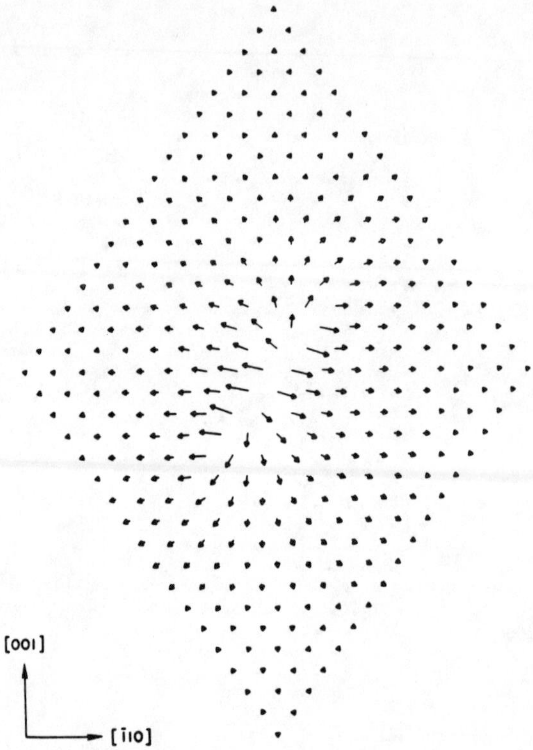

Fig. 3. Stress tensor field with the Baskes-Melius copper potential.

result of an intrinsic lattice or Peierls stress. In fcc metals, the dislocation dissociates into a planar fault, bounded by partial dislocations. In bcc metals, a number of atomistic simulations[26] have produced a quite different core geometry for screw versus edge dislocations. For a variety of central potentials, the edge dislocation dissociates into a narrow planar fault. In contrast, a screw dislocation dissociates into a narrow nonplanar configuration having threefold symmetry. A sample result for an atomistic calculation for a bcc metal is shown in Fig. 6. In that figure, a threefold symmetry in the core geometry is apparent. This figure uses a "differential displacement" map. The arrows joining nearest neighbors indicate the relative displacement of each pair in excess of the standard displacement found in a perfect (undislocated) bcc lattice. The potential used was a variation on the Johnson iron potential.[35] Duesbery and collaborators[36] have investigated the influence of shear stress on the atomic configuration, using a similar potential. That paper provides a further explanation of "differential displacement" maps as well as such figures for the dislocation core as the stress is increased. They

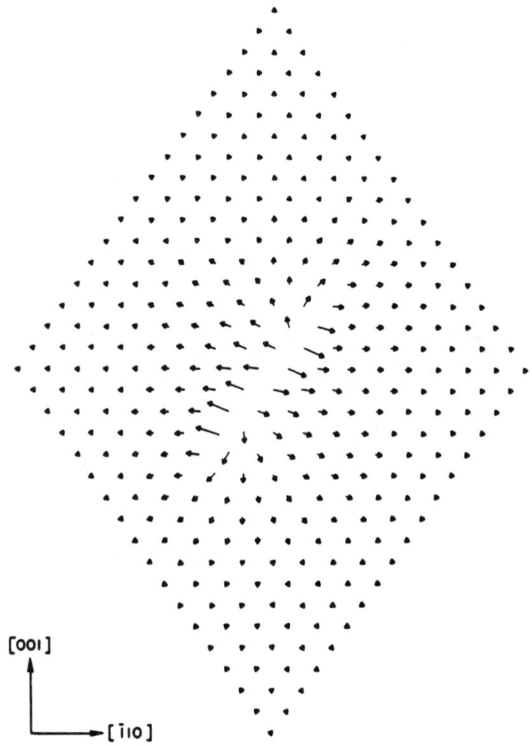

Fig. 4. Stress tensor field with the Baskes-Melius silver potential.

find successive slip on nonparallel {110} planes with net movement in a {112} plane.

These atomistic studies result in a Peierls stress for a screw dislocation well in excess of the Peierls stress for an edge dislocation. The latter tends to display a planar (rather than threefold) symmetry. Hence, these studies conclude that the slip process is controlled by the screw component of a dislocation. This is in agreement with electron microscope experiments.[37]

The strong temperature dependence in the yield stress that has been related to an activation energy due to the underlying large Peierls barrier can be understood from atomistic calculations. An elaboration of the temperature dependence of the yield stress is provided by Ref 38.

The Peierls stress in bcc metals displays a strong dependence on shear components which should not interact with a screw dislocation. Duesbery[39] has shown that the Peierls stress can vary from zero to the theoretical shear strength of the lattice, depending on the orientation of the lattice relative to the applied stress. This variation was shown to occur for a vari-

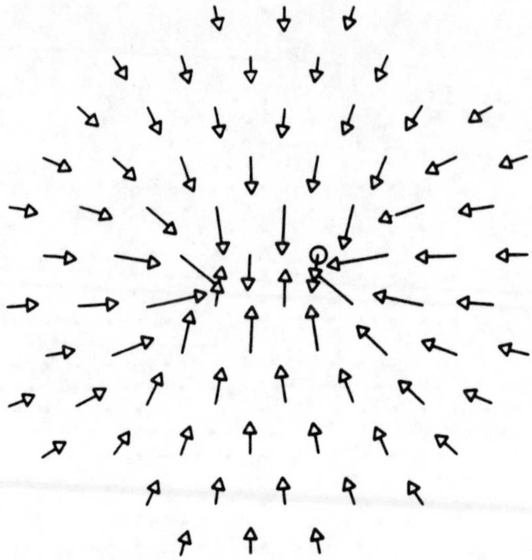

Fig. 5. Stress tensor field with the Moriarty aluminum potential.

Fig. 6. Differential displacement field for the Johnson-Wilson iron potential.

ety of interatomic potentials. Duesbery's atomistic simulations produced the important insight that in a crystal under stress, the core structure and corresponding Peierls stress of a screw dislocation depend most sensitively on the glide strain rather than the glide stress. Anisotropic elastic coupling effects produce a dependence of the glide strain on non-glide stress components—explaining a great deal about the anomalous orientation dependence of the yield stress in bcc metals.

Most atomistic simulations have been carried out with semiempirical central potentials which are of dubious validity in transition metals. For the simple metals (that is, sodium and potassium), the interatomic forces are well characterized by pseudopotential theory calculations which suggest a pair-potential model is quite valid for these metals. Basinski and Duesbery[(9)] carried out very detailed atomistic modeling in potassium and obtained an orientation dependence at variance with that found in the refractory bcc metals. After the initial modeling was performed, the orientation dependence was measured experimentally[(39)] and found to be in agreement with the calculations. In addition, the magnitude of the predicted Peierls stresses was in fair agreement with experiment.

All bcc atomistic calculations of a screw dislocation using central forces have produced a core configuration similar to that shown in Fig. 6. While this general geometry may indeed be the correct geometry, caution must be observed. Each of the atomistic calculations has used rather similar pair potentials, essentially Johnson potentials.[(35,7)] Takeuchi[(41)] and Sato and Masuda[(42)] have generated core geometry for a bcc screw dislocation with models that include noncentral forces. Alternate core geometries were generated. Esterling and co-workers[(15)] have investigated lattice stability and defect geometry in bcc lattices using two models—one with central, the other with noncentral forces—and obtained very different results. In the noncentral force model the interactions were obtained by a combination of contributions from the modified tight-binding recursion method described previously and a repulsive pair potential with parameters chosen such that the net forces led to the proper lattice spacing and bulk modulus in a perfect crystal. The contributions to the forces between atoms for each of these contributions were then calculated as a function of lattice spacing. In the central force model, the noncentral tight-binding terms were replaced by a pair potential which, together with the original repulsive pair potential, reproduced the same interatomic force versus lattice spacing. When each of these interatomic force models was used in an atomic simulation studying generalized stacking-fault behavior (essentially resistance to shear) and dislocation core geometry, very different results were obtained. Tight-binding parameters can be chosen which maintain stability against shear deformation.[(43)] The tight-binding parameters used in the above study were *ab initio*, obtained from an analysis of band-structure calculations. The points to be made are: (a) The *ab initio* calculations must be done accurately enough to produce tight-binding parameters which maintain lattice stability; and (b) noncentral force components can strongly influence defect geometries, in this case in a deleterious fashion.

4. Future Directions in Atomistic Modeling

In order to move atomistic modeling from an expository to a predictive capacity, we must have reliable interatomic forces. For the refractory bcc metals and alloys this almost certainly means the inclusion of noncentral forces. Baskes[10] has presented one mechanism which entails local "volume" contributions. The tight-binding recursion method offers a different approach. If both approaches are "correct," then there must be some equivalence. A study employing both techniques for the same localized defect would be of some interest. What is important is that there must be some way to systematically improve the model. In the past, this was done by using ever more data in fixing semiempirical potentials. The tight-binding method can be improved by better *ab initio* techniques to determine the tight-binding parameters (such as using quantum chemical cluster techniques to fit the parameters) and by putting all the electronic energy contribution (s, p, and d) on a common footing rather than treating the "non-d" term in an empirical fashion. There is no clear way to systematically improve the embedded atom central potential plus volume-dependent term models, since the volume dependence is again determined by empirical data rather than as a result of a systematic expansion such as is available in pseudopotential theory.

For the most part, atomistic calculations have dealt with isolated defects. In order to deal with real materials, we must accommodate alloy effects and complex defect structures. Except under very restricted conditions, these cannot be modeled by atomistic simulations because of the computational limits in current and any forseeable system. Rather, a link must be established via semicontinuum models, which are able to accommodate such complexities. The chapter by Arsenault in these proceedings (Chapter 8) carries on this discussion in the context of semicontinuum modeling.

References

1. J. P. Hirth and J. Lothe, *Theory of Dislocations*, (McGraw-Hill, New York, 1968).
2. R. J. Arsenault, Chapter 8 in these proceedings.
3. A. Englert and H. Tompa, J. Phys. Chem. Solids *21*, p. 306 (1961).
4. A. Englert and H. Tompa, J. Phys. Chem. Solids *24*, p. 1145 (1963).
5. P. C. Gehlen, *Interatomic Potentials and Simulation of Lattice Defects*, J. R. Beeler and R. I. Jaffee (Plenum Press, N.Y., 1972).
6. J. R. Beeler, Chapter 3 in these proceedings.
7. R. A. Johnson, Chapter 2 in these proceedings.
8. C. H. Woo and M. P. Puls, Phil. Mag. *35*, p. 1641 (1977).
9. Z. S. Basinski and M. S. Duesbery, *Proc. Int. Conf. on Dislocation Modelling of Physical Systems*, edited by M. F. Ashby, R. Bullough, C. S. Hartley, and J. P. Hirth, p. 273 (Pergamon Press, New York, 1980).

10. M. I. Baskes, Chapter 6 in these Proceedings.
11. J. A. Moriarty, Phys. Rev. Lett. *55*, p. 1502 (1985).
12. V. Heine, D. W. Bullett, R. Haydock, and M. J. Kelly, Solid State Physics *35*, edited by H. Ehrenreich, F. Seitz, and D. Turnbull (Academic Press, New York, 1982).
13. J. P. Gaspard and C. Lockmann, J. Phys. Chem. *6*, p. 3077 (1975).
14. I. M. Boswarva and D. M. Esterling, J. Phys. C *15*, L729 (1982).
15. D. M. Esterling, D. K. Som, A. K. Chaterjee, and I. M. Boswarva, J. Phys. F *17* (1987).
16. D. G. Pettifor, J. Phys. F *7*, p. 613 (1977).
17. D. G. Pettifor, J. Phys. F *8*, p. 219 (1978).
18. C. G. Gelatt, H. Ehrenreich, and R. E. Watson, Phys. Rev. *B15*, 1613 (1977).
19. M. P. Puls, in *The Structure and Properties of Crystal Defects*, edited by J. V. Paidar and L. Lejcek, p. 282-292 (Elsevier, Amsterdam, 1984).
20. I. L. F. Ray and D. J. H. Cockayne, Proc. Rog. Soc. *A325*, p. 543 (1971).
21. R. Jones, J. de Phys. *40*, p. C6 (1979).
22. S. Marklund, Phys. Stat. Sol. (B) *100*, p. 77 (1980).
23. *The Structure and Properties of Crystal Defects*, edited by V. Paidar and L. Lejcek (Elsevier, Amsterdam, 1984).
24. M. I. Baskes and C. F. Melius, Phys. Rev. *B20*, p. 3197 (1979).
25. D. M. Esterling, A. R. McGurn, U. M. Boswarva, and R. Arsenault, J. Mat. Sci. Eng. *68*, p. 97 (1984).
26. V. Vitek, Cryst. Latt. Defects *5*, p. 1 (1974).
27. P. A. Flinn, Tran. TMS-AIME, *218*, 145 (1960).
28. M. Yamaguchi, V. Paidar, D. P. Pope, and V. Vitek, Phil. Mag. *A45*, p. 867 (1982).
29. V. Paidar, M. Yamaguchi, D. P. Pope, and V. Vitek, Phil. Mag. *A45*, p. 883 (1982).
30. D. Caillard and J. L. Martin, Acta Met. *30*, p. 91 (1982).
31. J. A. Moriarty, Phys. Rev. *B10*, p. 3075 (1974).
32. D. M. Esterling and A. Swaroop, Phys. Status Sol. *B96*, p. 401 (1979).
33. D. J. Bacon in *The Structure and Properties of Crystal Defects*, edited by V. Paidar and L. Lejcek, p. 102-120 (Elsevier, Amsterdam, 1984).
34. I. P. Jones and W. B. Hutchinson, Acta Met. *29*, p. 951 (1981).
35. R. A. Johnson, Phys. Rev. *134A*, p. 1329 (1969).
36. M. S. Duesbery, V. Vitek, and D. K. Bowen, Proc. Rog. Soc. *A332*, p. 85 (1973).
37. H. Saka, K. Noda, and T. Imura, Cryst. Latt. Defects *4*, p. 45 (1973).
38. M. S. Duesbery, Proc. R. Soc. London *A392*, p. 145 (1984); *A392*, p. 175 (1984).
39. Z. S. Basinski, M. S. Duesbery, and G. S. Murty, Acta Met. *29*, p. 801 (1981).
40. S. Takeuchi, Phil. Mag. *A39*, p. 661 (1979).
41. A. Sato and K. Masuda, J. Phys. Soc. Japan, *50*, p. 519 (1981).
42. K. Masuda, N. Hamada, and K. Terakura, J. Phys. F *14*, p. 47 (1984).

8

Thermally Activated Motion of Dislocations

R. J. Arsenault

Engineering Materials Group
University of Maryland
College Park, MD 20742

Abstract

In this chapter the computer simulation of thermally activated dislocation is reviewed. This review involves the motion of a single dislocation (for the most part) on a slip plane in a lattice that contains a random array of barriers. These barriers can be discrete defects which are widely spaced so there is no interaction between the defects, or the barriers can be due to solute atoms in a concentrated solid solution, where there is an overlap of stress fields of the solute atoms. In all cases the interactions are determined by continuum elasticity theory.

The computer simulation modeling of dislocation motion in a dilute random array is fairly well understood and there is general overall consensus. In the case of concentrated solid solutions there have been very few investigations. However, the modeling as applied to fcc solid solutions, at low temperatures, agrees quite well with the experimental data, but for the bcc solid solutions there is no agreement. The modeling of high-temperature dislocation motion is very limited.

1. Introduction

The primary purposes of fundamental investigations of plastic deformation are to determine the barrier-to-dislocation motion and the mechanism by which the dislocation overcomes the barrier. The procedure generally followed is to measure first the activation parameters (that is, activation enthalpy ΔH, activation area A, and the appropriate Gibbs free energy ΔG). These experimentally determined activation parameters are then compared with the theoretically predicted values for a proposed mechanism. In this chapter the concern will be directed solely to determining how the theoretically predicted values can be obtained. Because of limited space, only some of the computer simulation modeling will be reviewed.

The discussion of the modeling will be broken down into the following sections:

- Dislocation motion through a dilute random array
- Multiple dislocation motion
- Solid-solution strengthening and weakening
- High-temperature creep

Each section begins with a brief introduction, followed by discussion of the model (in all cases, unless specifically stated, the modeling is that of Arsenault et al.), then a review of some of the results, and a conclusion.

2. Dislocation Motion Through a Random Array of Direct Barriers

2.1. Introduction

The critical stress required to move a dislocation through a random array of short-range barriers (srb) at *absolute zero temperature* has been determined by Kocks[1,2] and Foreman and Makin.[3] They observed that the stress required is less than that required to move a dislocation through a square array of the same concentration of srb. The lower stress required in the random array case is related to the fact that the stress required to overcome a line of srb is inversely related to the spacing between srb. In the random-array case it is always possible to find a spacing of srb greater than in the square-array case. Also, it was noted that the dislocation frequently moves by an "unzipping" mechanism—that is, a lateral or sideways motion of the dislocation. Dorn et al.[4] also investigated the motion or the configuration of a dislocation as it moved through a random array and again proved that "unzipping" does occur. Das, Guraraj, and Richman[5] have determined that "unzipping" also occurs if there is a viscous drag present.

Frost and Ashby[6] have investigated the motion of a dislocation through a square array of srb which have a square-shaped force-distance profile. They assumed the time necessary for the dislocation to overcome the barrier was a probability function which was dependent on temperature. However, this time was less than the time necessary for the dislocation to bow out between the srb. Once the dislocation had attained the equilibrium position, the dislocation had jumped the srb. Thus they were able to determine the conditions necessary for the motion of the dislocation at the drag velocity. They were not concerned directly with a configuration of the dislocation as it moved through the array. This analysis could be considered as a quasi-thermally activated motion of the dislocation.

2.2. Model

A random distribution of srb was placed on a plane, that is, the slip plane, and the internal stress was assumed to be zero. The minimum separation between srb was 1b. There were two different force-distance profiles assumed or used for the srb. One of the force-distance profiles (Fig. 1) used was that obtained by Gibbs[7] as follows:

$$F(x) = F_0\left\{\frac{(x/b)^2 + (1/\sqrt{2})(x/b) - 1}{((x/b)^2 + 1)^2}\right\} \tag{1}$$

where $F_0 = \Delta\epsilon b^2\mu/3.86$, $\Delta\epsilon$ is the tetragonal strain, μ the shear modulus. For vanadium the value of $\mu = 46.7$ GPa was used; b is the Burgers vector, and the value of b for vanadium $= 2.63 \times 10^{-8}$ cm was used. This force-distance relationship is zero at $x = \infty$. However, thermal fluctuations cannot occur over distances greater than 10b, so a cutoff of 10b was assumed. The second force-distance diagram used was a square-shaped function where the integrated area under the delta function is equal to the integrated area given by Eq 1, and the maximum force of the height of the square-shaped function is equal again to the maximum force defined by Eq 1.

Initially, a dislocation enters onto the slip plane and moves at a drag velocity defined by the following equation:

$$v_D = B\tau \tag{2}$$

where B is a drag coefficient and τ is the applied stress. Since the long-range internal stress is zero, $\tau = \tau^*$ (the effective stress). When the dislocation

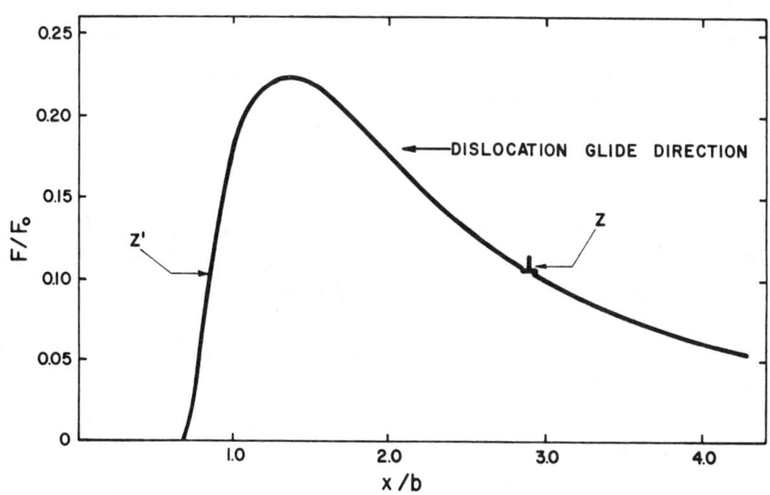

Fig. 1. Force-distance diagram for the tetragonal interaction.

makes contact with the srb, a force balance is determined by the following equation:[3]

$$\mu b^2 \cos\frac{\phi}{2} = F(x) \tag{3}$$

where ϕ is the included angle about the srb; and it is assumed that the line tension is $\mu b^2/2$, and F(x) is defined by Eq 1 or by the delta function. The bow out between the initial points of contact occurs as a circular arc and is defined as follows: $R = \mu b/2\tau$. When the dislocation is bowing out, a search is made to determine if contact is made with other srb. If contact is made, a new force balance has to be obtained at each point of contact with a srb. Then, again, the dislocation bows out between each srb where contact is made, and once more a search is made. This process is continued until an equilibrium force balance is obtained. The position of equilibrium is defined as Z′ (Fig. 1), and the second position of force equilibrium is defined as Z. When the square-shaped function is used to define the force-distance diagram, the positions of force equilibrium are defined directly by the square-shaped function. The main reason for using a delta function force-distance diagram was to reduce the computer time involved in making the calculations. The Gibbs force-distance profile was used initially for some of the investigations concerning the configuration of the dislocation as it moved across the slip plane. However, an immense amount of computer time was required to obtain a force balance because the force on a particular srb was dependent on the position of the dislocation on the neighboring srb.

If the applied force (F_A) is greater than the strength of the barrier (F_{max}), the dislocation is allowed to move over the barrier and the whole process is repeated. When a force balance is obtained and there are no points of contact where F_A is greater than F_{max}, the time required to overcome each srb where contact is made is determined by the following equation:

$$t_i = t_0 \exp\frac{\Delta G_i(\tau)}{kT} \tag{4}$$

where $t_0 = \ell/bv_D$, ℓ is the average spacing of the srb, v_D is the Debye frequency, and $\Delta G_i(\tau)$ is the activation energy of the ith barrier. The activation energy is determined by the following equation:

$$\Delta G_i(\tau) = \int_Z^{Z'} F(x)\,dx - F(x)|_z \Delta x \tag{5}$$

where the limits of integration are obtained from the force balance and F(x) is again defined by Eq 1 or the delta function.

The dislocation then is allowed to move over the srb that has the shortest jump time. In allowing the dislocation to jump only the srb with the shortest jump time, it is assumed implicitly that the probability of the dislocation jumping any other srb is very small during the time required to jump the srb with the shortest jump time. The shortest time then is stored after the jump has been made. This time also is accumulated at other points of contact and a linear relationship is assumed between the probability of overcoming a srb and time. The time accumulated by a dislocation at a srb reduces the time necessary to overcome the srb. For example, if the minimum time to overcome one of the srb (No. 1) is 1 s, and that of another srb (No. 2) is 10 s, then the dislocation has increased its probability of overcoming the srb No. 2 by 10% when it has jumped srb No. 1. Now, if as a result of the first jump and a new force balance, the jump time at srb No. 2 is decreased by 2 s (and this is the minimum jump time), then only 1.8 s is the time that is stored and not 2 s. This procedure is continued until the dislocation first emerges from the array. What is neglected is the possibility that the dislocation might jump the barrier during the time it is bowing out to the static equilibrium position.(6) However, the time involved in the bow out for the small stress case is approximately $\simeq 10^{-8}$ s, whereas the controlling jump time may be 10^3 s. For larger stress the bow out times are shorter, that is, 10^{-11} s, while the controlling jump times are still $\simeq 10^3$ s. Therefore, it was assumed that the dislocation would not jump the srb during the time required for the bow out to the static equilibrium position. The average velocity is determined as follows:

$$\bar{v} = \frac{X_L}{\sum t_i} \tag{6}$$

where X_L is the length of the slip plane in a direction of motion of the dislocation, and t_i is the time stored for each jump. Arsenault and Cadman(8) have examined four methods of selecting which srb the dislocation will jump, and they have shown that all of the methods give approximately the same result.

2.3. Results

Configuration of the Moving Dislocation. There are two possible means by which a dislocation can traverse the slip plane, depending on the concentration of srb, the strength of srb, applied stress and temperature. If the stress is large or the temperature is high, the dislocation moves at the drag velocity and traverses the slip plane as a straight line. If the stress is reduced or the temperature is lowered so that the time required to overcome *some* srb is not zero, then the dislocation line does not remain straight even though the dislocation motion is still at drag velocity. The nature of dislocation motion is not by an "unzipping" phenomenon; that is, it is not like the lateral motion of single kinks after a double kink has nucleated dislocations on a screw in bcc metals. Figure 2 is a plot of successive positions of a dislocation as it moves through a random array of medium-strength srb ($\Delta\epsilon = 1.2$). The x axis of Fig. 2 has been expanded to show the actual "bulge outs" that occur when the dislocation moves. Figure 3 is also a plot of successive positions of the dislocation as it moves across the slip plane. This is for higher-strength srb ($\Delta\epsilon = 12$). In this case, the dislocation moves by forming a

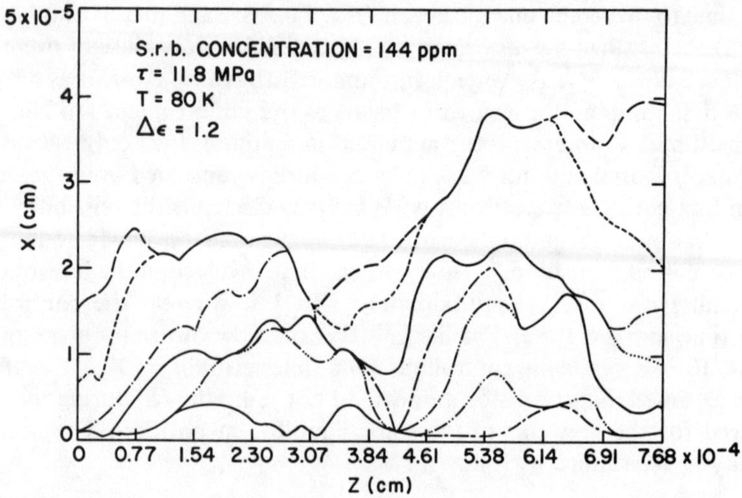

Fig. 2. The successive positions and configurations of the dislocation as it moves across the slip plane. The different types of lines represent the different locations of the dislocation.

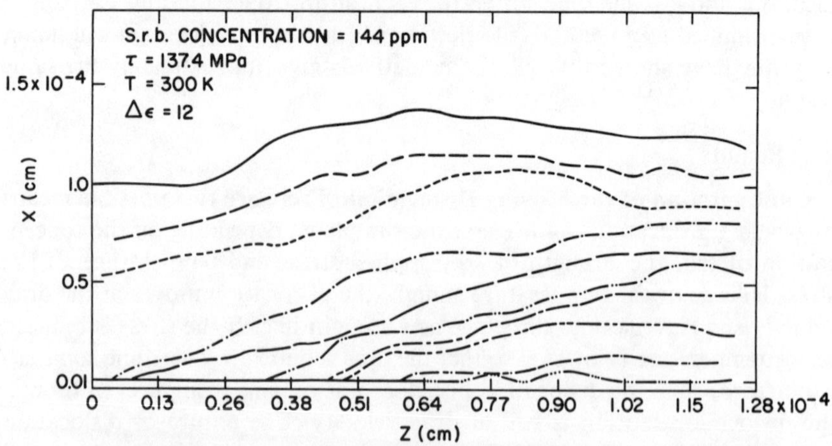

Fig. 3. The successive positions and configurations of the dislocation as it moves across the slip plane. The different types of lines represent the different locations of the dislocation.

small bulge; that is, the dislocation moves by what may be defined as "unzipping" over distances of five jumps or five srb in the z direction before it moves forward in the x direction. However, after the dislocation has made these five or so jumps in the z direction, it then makes another jump in the x direction; that is, a bulge begins to form. The dislocation bulges out a number of rows (where a row is defined as x_L/N_x, and N_x is the average number of srb in the x direction). The magnitude of the bulge out is defined as the number of rows between the minimum and maximum advance of the dislocation. After and during this bulge-out stage, the dislocation begins to move laterally, that is, "unzipping." However, the included angle ϕ between any three points of contact only rarely approaches the breaking angle as defined by Eq 3. As the "unzipping" is occurring, the portions of the dislocation which are moving more or less parallel to the x axis also advance slowly in the x direction. The lateral motion of the dislocation does not appear to be a unique function of the boundary conditions employed; but the magnitude of the bulge out is a function of these boundary conditions and parameters. The magnitude of the bulge out—that is, the number of rows advanced—can be a function of numerous parameters. However, one of the most important parameters is the effective force on the dislocation; that is, the ratio of the applied force to force or strength of the srb. If this ratio approaches one, then the magnitude of the bulge out is large. The parameters affecting the nature of dislocation motion are discussed in greater detail elsewhere.[9]

Steady State. Determination of the boundary conditions and parameters which result in a steady-state condition is *the critical factor* in determining the τ versus T relationship, and the activation parameters versus τ and T relationship.

The simplest manner by which the steady-state condition can be defined is as follows: the steady-state condition exists for a given temperature, stress, and concentration of srb; if increases in the size of the slip plane do not result in a change in the velocity, then an equilibrium size has been obtained. Figure 4 is a plot of dislocation velocity as a function of the size of the slip plane. A simple explanation as to why the width of the slip plane has a great effect on the velocity is based on one point: the probability of having a large spacing between points of contact increases as the width of the slip plane increases. As stated previously, the stress required to move a dislocation through an array of srb is inversely proportional to the spacing of srb. If the width of the slip plane increases, the probability of having a large spacing increases. Similarly, at a given stress the time required to overcome a srb decreases as the spacing increases, and as the width increases the probability of having a large spacing increases. Therefore the velocity increases as the width increases.

The conditions—that is, the width and length of the slip plane—required

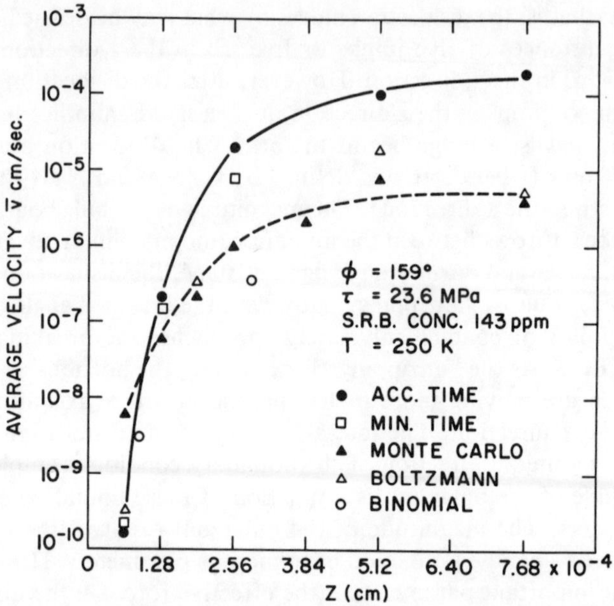

Fig. 4. The average velocity versus the width of the slip plane, i.e., length of dislocation line.

at a given temperature and stress do not necessarily hold for another temperature and stress condition. In general, as the stress is increased for a given concentration, the width and length have to be increased to obtain steady-state conditions.

The stress-versus-temperature relationship was obtained at approximately constant $\bar{v}$ conditions. Figure 5 is a plot of the stress required to obtain a $\simeq$ constant $\bar{v}$, a function temperature for two values of $\Delta\epsilon$, and three different concentrations. The stress at 0 K (τ_0) varies as $C^{1/2}$, and in all cases the values of τ_0 obtained by extrapolating to 0 K are in good agreement with those predicted by Foreman and Makin.[3]

2.4. Conclusions

The computer simulation modeling of the motion of a simple dislocation through a random array of discrete barriers is well understood, and the length of code required to undertake such modeling varies from 100 to 200 lines.

3. Multiple Dislocation Motion

3.1. Introduction

There have been several investigations of the velocity[10,11] and spacing[12,13] of a group of dislocations that have been generated by a source.

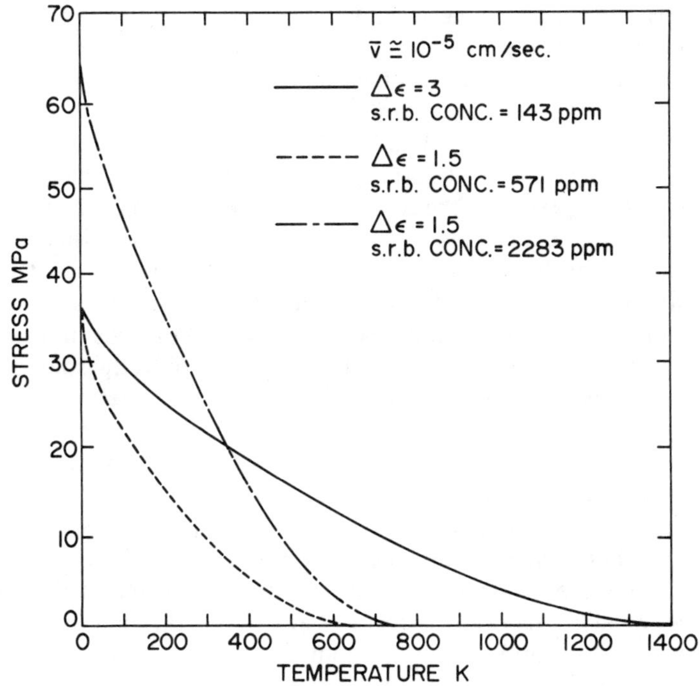

Fig. 5. Stress as a function of temperature at a given average velocity.

These investigations have assumed that the resistance to the dislocation motion is friction—that is, the dislocation does not make thermally activated fluctuations over discrete short-range barriers. These investigations will be defined as one-dimensional, for the dislocations remain parallel to each other.

3.2. Model

The procedure employed for the one-dimensional case is essentially the same as that developed in Section 1.2, with a few exceptions. It was found that the method used previously, that is, a minimum time criterion, was unsatisfactory. The minimum time criterion is based on determining the smallest activation energy (ΔG_i), and since ΔG_i is an increase function of the effective stress,

$$\tau_i^* = \tau_a \pm \tau_{ID}$$

where τ_a is the applied stress (which is the same on all dislocations) and τ_{ID} is the dislocation interaction stress. In the case of the first dislocation generated, $\tau_1^* = \tau_a + \tau_{ID}$, whereas for the second generated, $\tau_2^* = \tau_a - \tau_{ID}$. The net result is that ΔG_1 is always less than ΔG_2. This means that the first dislocation generated has to move completely off the slip plane before the second dislocation will move. Therefore, a binomial distribution method was developed to determine the time required to jump a given barrier.

The problem of source operation, that is, the generation of dislocation, has been considered before[14] and in this investigation two cases were considered. First, when the last dislocation has travelled 1000b (which is defined as the insert distance) from the source, another dislocation comes into existence at the source and can move. (It is possible that the first dislocation generated may approach the sound velocity when the second dislocation comes into existence at the source. The high-speed dislocation interaction effects have been ignored.) In the second case, the dislocation came into existence as the source when the last dislocation was 10,000b from the source. These differences in insert distance are equivalent to arbitrarily defining the stress required to operate the source. Changing the stress to operate the source has only a small effect on the dislocation velocity-distance relationships.

The motion of a single dislocation on a slip plane containing a two-dimensional array of short-range barriers, which can be jumped by thermal fluctuations, has been described in detail elsewhere.[8,9,15] When considering the motion of more than one dislocation on this two-dimensional slip plane, there is a possibility that the dislocations are not parallel and that force per unit length of dislocation is not constant along the dislocation line. If the assumption that was employed in the two-dimensional single dislocation case—that is, that the dislocation bows out as the arc of a circle between two adjacent short-range barriers—is employed herein, it is necessary to determine the average effect stress ($\bar{\tau}_{eff}$) on the segment of dislocation between the two adjacent short-range barriers which are in contact with the dislocation. The radius of the arc is equal to $\mu b/2\bar{\tau}_{eff}$, assuming the line tension is $\mu b^2/2$. The first method involved a detailed calculation of the interaction stresses, by dividing the dislocations into short segments.[16] The second method was to take the midpoint on the dislocation line between two points of contact and then to determine the average distance to the other dislocations. These dislocations were treated as a group of parallel dislocations. If the dislocations have a separation of 2000b or greater then the difference in $\bar{\tau}_{eff}$ between the two methods is $<4\%$.[17]

After determining $\bar{\tau}_{eff}$ and R the applied force on the barrier was determined by Eq 3. The activation energy to jump the ith barrier was determined by Eq 5.

3.3. Results

One-dimensional Parallel Group. The motion of multiple parallel dislocations (that is, one-dimensional cases) through a random array of short-range barriers was examined for three different types of short-range barriers.

One of the barriers was chosen to represent the dislocation etch pit data obtained from Fe-Si.[18] From these data it is possible to obtain a value for τ_i.

The long-short dashed line in Fig. 6 is the steady-state velocity of the first of 11 dislocations generated. The velocity of this first dislocation generated is approximately 1.65 times faster than that of a single dislocation traversing the same short-range barrier array at 298 K, which is in agreement with the results of Yokobori et al.,[19,20] and 2.80 times faster at 77 K. The velocity of any dislocation in the group appears to have reached an average velocity when it had traveled two or three times the insert distances. Figure 7 is a plot of instantaneous velocity versus distance of the first dislocation generated. The instantaneous velocity is defined as the distance between 10 short-range barriers divided by the time required to traverse these short-range barriers. There is a local maximum in the instantaneous velocity of the first dislocation after it has traveled a distance of 10^4, which corresponds to the fact that at this distance the second dislocation came into existence at the source. However, when the third dislocation came into existence at the

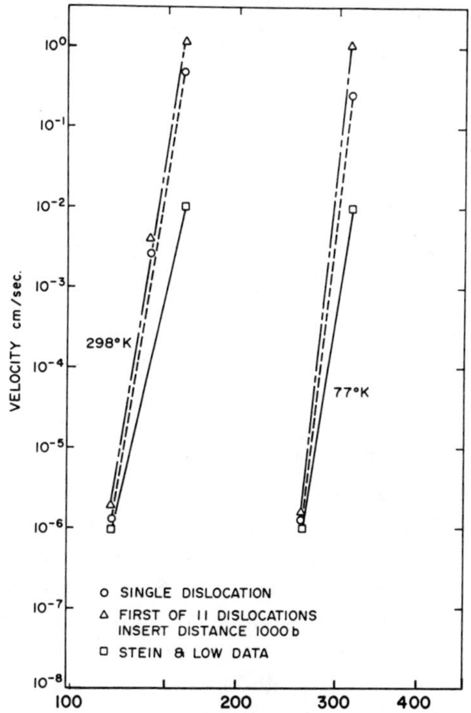

Fig. 6. The average velocity of the first dislocation generating versus stress, and the experimental results for Fe-Si.

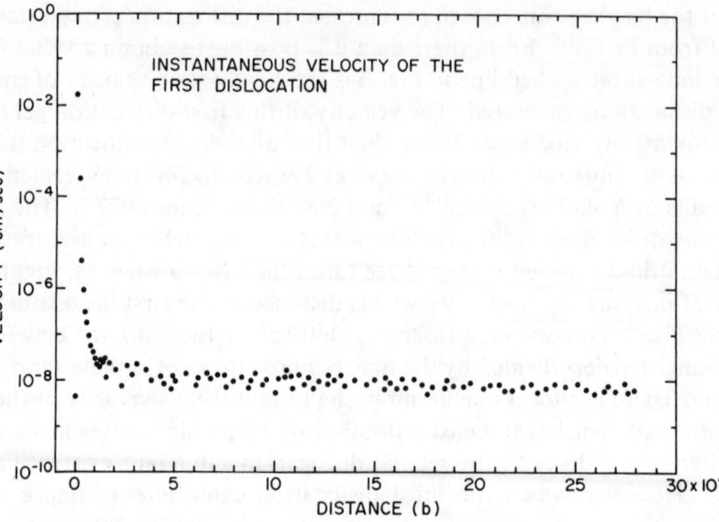

Fig. 7. The instantaneous velocity versus distance of the first dislocation generated from a continuously operating source as a function of distance for the source.

source, there was no discernible local maximum in the instantaneous velocity of the first dislocation.

Two-dimensional Coplanar Group. Two different-size two-dimensional arrays were chosen for this investigation. A slip plane width of 4×10^{-4} cm was chosen to determine whatever effect the other dislocations on the slip plane had on the configuration of the first dislocation generated. In this particular part of the investigation, the more detailed method was used to determine the interaction stresses between the dislocations.(16) After the dislocation had progressed two to three insert distances on the slip plane, there were no detectable differences in the configuration of this first dislocation and of a single dislocation traversing the same area of the slip plane. Ideally, this wide slip plane should be used to determine how a group of dislocations affects the velocities of the first dislocation of this group, but the computer time required is prohibitive. Therefore, a slip plane that was 7.3×10^{-5} cm wide was employed to determine the velocity differences. Figure 8 is a plot of the positions of four dislocations that have been inserted on the slip plane, and from Fig. 8 it is evident that the dislocation group resembles an inverted pileup. This inverted pileup resembles the transmission electron micrographs of inverted pileups generated by grain-boundary sources.(21) The velocity of the first dislocation was about two times greater than that of a single dislocation traversing the same area of the slip plane. Also, the velocity of the first dislocation traveling in this two-dimensional slip plane is within a fac-

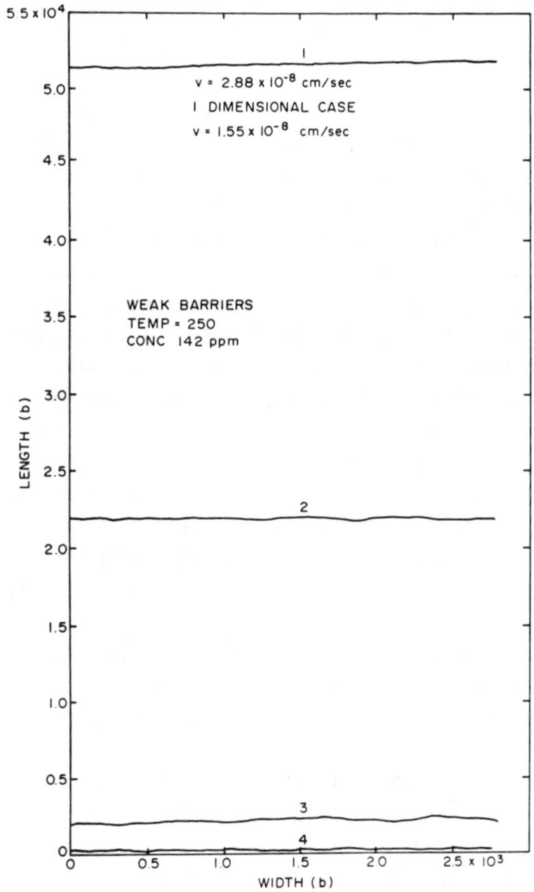

Fig. 8. The position of four dislocations on a slip plane.

tor of two of that obtained in the one-dimensional case. This small difference is due to the fact that the empirical expression of the ΔH used for the one-dimensional case is not exact.

Conclusions. From an investigation of the discrete short-range barriers that can be overcome by thermal fluctuations, it was evident that the velocity of the first dislocation of a group of dislocations emitted by a continuously operating source was not significantly different from that of a single dislocation traversing the same area of the slip plane. Also, the results obtained from an investigation of a one-dimensional coplanar model were in very good agreement with the two-dimensional nonparallel model. Therefore, it is not necessary to consider the more complicated or computer-time-consuming case of the two-dimensional problem.

4. Strengthening of fcc Solid Solutions

4.1. Introduction

There have been several computer simulation investigations that have dealt with concentrated solid solutions. Ono[22] investigated the motion of a dislocation by a computer simulation technique in a concentrated solid solution. However, the dislocation motion was restricted to plus or minus a half lattice spacing, which essentially means that the dislocation line remains straight and only a very small fraction of the solute atom arrangements was sampled. Kuo and Arsenault investigated the strengthening of concentrated solid solutions based on the condition that the dislocation line remains straight.[23] However, they allowed the dislocation to move large distances and took into account the three-dimensional nature of the interaction between the solute atoms and the dislocation line. From the results of Kuo and Arsenault it is evident that the maximum interaction force (that is, maximum degree) of solid-solution strengthening occurs from the interaction of strain field of an edge dislocation with the strain field of a solute atom induced by the size misfit effect. This was what was to be expected. Surprisingly, by using a dislocation line length of about 3 μm, there was very good agreement between calculated yield stress and experimental values of yield stress at 4.2 K. The dislocation line length of 3 μm corresponds to a size of dislocation cell structure in solid solutions. If the line length required to obtain agreement were much smaller or larger, then the concept of a straight-line approximation would be of no value.

The data obtained by Kuo and Arsenault is at variance with the hypothesis of Fleischer,[24-26] who suggested that yield stress in fcc solid solutions is controlled by the motion of screw dislocations. Also, Haasen[27] has suggested that in certain fcc alloy systems, the modulus effect is a major factor contributing to strengthening, whereas Kuo and Arsenault found the modulus effect to be small.

4.2. Model

An fcc lattice was generated and substitutional solute atoms were randomly distributed in the lattice in the concentration range of 0.1% to 10%. As mentioned in the Introduction, the maximum interaction force was found to be due to the interactions of the solute-atom-size misfit strain and an edge dislocation strain field. Hence, the data generated only for an edge dislocation will be considered. (The modulus effect was also considered, but will not be discussed.) In the case of fcc metals and their solid solutions, the dislocations generally exist as extended dislocations. The partial dislocations are separated from a few b to 50b or more. Ideally, the interaction of the solute atoms should be considered with that of the partials, and the motion of the extended dislocation should also be considered. However, the computer time necessary for these calculations was prohibitive. The extended edge dislocation was simulated as a perfect dislocation, but the Burgers vector was reduced to that of the edge component of the partial. The screw component was neglected because the interaction was found to be <10% of that of the edge.

An initially straight edge dislocation was introduced into the lattice with the dislocation line oriented along the z direction in the slip plane. The x direction was taken to be the direction of the dislocation motion (Fig. 9).

The lattice repeat distance along the x axis is the Burgers vector b. The interaction force f(i) between a given length (b) of dislocation and a solute atom for the size effect was determined by the following equations:[28]

$$f(i)_{size} = \int_{z-b/2}^{z+b/2} \sigma(i)_{size}\, dz \tag{7}$$

$$\sigma(i)_{size} = 6\mu\epsilon r_0^3 xy[x^2 + y^2 + z^2]^{-5/2} \tag{8}$$

In the equations, $f(i)_{size}$ is the component in the slip direction of the interaction force due to the ith solute atom, μ is the shear modulus of the solvent, r_0 is the radius of the solvent atom, and ϵ is the misfit strain due to the size difference between the solute and solvent atoms. Specifically, $\epsilon = (1/a)/(da/dc)$, where a is the lattice parameter of the matrix, and da/dc is the change in lattice parameter with concentration. For the purpose of evaluating the forces, the origin of the coordinate system is placed at the site of each solute atom as shown in Fig. 9.

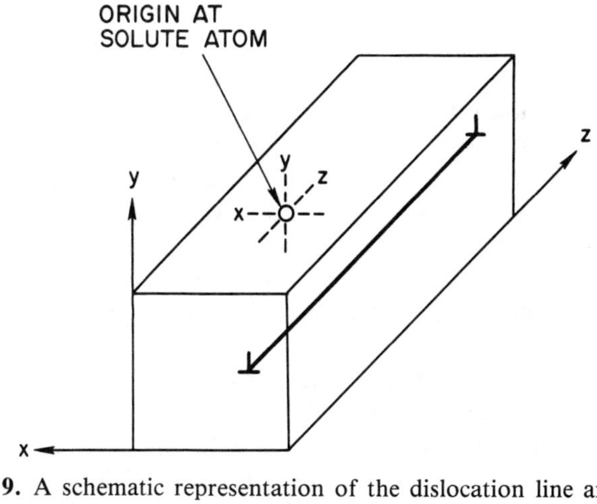

Fig. 9. A schematic representation of the dislocation line and the solute atom.

A similar expression can be obtained for the modulus effect. The modulus interaction between a spherical inclusion in a straight edge dislocation was derived by Cominou and Dunders.[29,30] This interaction will be contrasted, in both magnitude and geometric dependence, with the size interaction due to the misfit strain from a spherical inclusion which is the basis of most theories of solid-solution strengthening. The derivation of Cominou and Dunders produced an interaction force between the solute atom and the entire dislocation line. This approach was modified to obtain the interaction force with a given segment of the dislocation line. The interaction force, $f(i)_{modulus}$, between an edge dislocation and a solute atom as the result of a modulus difference between a solute and solvent atoms is defined elsewhere.[23]

The total force in the direction of motion (along the x axis) on a given segment of dislocation will be the sum from all the solute atoms. As already noted, in evaluating each contribution, the origin is shifted to each solute atom. The net result will refer to a dislocation line segment at position (x,z) relative to an absolute origin in Fig. 9. This total force, F_s, will be

$$F_s(x,z) = \sum_{i=1}^{N} f(i) \tag{9}$$

where N is the number of solute atoms in a search volume around the dislocation line segment, and the interaction force, f(i), is due either to the size effect (Eq 7) or to the modulus effect.[23] As emphasized in Eq 9, F_s depends on the location of the dislocation line segment on the slip plane. The size of the search volume depends on the range of the interaction as discussed below. On the given segment of the dislocation line, the sum in Eq 9 should be over every solute atom in the lattice. However, this was not necessary. Only atoms within an outer cutoff radius were included in the sum. This radius was determined by calculating the total force on a given segment of dislocation line as a function of search radius from 2b to 50b. We found that, if the cutoff radius was set to 4b, the force was within 10% of the asymptotic value determined at distances up to 50b. In addition to this outer cutoff radius, an inner cutoff radius was imposed as well. At short range, the elastic formulation is not valid. Interactions at or inside this inner radius were neglected as well. Little guidance exists, at present, for the interactions at short ranges so the inner radius was chosen to exclude the nearest-neighbor atoms.

For the future atomistic consideration, the interactions at close range, that is, within the core, can be accounted for.

If a straight dislocation line is placed on a slip plane, its shape will change as the result of the interaction forces between the dislocation and solute atoms. The detailed configuration of the dislocation (if the local change in the radius of curvature is small) can be obtained by considering the force balance of each 1b-long segment of dislocation line:

$$\Gamma \frac{d^2x}{dz^2} + \frac{F_s(x,z)}{b} + \tau b = 0 \tag{10}$$

where τ is the externally applied shear stress, Γ is the line tension, and $F_s(x,z)$ is the total interaction force (size or modulus) due to the solute atoms on the dislocation segment of interest. In this work, the line tension Γ was chosen as 0.5 μb^2. There are several possible methods of numerically solving Eq 10, one of which is the shooting technique.[31] However, this technique and most others are unsatisfactory.

The difficulty in obtaining a solution is due to the function F_s, which is a random function in both amplitude and period. Figure 10 is a typical plot of $F_s(x,z)/b$ as a function z for a given value of x on the slip plane. In the shooting technique, two boundary conditions are imposed and a solution is attempted. For example, we could choose the dislocation line length L, which is set $x(z = 0)$ as the same value as $x(z = L)$. In attempting to match this condition, we vary the slope dx/dz at $z = 0$ and then integrate Eq 10

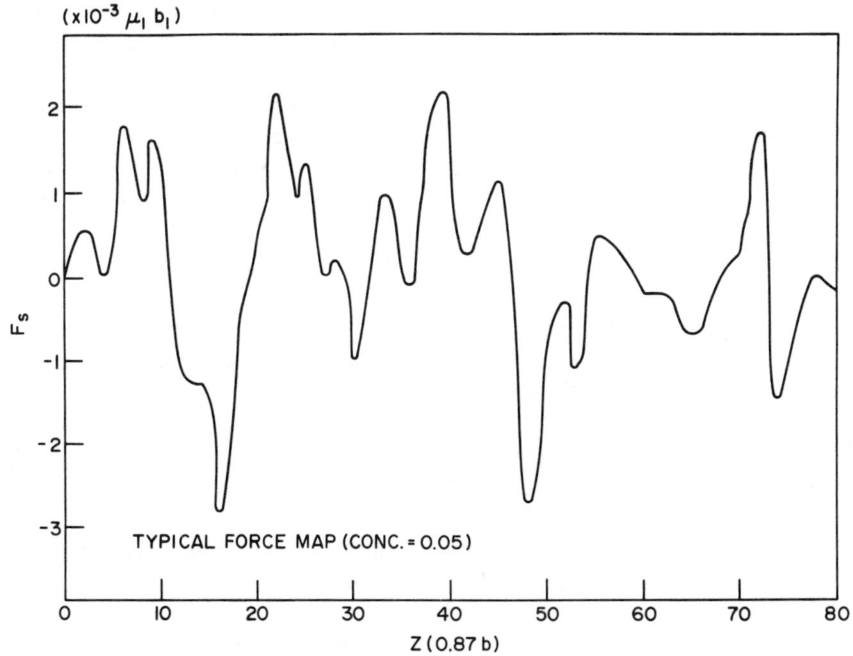

Fig. 10. A typical sample of the force per unit b as a function of distance in the z direction for a given value of x.

numerically to obtain ("shoot") a solution across to $z = L$. The slope is adjusted until $x(z = 0)$ is the same as $x(z = L)$. In practice, for moderate to large L and for typical random solute forces, we find the solution to behave poorly. That is, the solution is extremely sensitive to the change in the initial slope and the solution is multivalued even for a specified slope. In addition, there is no guarantee that the solution obtained is physically correct. By this we mean that it may not be possible to reach a particular configuration, even though it satisfies Eq 10, from some initial configuration for the dislocation. There may exist a barrier (due to the random solute forces) which prevents the dislocation from moving from an initial to a current configuration, even though both the initial and final configurations are solutions to Eq 10. There are other difficulties as well (such as how to define further motion of the dislocation), and so a different approach was taken.

In our numerical simulation study, focus was on a dislocation line segment of length L. This segment was embedded in an fcc lattice block containing the segment and its slip plane. As is typical in simulation studies, the system must be finite. The chosen lattice block was large enough to avoid spurious edge effects in the y and z directions of Fig. 9. The chosen size of the block in the x or slope direction typically was about 100 to 200b. This length was found to be sufficiently large to define the onset of the yield condition. An effectively infinite dislocation line segment was modeled by imposing minor boundary conditions at the ends of the dislocation line segment (at $z = 0$ and $z = L$). Because of our relaxation process (described below), this resulted in the dislocation line being parallel to the z axis at $z = 0$ and $z = L$.

An initial dislocation configuration was obtained by relaxing a straight dislocation at zero external stress. The stress was increased in small increments, with an appropriate relaxation at each stress level, until equilibrium or a force-balance condition could not be maintained. At that stress level, the dislocation line segment moved rapidly to the uppermost x value in the block. This defined our yield stress. We now turn to the details of this process.

The net force df on any 1b-long segment of the dislocation line can be obtained by evaluating the residual in Eq 10; that is,

$$df = \left[\Gamma \frac{d^2x}{dz^2} + \frac{F(s,x)}{b} + \tau b\right] b \tag{11}$$

Note that df can be positive or negative and, as such, can move the dislocation in the positive or negative x direction. However, because of the random nature of the solute interaction forces and the elastic nature of a dislocation line, the force may vary substantially from one segment to the next, and the motion of a segment of the dislocation line definitely affects that of the near segments. Hence, the movement of these segments must be coupled. In recognition of this, we defined an adjusted force on the nth 1b segment as

$$DF_n = df_n + \frac{4}{5}(df_{n-1} + df_{n+1}) + \frac{3}{5}(df_{n-2} + df_{n+2}) + \frac{2}{5}(df_{n-3} + df_{n+3}) + \frac{1}{5}(df_{n-4} + df_{n+4}) \quad (12)$$

and suppose the displacement of the nth segment DX_n is proportional to DF_n.

Taking into account the elastic nature of the dislocation line greatly reduced the computational time for relaxation of the entire dislocation segment. Several *different* methods of adjusting for the neighboring segment forces, similar to that in Eq 12, were tried and little effect was noted on the end result.

The relaxation process itself involved locating the 1b segment which was subjected to the maximum adjusted force (DX_{max}). This segment was moved 0.005b in either the positive or negative x direction, depending on the sign of DF_{max}. The other segments of the dislocation line were also moved, but in an amount 0.005b (DF/DF_{max}), where DF is the (signed) adjusted force for that particular 1b segment. This process was iterated until the DF for every segment was suitably small.

To recap, the dislocation is placed in the fcc lattice block as a straight line at x = 5b. Then it is allowed to relax to zero external stress. When the dislocation line is straight (in the initial position), the only force on it is F(s,x)/b. After the first increment of motion or relaxation, the line tension component gives rise to an additional force on the dislocation. The dislocation is moved incrementally as described above until it reaches the condition such that the force balance is equal to zero (Eq 10 is satisfied).

At this point the dislocation line has changed its shape. The line labeled 1 in Fig. 11 is the relaxed shape of the dislocation line. Then the external stress x is increased in small steps (0.0001 μ) and at each stress level a new force balance is achieved. The dislocation lines labeled 2 and 3 in Fig. 4 correspond to intermediate stress levels. The process of incrementally increasing τ is continued until at some stress level a force balance cannot be obtained. At this stress level, the relaxation process produces a dislocation line which rapidly extends up to the maximum value of the lattice in x direction for the simulation block. The stress level at which this occurs, that is, a force balance cannot be obtained, is defined as the yield stress of that particular lattice chosen. Figure 12 is a typical graph of how the stress is incremented as a function of distance.

As noted earlier, the solute force distribution is highly complex and random. Hence, for a specified boundary condition, there exists a multitude of

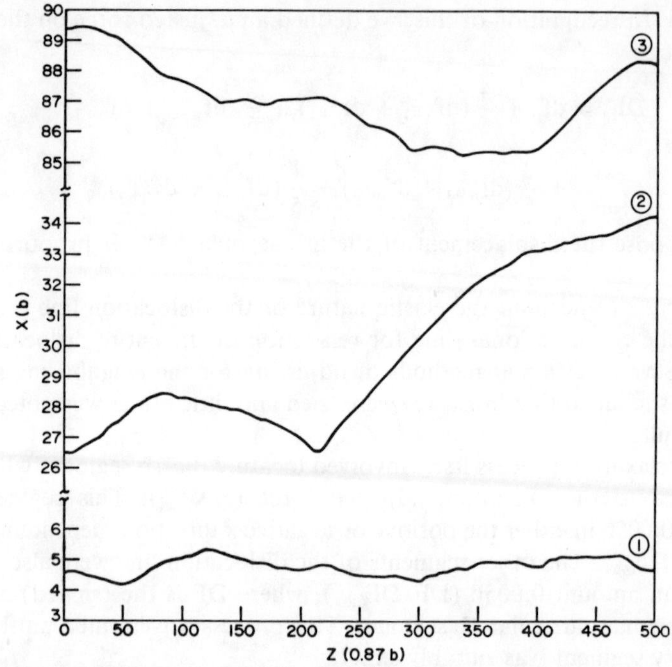

Fig. 11. The configuration of the dislocation line at various positions on a slip plane and subjected to various values of stress. The line labeled 1 is that of the relaxed dislocation shape configuration at zero applied stress. The line labeled 2 is the equilibrium position of the dislocation line at a higher value of stress, and the line labeled 3 is the equilibrium position of the dislocation line at a still higher value of stress.

equilibrium dislocation configurations at any stress level. The process of increasing the external stress in small steps and then perturbing the 1b line segments to obtain equilibrium at each stress level provides a succession of dislocation configurations as shown, for example, in Fig. 12. This perturbational approach generates a related sequence of configurations which we characterize as having "physical continuity" with the original, stress-free configuration.

The major requirement in the consideration of thermally activated motion of the dislocation in a concentrated solid solution is a determination of the force-distance diagram for the incremental motion of a segment of the dislocation line.

From the interaction due to size effect difference of the solute atom and the edge dislocation, it is possible to produce a "force map." A force map is the interaction force present on the slip plane at each location on the slip

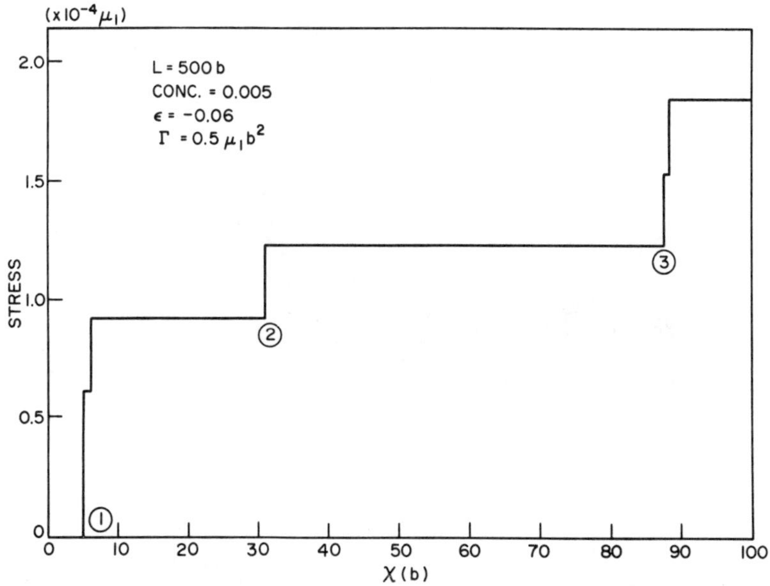

Fig. 12. A stress versus distance diagram. This stress is that required to move the dislocation forward across the slip plane, as a function of the position of the dislocation line with respect to its forward motion along the x direction.

plane, which means that if a 1b segment of an edge dislocation were at a given point, the force would be obtained from this force map. The vertical axis of the plot in Fig. 13 is the interaction force, and it can be seen that the interaction force can be positive or negative. A positive interaction force opposes the motion of the edge dislocation in the plus x direction, and a negative interaction force aids the motion of the dislocation in the plus x direction. From an examination of the two parts of Fig. 13, it is possible to consider the force map as consisting of a group of "hills" and "valleys." The hills oppose the forward motion of the dislocation. Therefore, if a straight dislocation were randomly placed on the slip plane, it would be subjected to positive and negative forces and "relax" into a curved configuration. If it were placed as a straight line or curved line in another position, it would relax to a different configuration. Also, if the initial relaxed position is considered as the starting point and if a given segment were placed on the other side of a "hill," the dislocation would have a different relaxed position. These two different positions could represent the thermally activated "jump" of the dislocation. Following is a discussion of the details of the procedure of determining the activation energy to jump the hills.

Initially, the dislocation line was placed into the lattice as a straight line

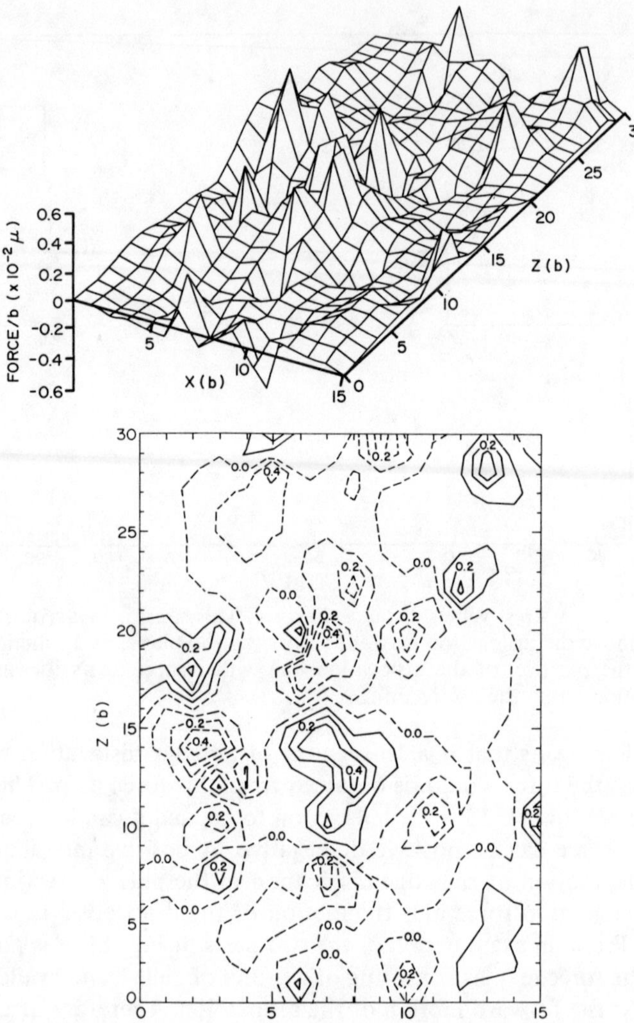

Fig. 13. (Top) A force map where vertical axis is the force which can act upon an edge dislocation at that position. (Bottom) A contour map of equiforce lines corresponding to the three-dimensional force map (top). The solid lines represent positive force values, and the dashed lines represent negative force values.

and allowed to relax under zero τ. (The applied stress and the effective stress are the same; that is, there was zero long-range internal stress in the lattice, other than that which may be developed by the solid atoms.) Then a predetermined τ was applied and the forward dislocation motion was the same as that described above. After the dislocation reached static equilibrium, the entire dislocation line was given a virtual displacement forward of 0.5b (as shown by the dashed line in Fig. 14), and the dislocation was allowed to relax. Two possible relaxation configurations could occur:

1. The dislocation line could relax to its original position, that is, the solid line.
2. The dislocation line could form one or multiple loops as shown by the solid and dashed lines in Fig. 14.

When the first case arose (that is, no loops formed), the dislocation was virtually displaced 1.0b forward and the relaxation process repeated. The virtual displacement with additions 0.5b (for example, the third displacement was 1.5b and the relaxation process repeated) when one or more loops had formed, then the force-distance diagram for each loop was determined. This was accomplished by tracking the virtual motion of the dislocation through the loop in very small increments and calculating the solute interaction force on the dislocation at each position. After the force-distance diagram had been determined, an integration was performed to determine the activation energy, ΔG_i, associated with that particular loop. If more than one loop

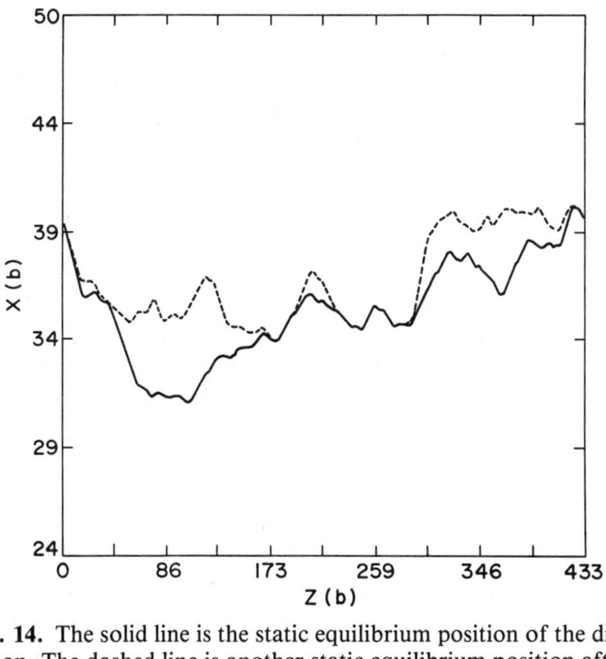

Fig. 14. The solid line is the static equilibrium position of the dislocation. The dashed line is another static equilibrium position after a virtual displacement.

formed, as shown in Fig. 14, then a force-distance diagram was calculated for each loop and the smallest ΔG_i was found for all of the potential loops. This calculation is only for the loops that have formed as a result of a given virtual displacement. The smallest ΔG_i is *not* found for the entire slip plane. Dislocation was allowed to "jump" with the loop which had the smallest ΔG_i. The time required to "jump" was calculated by Eq 4.

After the loop with the minimum ΔG has jumped, the dislocation is allowed to relax again. If there are still any other potential loops, the loop with the smallest ΔG is allowed to jump and the relaxation process is allowed to occur after the jump. When all the potential loops have jumped and the dislocation is allowed to relax to a new configuration, the process is repeated—that is, the entire dislocation line is virtually moved forward 0.5b as shown in Fig. 1. The process is continuously repeated until the dislocation advances completely across the slip plane. The time, t_i, of each jump is stored and the average velocity of the dislocation is determined by Eq 6.

In order to obtain and average the $\bar{v}$ of 1.28×10^{-5} cm/s at a given T, the τ was incrementally decreased or increased until the $\bar{v}$ was obtained with 10% of 1.28×10^{-5} cm/s. This process was repeated in the temperature range from 0.0 to 1000 K for various solute atom size differences, that is, misfit strains and configurations.

The ΔG versus τ relationship was obtained by two different procedures. In the first procedure, the local maximum activation energy was determined for each 1b strip in the x direction. In general, there were numerous ΔG's determined in each wide strip. The location of a given loop was determined by its central line position. For example, in Fig. 14 the dislocation of the loop "A" is at 34b. All the ΔG's of loops on center line 13b were determined and the largest ΔG was defined as $\Delta G_{LOC\,MAX}$. Then a plot of $\Delta G_{LOC\,MAX}$ was constructed as shown in Fig. 15; this was for a given misfit strain concentration and stress. Then the largest $\Delta G_{LOC\,MAX}$ was determined and the process repeated at various values of τ for the same misfit strain and concentration in order to produce a plot of ΔG versus τ.

The second procedure, which consumes more computer time but is more rigorous, involves determining the difference in $\bar{v}$ at various temperatures for a constant τ, ϵ, and C. Then the following equation can be employed:

$$\Delta G = k \left[\frac{\ln \bar{v}_1/\bar{v}_2}{\left(\dfrac{1}{T_2} - \dfrac{1}{T_1} \right)} \right]_\tau \tag{13}$$

where $\bar{v}_1$ is the average dislocation velocity at temperature T_1, and $\bar{v}_2$ is the average dislocation velocity at temperature T_2, and k is again Boltzmann's constant. The process is then repeated at various levels of τ, and ΔG can be determined as a function of τ.

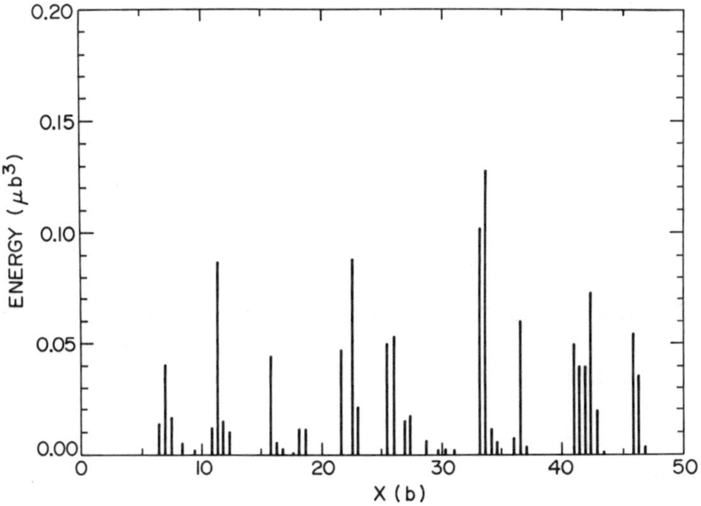

Fig. 15. A plot of the maximum value of ΔG at a given value of x as a function of the distance the dislocation moves in the x direction.

The activation volume (v^*) was determined in the following manner. The $\bar{v}$ was determined at two different stress levels at a constant T and ϵ, and the following equation was used:

$$v^* = kT\left[\frac{\ln(\bar{v}_1/\bar{v}_2)}{\tau_1 - \tau_2}\right]_T \tag{14}$$

where $\bar{v}_1$ is the average velocity at τ_1, and $\bar{v}_2$ is the average velocity at τ_2, and then v^* can be determined as a function of τ.

The self-consistency of the model was determined by analytically calculating the force-distance diagram for the interaction of a single solute atom (1b above the slip plane) and a dislocation, and then calculating ΔG_0, the activation energy at zero τ. Next, a single solute atom was placed into the fcc lattice (1b above the slip plane), and the dislocation was moved through the lattice at zero stress by making 0.3b virtual displacements of the dislocation. When the dislocation encountered the solute atom, the dislocation line still remained almost perfectly straight. (In an analytical solution, an assumption is made that the dislocation *does* remain perfectly straight.) The dislocation makes virtual displacements as described above, then a force-distance diagram can be obtained, and again the activation energy can be determined. The analytical result and the computer result are in very good agreement. The analytical ΔG_0 was 0.1 eV, and the computer simulation result was 0.098 eV.

4.3. Results

Size of Lattice. As previously determined in a dilute random-array computer simulation investigation, the stress required to move a dislocation across the slip plane decreases as the length of the dislocation line segment L increases. This stress decreases to some asymptotic value, where it remains essentially constant as the value of L is further increased. Similar results were obtained in the present investigation (Fig. 16) for absolute zero temperature. The result was that the yield stress reached asymptotic value at a relatively small L value of approximately 250b. As observed previously, the value of L at which the stress reaches an asymptotic value increases as the concentration increases.

A further check was carried out to determine if, indeed, the yield stress was insensitive to the dislocation line segment length L for L larger than about 250b. A variety of random arrays of solute atoms (at concentrations of 0.5, 5, and 10%) were used. For small L (L < 250b), there was a distinct relation between L and the yield stress for each of the random arrays. However for L ~ 250b and larger, the yield stress became independent of L (as shown in Fig. 16) for a particular solute atom distribution.

As a further determination of whether the sample was sufficiently large, several different random arrays were used at a given concentration (5%) as a function of dislocation line length. If the line length is small (25b), then a change in the random array has a large effect (that is, ±14%) on the yield stress. If the length of the segment is increased to 250b, then the yield stress varies by 9% as the result of a change in the random array. However, if the

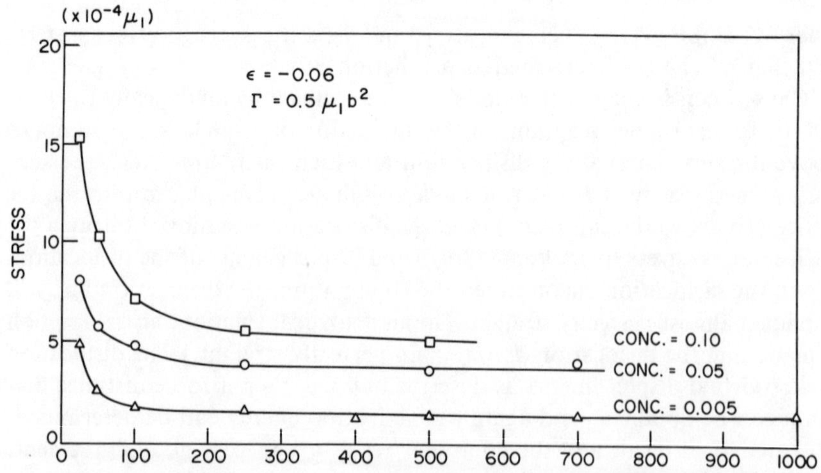

Fig. 16. Yield stress as a function of the length of the dislocation for various solute concentrations.

length of the segment is increased to 500b, there is no change in the yield stress as the result of a change in the random array.

The present computer simulation results indicate that the contribution of the size effect is much greater than that due to the modulus effect for physically reasonable values. Therefore, only the strengthening due to the interaction of the size effect and an edge dislocation will be considered in the comparison with experimental results. The computer simulation results were obtained for the motion of the dislocation at 0 K.

Hence, a comparison can be made with the experimentally determined yield stress at 4.2 K. The solute concentration of the alloys chosen for comparison was moderate (<10%) because the computer simulation results were obtained assuming a random array of solute atoms. For real alloys the degree of short-range ordering increases as the solute concentration increases.

In Fig. 17, the experimental data were plotted as a series of points, which were labeled, and only the size effect was considered. The experimental data represent a fairly extensive range of size differences and concentrations. The solid line was obtained from the simulation results, and it is in fair agreement with the experimental results. However, there is an exception, the case of Ag-Al; Riddhagni and Asimow[32] have put forward agreements why the data in the Ag-Al system may be wrong.

Macroscopic yield is assumed to occur as a result of long-range motions

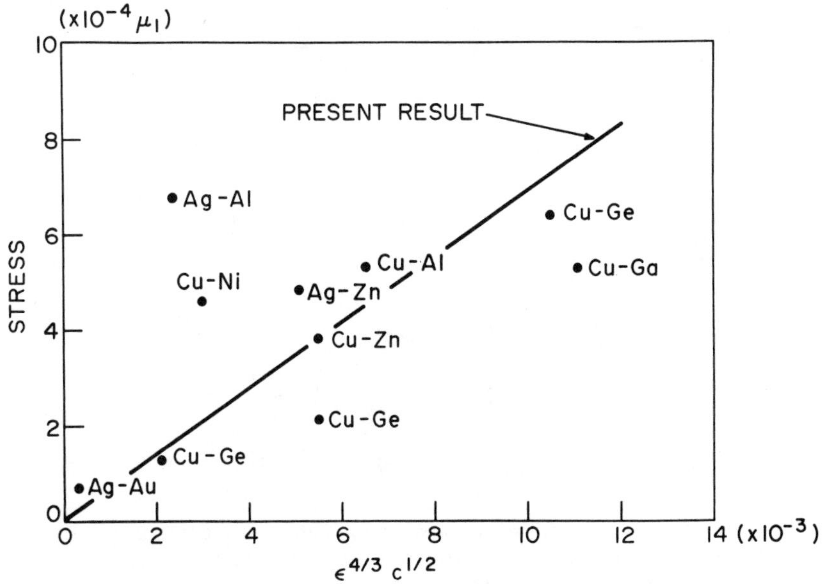

Fig. 17. The yield stress versus size effect and concentration for various alloy systems. The experimental results are taken from Ref 31.

of dislocations. The stress required to produce long-range motion of dislocations is defined as the yield stress. The applied stress under consideration in this investigation is the stress required to move the dislocation through the lattice at a given temperature and a given average velocity, and therefore, this stress will be equated to the experimental yield stress. As stated in a previous section, in constructing a plot of τ versus T, it is necessary that the $\bar{v}$ be held constant. A portion of the data obtained is plotted in Fig. 18(a). As

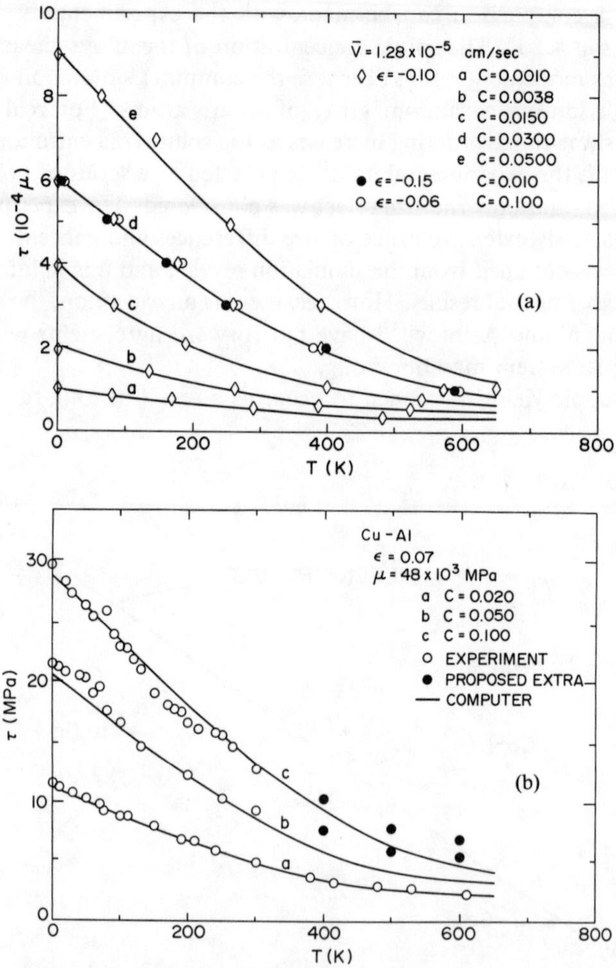

Fig. 18. (a) The yield stress versus temperature at a constant average dislocation velocity for various misfit strains and concentrations. (b) The yield stress versus temperature at a constant average dislocation velocity. The open circles are actual experimental data points of Suzuki;[32] the solid points are extrapolations proposed by Suzuki.[32]

can be seen in Fig. 18(a), for higher concentrations and the large misfit strains, there seems to be an athermal region which would appear to be due to a long-range internal stress. However, as a reminder, a long-range internal stress was not incorporated into the model.

In Fig. 18(b), a specific comparison is made between the computer simulation results and the experimental results obtained from various Cu-Al alloys.(33) In order to make this specific comparison, it is necessary to choose the misfit strain of the Al solute atoms; $\epsilon = 0.07$ has been accepted by previous investigations, and therefore $\epsilon = 0.07$ was chosen for Al in Cu. The open circles in Fig. 18(b) are experimental data points given by Suzuki,(33) the solid circles are the extrapolations as proposed by Suzuki.(33) The solid lines are the results obtained by computer simulation. In general, the comparison between the computer simulation and the experimental data is very good. However, the fit to proposed extrapolations and the computer data results in some discrepancies.

The terminology of stress equivalency was put forward by Basinski et al.(34) The phenomena of stress equivalency can be expressed in several different ways. A straightforward way of demonstrating whether stress equivalency exists is to consider several different alloys which have different compositions but the same yield stress at absolute zero. For example, consider three alloys with the following misfit strain and composition:

Alloy 1	$\epsilon = -0.10$	$c = 0.03$
Alloy 2	$\epsilon = -0.15$	$c = 0.01$
Alloy 3	$\epsilon = -0.06$	$c = 0.10$

If the yield stress versus temperature plot for these three alloys is the same, then this would be a manifestation of stress equivalency. These three alloys indeed can be plotted on one line as shown in Fig. 18(a). Another way is in a plot of v^* versus τ at constant temperature. If numerous alloys are considered and these alloys represent various misfit strains and concentrations, the data obtained at a given temperature define a single line. If a different temperature is considered, a different line is obtained, as shown in Fig. 19. The data points were obtained by Basinski et al., and the lines were obtained from the computer modeling, in the same manner as the experimental data of Basinski et al.

4.4. Conclusions

In several recent reviews of solid-solution strengthening in fcc alloys, it has been strongly argued that the resistance to dislocation motion due to a random array of solute atoms could not account for the temperature dependency of the yield stress of fcc alloys. This discrepancy could be stated in another manner: the activation energy at zero stress which is predicted by the interaction of a single atom and a moving edge dislocation is at least five

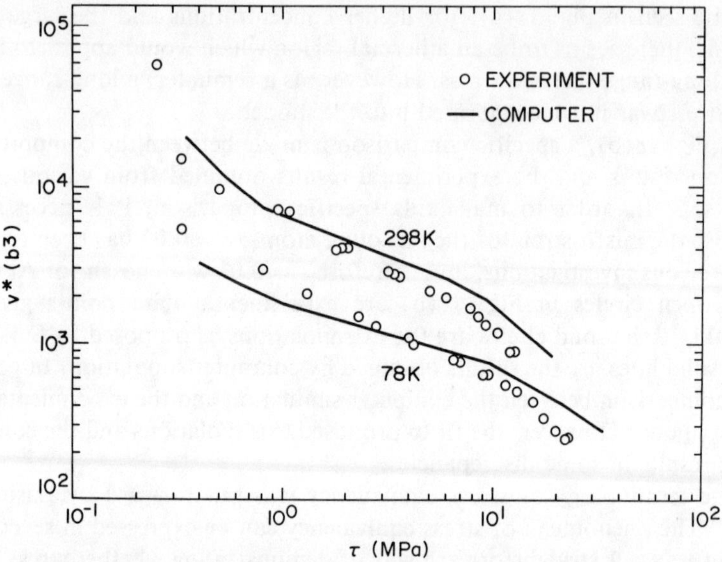

Fig. 19. A plot of the activation volume as a function of stress at two different temperatures for numerous different solute concentrations.

times smaller than that observed experimentally, and this type of interaction would lead to the conclusion that there would be zero solid-solution strengthening above approximately 77 K.

The computer simulation results are in very good agreement with the analytical solution of the activation energy of a single solute atom and an edge dislocation. However, the strengthening does not occur as a result of the interaction of a single solute atom and edge dislocation, but is the result of the interaction of numerous (5 to 10) net solute atoms. In general, the stress field of one solute atom is canceled by the stress field of another. However, even though the solute atoms are located in the lattice at random locations, there are regions where there are several solute atoms which oppose the motion of dislocation and there is no canceling effect by any neighboring solute atoms. The net result is that ΔG_0 can be as large as 1.4 eV. A ΔG_0 of 1.4 eV results in solid-solution strengthening up to very high temperatures, such as 1000 K.

The high ΔG_0 has implications in determining the τ^* from solid solutions. If the tests were not conducted at high enough temperatures, there would be a problem. Even if the tests were conducted at high temperatures, serrated yield occurs in numerous fcc solid solutions. The presence of serrated yielding would obscure any determination of the athermal region. The present computer simulation results suggest that previous determinations of τ^* from the experimental results are highly questionable.

A strengthening mechanism based on a thermally activated motion of an edge dislocation through a random array of solute atoms will result in satisfying all the requirements of the stress equivalency phenomena.

5. Solid-Solution Weakening

5.1. Introduction and Model

Many bcc alloys display anomalous mechanical properties.[35-37] In particular, it is possible for the yield stress of the alloy to be lower than that of the pure metal at low temperatures (solid-solution weakening). A model which may provide an explanation for this phenomenon—at least for certain metals—is concerned with dislocation double-kink formation and lateral motion of single kinks[38-44] in the presence of solute atoms. These atoms occupy substitutional positions in a bcc lattice with a given Peierls stress. The deformation is assumed to be controlled by the motion of screw dislocations. The solute atoms are assumed to interact with the dislocation via a classical size[28] or modulus[29,30] effect interaction. Hence, this semicontinuum model neglects such important effects as the dislocation core structure, the influence of solute atoms on the core structure and the associated Peierls barrier, and the modification of the defect-dislocation interaction at short range.

Substitutional atoms can influence both the formation of a double kink (DK) and the lateral propagation of a single kink. In a random alloy, the internal stresses due to the impurities can either aid or oppose both processes. An initially straight dislocation (parallel to a Peierls valley) will form a DK in the region where the impurities provide the greatest "assist." Hence, the critical nucleation stress is lowered by the presence of solute atoms, which can lead to weakening. For the dislocation to move, each link must propagate laterally (either to the assumed pinning points or to be annihilated by an opposite kink). In a pure metal there is only a small barrier to lateral motion. In an alloy the lateral motion will be controlled by the maximum opposing stress arising from the solute atoms. This leads to a hardening against kink motion. Therefore, the observation of weakening will result from solute atoms assisting in the formation of a DK, whereas strengthening will be due to the solute atoms opposing lateral motion.

Earlier studies in this area have either limited the effect of solute/interstitial atoms to the role of pinning points[38-40] or included only one[41,42] or at most two[43] isolated solute atoms interacting with the dislocation via the size effect alone. An important exception is the statistical treatment by Suzuki.[44] In the latter work, the core is considered in an atomistic fashion, but is not relaxed from the linear elastic solution. Detailed superposition effects cannot be accounted for in this statistical approach. These effects are important since Suzuki requires statistical fluctuations consisting of linear arrays of about 10 to 15 solute atoms in order to get physically significant results. Further, these fluctuations will be shown unlikely even in large crystals. Finally, Suzuki focuses on the modulus effect, including only a second-order size effect due to nonlinear elasticity.[45]

The screw dislocation was placed on the (101) slip plane with Burgers vector in the z or [111] direction and slip in the x or [121] direction (Fig. 20). The shape of the dislocation was determined by the solution of the force balance equation for each element of the dislocation line.[39]

$$\Gamma \frac{d^2x}{dz^2} + \tau b - \tau_p b \sin\left(\frac{2\pi x}{a}\right) + F_s(x,z)/b = 0 \qquad (15)$$

This equation neglects certain terms proportional to $(dx/dz)^2$ which are on the order of 1% or less for our line shapes. Here Γ is the line tension, τ is the external stress, τ_p is the assumed Peierls stress, b is the Burgers vector, a is the repeat distance in the x direction (~0.95b), and $F_s(x,z)$ is the total solute force at position (x,z) on the slip plane. In some of the calculations to be described, we followed the formation and propagation of a DK with the shape constrained to the solute-free ($F_s = 0$) solution. However, given the dislocation shape(s), the calculation of the activation energies for each process still included contributions from the internal solute atom stress fields.

In this study, τ_p was taken as 0.006 μ where μ is the shear modulus of the pure metal. This is a representative value for τ_p/μ. The size misfit parameter (ϵ in Eq 6.10 of Ref 35) was taken as 0.10. The solute/solvent modulus ratio was taken as 2.0. These parameters were chosen to be large to

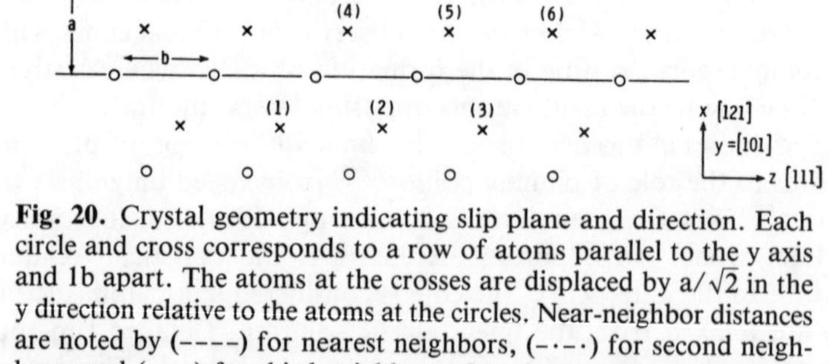

Fig. 20. Crystal geometry indicating slip plane and direction. Each circle and cross corresponds to a row of atoms parallel to the y axis and 1b apart. The atoms at the crosses are displaced by $a/\sqrt{2}$ in the y direction relative to the atoms at the circles. Near-neighbor distances are noted by (----) for nearest neighbors, (-··-·) for second neighbors, and (------) for third neighbors. Certain "cross"-type rows are denoted by numbers in defining clusters in the text.

provide general bounds on the magnitude of the respective effects. The line tension was $\mu b/2$.

Early studies focused only on the magnitude of the single defect-dislocation interaction in assessing its importance. However, the geometrical arrangement of solute atoms is crucial in determining whether the net stress field from a cluster of atoms is additive or tends to cancel. For example, with our dislocation geometry, the individual atom-dislocation size effect interact is even in the x coordinate and odd in the z coordinate (with the solute atom placed at the origin). Conversely, the modulus effect interaction is odd in the x and even in the z coordinate. Hence, a line of impurity atoms parallel to the z axis will have an additive modulus effect, but there will be a canceling among the size effect contributions. As will be shown, this leads to a saturation in the hardening effect on lateral motion due to the size effect when successively larger linear clusters (parallel to the z axis) of impurity atoms are considered. The modulus contribution does not saturate for similar arrays parallel to the z axis, but would involve canceling for arrays parallel to the x axis.

The nucleation energies were determined in a manner analogous to the saddle point calculation described by Arsenault[39] and modified to include forces due to the solute atoms. For the pure metal, formation energies became insensitive to the separation between individual kinks when the full DK width exceeded about 50b. This criterion (constraint on minimum DK width) was maintained in the presence of solutes. Only single step kinks were considered in this particular study. The activation energies for lateral motion of the single kinks were obtained by calculating the total solute atom interaction force on the kink as a function of the position of the kink along the z axis. The resulting force-distance curve was evaluated to obtain the maximum activation energy. Similarly, the formation energy as a function of the center of the DK was evaluated to obtain the minimum nucleation energy. These activation energies were, in turn, evaluated as a function of applied stress.

It would clearly be preferable to follow the change in the dislocation line shape as the DK is nucleated or a single kink propagates laterally. However, difficulties were encountered with the solution of Eq 15 in modeling this process, particularly for lateral motion, as now described. The movement of the (single) kink was controlled by the boundary conditions imposed on Eq 15. The top of the kink was taken to have zero slope at a specified value of the z coordinate (z_{top}). The x coordinate at this point was varied to control the half-length (fixed at 25b in our case). In the absence of impurity atoms, the line shape was sigmoidal in form. Being relatively flat for about 10 to 13b units along the x axis, it was followed by a sloping region of about 10b (the actual width) and a final region with a small slope of width about 2b.

As the kink was forced to approach a hard spot (a region where the stress field of the solute atoms opposed the motion of the kink), the final small slope region tended to widen; that is, the motion of the sloping region stalled out. Upon further increment in z_{top}, two possibilities occurred. Either the line solution would then "pop" past the obstacle for narrow moderately weak regions, or no solution to Eq 15 could be found for the given boundary conditions. In both such cases the system avoided the saddle point configuration. Lacking a straightforward way to obtain such a configuration, we imposed a rigid shape (pure metal shape) on the dislocation kink. The flexible solution displayed only a small variation in its form in the sloping region when a solution existed, and hence the rigid solution should yield a reasonable estimate of the activation energy for lateral motion. The alternative (using a flexible line shape) led either to an underestimate of the barrier—for a narrow region—or to no solution at all.

The use of a rigid line shape overestimates the true (saddle point configuration) lateral energy barrier. In our conclusion, we will contend that the model lateral energy barriers are much too small to explain the experimentally observed strengthening. Hence, the significant conclusions from our model cluster study would be altered only quantitatively, but not qualitatively, with the use of a flexible line solution that included the correct saddle point configuration.

The nucleation process required finding the first energy minimum and the first energy maximum for a DK (fixed at a total width of 50b). Although there were cases where no solution (subject to our particular boundary conditions) could be found for the nucleation process, there was less difficulty than with the lateral motion process. We will not report many of the details of the nucleation process with a flexible line shape since, where a solution existed, resulting activation energies were rather similar to the rigid line shape results with the preferred nucleation center at similar locations.

Very little strengthening was obtained where computer modeling used a random array of solute atoms. In order to assess the importance of various configurations of solute atoms, certain cluster geometries of solute atoms were chosen. These clusters provide insight into statistical fluctuations in a random array. The cluster geometries were chosen to maximize the solid solution effects on double-kink nucleation and lateral motion of single kinks.

Eight cluster geometries were considered, all of which involved the solute atom(s) placed $a/\sqrt{2}$ above the slip plane: (1) a single impurity atom (cluster 1); (2) three, five, ten, or twenty atoms in a linear array parallel to the z or [111] axis (clusters 3A, 5A, 10A, and 20A); (3) three or five atoms in a linear array parallel to the [111] axis, hence inclined to the x axis (clusters 3D and 5D); and (4) rhombus-shaped clusters of four or nine atoms with edges parallel to the [111] or [111] axis (clusters 2S and 3S). Clusters 3A and

3D would correspond to placement of solute atoms $a/\sqrt{2}$ above the slip plane in rows (1, 2, 3) or in rows (1, 4, 7), respectively, in Fig. 20. Clusters 5A–20A and 5D–20D are obvious extensions. Cluster 2S corresponds to placement of solute atoms above the slip plane at rows (1, 2, 4, 5) in Fig. 20. Cluster 3S corresponds to placement of solute atoms above the slip plane at rows (1 to 9) in Fig. 20.

5.2. Results

In general, the activation energy of DK formation dominates that of lateral motion and, in the presence of the solute atom clusters, is less than that of the pure metal. Therefore, solid-solution weakening occurs, but it is possible to have strengthening in a few unique cases.

The exact magnitude of the strengthening or weakening was not determined, for this would require a computer simulation investigation at a constant average dislocation velocity as a function of temperature and composition, with the dislocation traveling in a three-dimensional sample. However, the general aspects of strengthening and weakening can be considered from the stress dependence of the activation energy, assuming the following relation between activation energy (ΔH) and average dislocation velocity ($\bar{v}$):

$$\Delta H(\tau) = kT \ln \frac{\bar{\nu}_0}{\bar{v}} \tag{16}$$

where $\bar{\nu}_0$ is related to the attempt frequency and the distance traveled by the dislocation in a successful thermal fluctuation, T is the temperature, and k is Boltzmann's constant. This equation will be used to generate an estimate of the yield stress (τ) as a function of the temperature. In so doing, we will fix $\bar{v}$ at some appropriate constant value. That value will be subsumed into the constant T_c defined below.

Prior to using the activation energy versus stress data it is necessary to take into consideration that the process with the larger activation energy at a given stress will be the rate-controlling process. It is also possible that one process may be rate controlling in the low stress range with the other controlling in the high-stress range. From the data of activation energy versus stress and by using Eq 16, it is possible to plot τ/τ_p versus T/T_c where T_c is the temperature at which Eq 16 is satisfied for $\tau = 0$ in a pure metal. From data shown in Fig. 21, it is obvious that softening is the general result for the size effect with the exception of cluster 3S, where strengthening occurs at low temperatures and weakening at high temperatures. However, this is indirect conflict with experimentally observed weakening which displays an opposite temperature dependence.

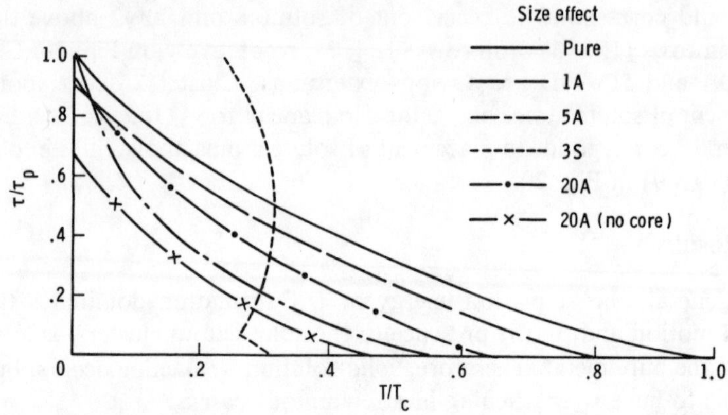

Fig. 21. Estimated yield stress versus temperature for various clusters interacting with the dislocation due to size effect.

5.3. Conclusions

From the results obtained in this investigation we may conclude that a continuum interaction model based on size and modulus differences between solute atoms and the matrix is incapable of explaining the observed strengthening of bcc alloys. While the observed weakening could be explained within such a model, a satisfactory description must account for the full temperature-dependent strengthening and weakening. Particularly in view of the very short-range interactions required, this suggests that an atomistic model including core and relaxation effects will be necessary.

Also, this investigation is a clear indication of a case where there is a significant reduction in time (computer and man time) if special cases are considered first.

6. High-Temperature Creep of Pure Metals

6.1. Introduction

The phenomenon of high-temperature creep has been investigated experimentally and theoretically for many years.[46-49]

In the previous analytical models that have been developed to explain high-temperature creep, numerous assumptions have been made explicitly and implicitly. An important advantage of computer simulation experiments or modeling is that the number of assumptions required to obtain a given result is drastically reduced. However, the major advantage of computer simulation experiments is that greater insights into the physical mechanisms can be obtained; the specific numerical results are of secondary importance.

A computer simulation experiment was undertaken to determine whether steady-state creep could be modeled in a way that would produce a fifth-power stress dependence. A basic assumption of this model is that the dislocations are generated at subgrain boundaries and that the climb of edge dislocations is the rate-controlling mechanism.

6.2. Models

A brief listing of the basic parameters of the two models follows.

1. The rate-controlling step in the two models is the climb of the edge of dislocations.
2. Aluminum was chosen as the model material; so all diffusion coefficients, etc., were those corresponding to aluminum.
3. In steady-state creep a well-defined subgrain structure exists.
4. The dislocation sources are at the subgrain boundary.
5. Each dislocation source generates more than one dislocation. The number generated depends on the stress, or some arbitrary exhaustion was imposed.
6. The internal stress, both glide and climb, does not change with time or strain. In other words, "recovery" (such as that proposed by McLean[(50)]) does not occur by climb or cross-slip so that the magnitude of the internal stress, either locally or on average, is not changed.
7. The annihilation process is due to encounters of edge dislocations of opposite sign.
8. The sample is divided into two regions. In the interior region, the image forces from the surface are very small and are considered to be zero. In the surface region, the image forces contribute to the force of the moving dislocations.

The difference between the models is related to the distance the edge component of the dislocation line traverses before it is annihilated. In the first model, the edge dislocation moves from one side of the subgrain (that is, where it was generated), climbs a given distance, and is annihilated. In this model the internal stress at the subgrain boundary is large, preventing the edge dislocation from gliding through the subgrain boundary. In the second model the edge dislocation can glide through the boundary after it has climbed a given distance and traversed several subgrains before it is annihilated.

In the second model the internal stress varies not only in the x direction (the x-z plane is the slip plane) but also in the y direction. This means that in the second model the edge dislocations can climb to a position such that the applied stress is equal to the glide internal stress and it can continue to glide.

The first model, which is defined as the single-subgrain model, is very similar to a model proposed by Weertman[(51)] and derived from the work of Caillard and Martin.[(52-54)] In this model the dislocation sources are located within the subgrain walls and the edge component of the dislocation loop only moves from one subgrain wall across the subgrain to the opposite wall. Figure 22 is a schematic diagram of subgrains near the surface of the sample. The positive edge dislocations move towards the interior, whereas the negative edge dislocations move toward the surface. Because of the image effect, it is necessary to take into account the difference between the signs of the edge dislocations. In the interior it is not necessary to make this distinction. Since the dislocations are being generated at subgrain walls, the edge dislocations are either negative or positive. In contrast, dislocations generated by a Frank-Read source form a complete loop; so one side of the loop is negative and the opposite side is positive. The dislocations generated from a subgrain wall do not form a complete loop, as shown in Fig. 22(b). Therefore the edge component is either positive or negative. The edge component of the dislocation does not leave the subgrain in which it was generated and is annihilated by climbing in the vicinity of the opposite subgrain wall.

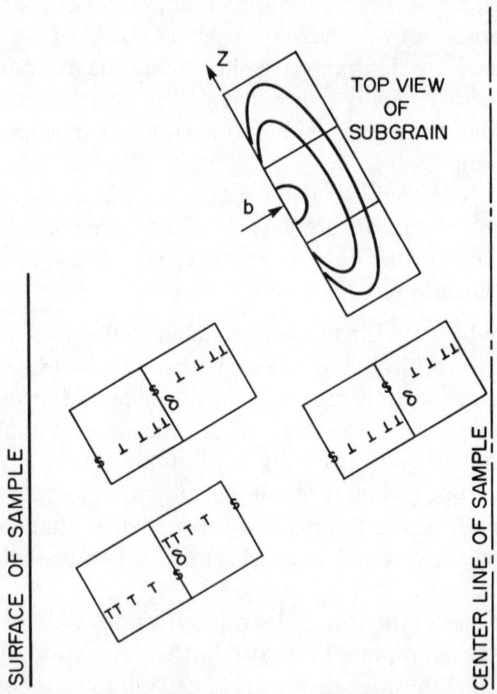

Fig. 22. A schematic diagram of two subgrains in the near-surface region. The near-surface region has both types of source, positive and negative. The annihilation distance is δ. Also shown is a schematic diagram of two subgrains in the interior. The annihilation distance is δ.

The width of the subgrain is inversely related to the applied stress[46] and is taken from the experimental relationship

$$\lambda = \frac{K}{\sigma_a} \tag{17}$$

where K is a constant.

In the first variation of this model, τ_{IG} can be equal to zero over the interior of the subgrain, but τ_{IG}^0 (the maximum value of glide internal stress) has to be large enough at the subgrain wall that the dislocations do not penetrate the subgrain wall. Another variation is that there is a finite τ_{IG} in the interior of the subgrain but that τ_{IG}^0 in the interior of the subgrain is less than the glide component applied stress τ_a. The applied stress has two components that are important to the motion of the edge dislocations, namely, the glide applied stress τ_a and the stress σ_{ac} that promotes climb. The orientation of the sample is such that $\tau_a = \sigma_{ac}$.

σ_{IC} is either zero in the interior of the subgrain and at the subgrain wall, or σ_{IC} has a finite positive or negative value at the subgrain wall and is zero in the subgrain interior.

There are forest dislocations within the subgrain which can be overcome (cut) by thermally activated "jumping." The stress required for source operation was an arbitrarily chosen parameter;[15] changes in the stress for source operation had very little effect on the end results.

If the sample is loaded initially with a stress greater than the maximum source operating stress, then all sources operate. It is assumed that the near-surface sources operate at a stress lower than that in the interior; in general, the near-surface operating stress was one-half that of the interior sources. The near-surface is the region from the surface up to a depth of 100 μm from the free surface of the sample, and obviously the interior includes the remainder of the sample. The position of the source in a subgrain wall is at an arbitrary distance from the slip plane of the adjacent subgrain, as shown in Fig. 22. Within a very short period of time a pileup occurs, stopping the sources because of the back stress of the pileup, but the pileup can aid in operating the source in the subgrain wall at which the pileup occurs. Figure 22 presents schematic diagrams of the two subgrains in the near-surface and the interior of the sample after the initial loading. These dislocations initially are coplanar but do not remain coplanar, for the motion of the edge dislocation can be both climb and glide. This means that the glide effective stress τ_i^* on each dislocation as well as the climb effective stress σ_{xxi}^* have to be determined by the following equation:

$$\tau_i^* = \tau_a \pm \tau_{IM} \pm \tau_{IG} \pm \tau_D \tag{18}$$

where τ_a is the externally applied glide stress, τ_{IM} is the glide image stress[12] (for dislocations in the interior, the image stress is zero), τ_{IG} is the long-range internal stress and is assumed not to change with time after the sample has been loaded, and τ_D is the interaction stress due to all the mobile dislocations. The climb effective stress was determined from the following equation:

$$\sigma^*_{xxi} = \sigma_{ac} + \sigma_{xxIM} \pm \sigma_{xxD} \pm \sigma_{IC} \tag{19}$$

where σ_{ac} is the external applied climb stress, σ_{xxIM} is due to the image stress[12] (again, for dislocations in the interior, the image stress is zero), σ_{xxD} is due to all the mobile dislocations, and σ_{IC} is the climb internal stress which is assumed to be independent of time. There are two possible variations in σ_{IC}. The first is that σ_{IC} is zero where the slip plane intersects the subgrain boundary and increases to a maximum at the top and bottom of the subgrain, and σ_{IC} can be either positive or negative going up or down. The second variation assumes that σ_{IC} is zero at the bottom and increases to a maximum at the top of the subgrain and can be either positive or negative; the sign of σ_{IC}, however, changes from one subgrain boundary to the next.

The generation and motion of each individual dislocation were determined and the strain was obtained from their cumulative motion. The glide motion is controlled by two components: the first is due to viscous drag and the second is due to thermally activated jumping of random short-range barriers. The time required for glide due to viscous drag is

$$t_{Di} = \frac{X_f}{B\tau^*_i} \tag{20}$$

where X_f is the distance between short-range barriers or the distance between the short-range barrier and the long-range barrier, B is constant, and τ^*_i (defined in Eq 18) is the effective glide stress on the ith dislocation. The thermally activated barriers are assumed to have a nature similar to that of forest dislocations. The time required for thermally activated jumping is given by Eq 4.

The time required for the climb of the edge dislocations[55] in the y direction is given by the following equation:

$$t_{Ci} = \frac{b}{\{C\exp(-Q_{sD}/kT\}\{\exp(\sigma^*_{xxi}\Omega/kT) - 1\}} \tag{21}$$

where C is constant, Q_{sD} is the activation energy for self-diffusion, σ^*_{xxi} is given in Eq 19, $\Omega \approx 0.7b^3$, k is Boltzmann's constant, and T is the temperature.

The motion of a given dislocation can be in either the positive direction or the negative direction, depending on the sign (that is, the direction) of the effective stress.

On loading the sample, the τ^*_i and σ^*_{xxi} values are determined for each dislocation generated, and then in a given time increment the possible motion of all dislocations is considered. The computational time required for this procedure is very long, but, as expected, the times for various types of motion are vastly different. The motion can be divided into three groups: (i) viscous drag, (ii) thermally activated jumping, and (iii) climb. In other words, initially all the dislocations that could move by viscous glide are allowed to move, and the time required is stored. Thermally activated jumping then follows the viscous glide and again the time required is stored. This process is repeated until all dislocations reach a static equilibrium. Figure 23 is a schematic representation of the dislocation configuration before there is any possibility of climb. The configuration is that of a classical dislocation pileup at a barrier. The time involved in this process is negligible compared with the time required to climb a distance of b. The time required to climb the distance b is then determined and stored. Next it is determined whether glide would occur. If glide occurs, the time of glide is not stored because it is approximately five orders of magnitude less than the time for a climb of b. The dislocations climb an arbitrary distance (generally some fraction of the subgrain size) and then are annihilated. (The distance climbed for annihilation affects the strain rate but has no effect on the stress exponents.) Figure 23(b) is a schematic representation of the dislocation configuration when the first dislocation is about to be annihilated. In Fig. 23(a), all the dislo-

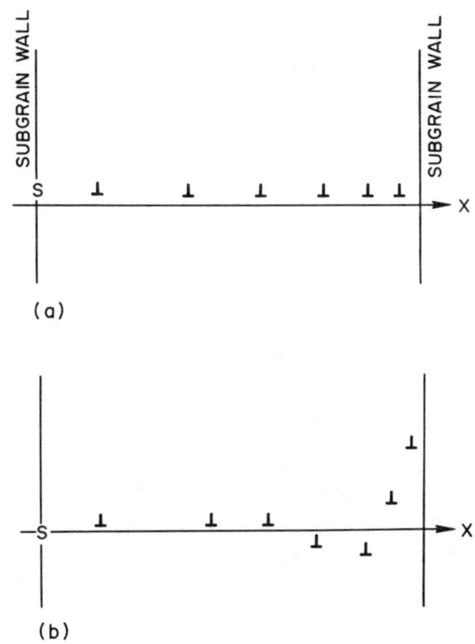

Fig. 23. Schematic representations of (a) static equilibrium positions of dislocations, in which no climb has taken place, and (b) the dislocation arrangement after a finite amount of climb has taken place.

cations are coplanar, but now the first dislocations generated are on several planes above and below the original slip plane.

The following equation was used to obtain the strain versus time relationship:

$$\epsilon(t) = \frac{\sum_{i=1}^{N} \{\Delta X_{Ei}(t) \Delta Z_{Si}\} b}{h(n\lambda)L} \tag{22}$$

where $X_{Ei}(t)$ is the distance traveled by the edge component of the ith dislocation loop at the given time, ΔZ_{Si} is the distance traveled by the screw component of the *i*th dislocation loop in the first model, h is equal to the width λ of the subgrain as defined in Eq 17 (it is assumed that the subgrains are cuboidal in shape), and L is the maximum glide distance of the edge component of the first dislocation generated, which is summed over all the N moving dislocations. In the single-subgrain model the maximum distance L traveled by the edge component is equal to λ, and the screw component is also equal to λ.

A variation of the first model is based on the experimental observations of Caillard and Martin.[52-54] They observed that the screw component of the generated dislocation loop cross slips and expands into several neighboring subgrains; that is, it travels a distance of nλ, and the extent of travel is linearly related to the applied stress. Also, it is assumed that the density of the dislocation sources is related to the subgrain boundary area. The net result is that the density of sources increases as the stress increases.

The second model, defined as the multiple-subgrain model, has many similarities to the first model, but there is one distinct difference. An edge dislocation generated from a source can travel through several subgrains before it is annihilated, and the annihilation step is not controlled by climb; the edge dislocation glides into an edge dislocation of an opposite sign. The dislocations climb until $\tau_i^* = \tau_{IG}$ and then they can move by glide, which means that $\tau_{IG}^0 > \tau_a$. This means that the glide internal stress depends not only on x (the direction of slip) but also on y (the direction of climb). Figure 24 is a schematic representation of the dislocation configuration after N dislocations have been generated.

In this model, it becomes increasingly difficult to account for the location of each individual dislocation. In this case, it is not only possible but very probable that the n + 1 dislocation may advance further than the nth dislocation generated, and the dislocation may climb up or down several subgrain heights. Also, since the dislocations travel through several subgrains, the image dislocations from the near-surface source that generated dislocations have to be accounted for in the same manner. However, the stress and time required for each individual step or jump are the same as those given in Eq 4 and 21.

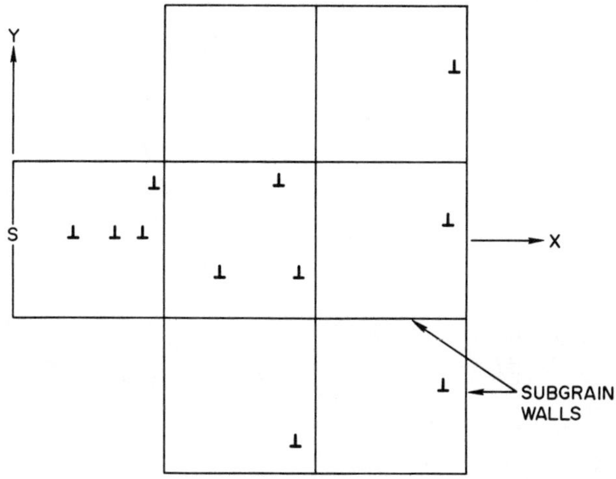

Fig. 24. A schematic representation of the dislocation configuration after N dislocations have been generated, for the second model.

The strain is again determined by Eq 22, but, in this model, h is the maximum distance in the y direction that the dislocations have climbed up and down (that is, the maximum vertical spread of the dislocations) and L is the annihilation distance.

In the second model, the variation in cross-slip of the screw component and the density of sources within the subgrain boundary area were also considered.

The relationship to the macrosample is obtained by combining the subgrains in the interior with those in the surface region and assuming that this combination is representative of the macrosample. In order to maintain plastic strain compatibility, the difference between the strain developed in the surface region and that in interior subgrains cannot exceed the elastic strain. If the strain in the two regions is significantly different, then further adjustments must be made in the variations, such as increasing or decreasing the magnitude of the internal stress.

6.3. Results

The simulation results obtained from both models are essentially the same; therefore, only the results from the first model will be discussed.

The screw component cross slips into multiple subgrains and the density of sources is a linear function of the subgrain boundary area, and the stress exponent is nearly 5 (Fig. 25). Again, changes in τ_{IG} to $\tau_{IG} \neq 0$ and in σ_{IC} to $\sigma_{IC} \neq 0$ had very little effect on the stress exponent.

Another variation was to assume arbitrarily a dislocation source exhaustion. (The purpose of this was to determine whether small values of τ_{IG}^0

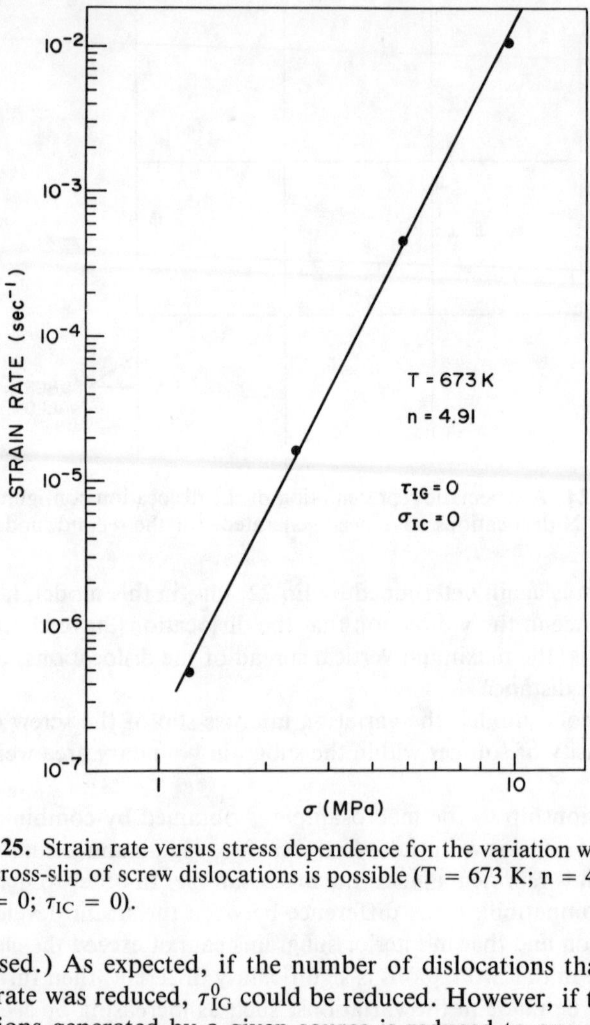

Fig. 25. Strain rate versus stress dependence for the variation where the cross-slip of screw dislocations is possible (T = 673 K; n = 4.91; $\tau_{IG} = 0$; $\tau_{IC} = 0$).

could be used.) As expected, if the number of dislocations that a source could generate was reduced, τ_{IG}^0 could be reduced. However, if the number of dislocations generated by a given source is reduced to one, difficulties arise, which are not obvious. If the number of dislocations generated was greater than 5, the stress exponent was unaffected.

6.4. Conclusions

Two models were considered in these computer experiments of high-temperature creep of pure metals in the power-law range and the following conclusions can be drawn:

1. Steady state creep can be obtained.
2. A stress exponent of about 5 can be obtained.

3. The activation energy is that of self-diffusion (climb is the rate-controlling step) at low values of applied stress and the activation energy decreases as the stress increases.
4. If climb is the rate-controlling step in high-temperature creep, then τ_{IG}^{0} (the maximum value of glide internal stress) has to be greater than τ_a.

7. General Conclusions

The computer simulation of dislocation motion in dilute random arrays is well understood.

The temperature dependency and the magnitude of the strengthening that is obtained from computer simulation modeling of fcc solid solutions at low temperature is in extremely good agreement with the experimental results. The phenomenon of "stress equivalency" is completely described.

However, in the case of bcc solid solutions, at low temperatures, the solid-solution weakening can be accounted for, but the strengthening cannot be described, with a semicontinuum using the classical defects instruction model.

The computer simulation modeling of high-temperature deformation is in its infancy, but in simple cases, the modeling was able to produce a steady-state creep rate and the correct power-law dependency.

Acknowledgement

The author wishes to acknowledge the generous support provided by the Computer Science Center of the University of Maryland.

References

1. U. F. Kocks, Phil. Mag., 1966, *13*, p. 541.
2. U. F. Kocks, Can. J. Phys., 1967, *45*, p. 737.
3. A. J. E. Foreman and M. J. Makin, Phil. Mag., 1966, *14*, p. 911.
4. J. E. Dorn, P. Guyot, and T. Stefansky, *Phys. of Strength and Plasticity*, edited by A. S. Argon, M.I.T. Press, 1969, p. 133.
5. E. S. P. Das, P. M. Gururaj, and M. H. Richman, J.A.P., 1972, *43*, p. 2210.
6. H. J. Frost and M. F. Ashby, J.A.P., 1971, *42*, p. 5273.
7. G. B. Gibbs, Phil. Mag., 1969, *20*, p. 611.
8. R. J. Arsenault and T. W. Cadman, Scripta Met., 1973, *7*, p. 631.
9. T. Cadman and R. J. Arsenault, Scripta Met., 1972, *6*, p. 593.
10. R. J. Arsenault and C. K. Kuo, Met. Trans., 1978, *9A*, p. 459.
11. A. R. Rosenfield and G. T. Hahn, Dislocation Dynamics, edited by Rosenfield, Hahn, Bement, and Jaffee (New York, McGraw-Hill), 1968, p. 255.
12. A. K. Head, Phil. Mag., 1972, *26*, p. 43; ibid., 1972, *26*, p. 55; ibid., 1973, *26*, p. 65; ibid., 1973, *27*, p. 531.

13. A. K. Head and W. W. Wood, Phil. Mag., 1973a, *27*, p. 505; ibid., 1973b, *27*, p. 519.
14. R. J. Arsenault, Phil. Mag., 1971, *24*, p. 259.
15. R. J. Arsenault and T. W. Cadman, *Rate Processes in Plastic Deformation of Materials*, edited by Li and Mukherjee (Metals Park: American Society for Metals), 1975, p. 102.
16. D. J. Bacon, U. F. Kocks, and R. O. Scattergood, Phil. Mag., 1973, *28*, p. 1241.
17. R. J. Arsenault and T. W. Cadman, Scripta Met., 1978, *12*, p. 633.
18. D. Stein and J. Low, J. Appl. Phys., 1960, *31*, p. 362.
19. A. T. Yokobori, Jr., T. Yokobori, and A. Kamei, J. Appl. Phys., 1975, *46*, p. 3720.
20. T. Yokobori, A. T. Yokobori, Sr., and A. Kamei, Phil. Mag., 1974, *30*, p. 367.
21. L. E. Murr, Met. Trans., 1975, *6A*, p. 505.
22. K. Ono, Proceedings of the 3rd International Conference on the Strength of Metals and Alloys, Cambridge, England, 1973.
23. C. T. K. Kuo and R. J. Arsenault, Mat. Sci. Eng., 1976, *27*, p. 11, and 1977, *30*, p. 65.
24. R. L. Fleischer, Acta Met., 1961, *9*, p. 996,and 1963, *11*, p. 203.
25. R. L. Fleischer, Acta Met., 1962, *10*, p. 855.
26. R. L. Fleischer, *The Strengthening of Metals*, edited by D. Peckner, Reinhold Publ. Co., New York, 1964, p. 93.
27. P. Haasen, *Physical Metallurgy*, Cambridge University, Cambridge, 1978.
28. J. Weertman and J. R. Weertman, *Elementary Dislocation Theory*, Macmillan Co., New York, 3rd ed., 1967, p. 173.
29. M. Cominou and J. Dundurs, J. Appl. Phys., 1972, *43*, p. 2461.
30. M. Cominou, M.S. Thesis, Northwestern University, August 1971.
31. R. J. Arsenault, R. Hsu, and D. Esterling, Scripta Met., 1981, *15*, p. 567.
32. B. R. Riddhagni and R. M. Asimow, J. Appl. Phys., 1968, *39*, p. 4144, and 1968, *39*, p. 6159.
33. H. Suzuki, *Strength of Metals and Alloys*, edited by H. J. McQuean et al., ICSMA7, 1985, p. 1727.
34. Z. S. Basinski, R. A. Foxall, and R. Pascual, Scripta Met., 1972, *6*, p. 807.
35. E. Pink and R. J. Arsenault, Progress in Materials Science, 1979, *24*, p. 1.
36. V. A. A. Brathwaite, B. L. Mordike, and K. D. Rogausch, Zeitscaift fur Metallkunde, 1971, *62*, p. 596.
37. J. R. Stephens and W. R. Witzke, J. of Less-Common Metals, 1976, *48*, p. 285.
38. J. Dorn and S. Rajnak, Trans. TMS-AIME, 1964, *230*, p. 1052.
39. R. J. Arsenault, Acta Met., 1966, *14*, p. 831.
40. K. Ono and A. W. Sommer, Metall. Trans., 1970, *1*, p. 877.
41. A. Urakami, Ph.D. Dissertation, Northwestern University, Evanston, IL, 1970.
42. A. Urakami and M. E. Fine, Proceedings of the First International Conference on the Mechanical Behavior of Materials, Kyoto, Japan, 1971.
43. A. Sato and M. Meshii, Acta Met., 1973, *21*, p. 753.
44. H. Suzuki, Chapter 15, part 2 in *Dislocations in Solids*, edited by F. R. N. Nabarro, North-Holland Publishing Co., 1979.
45. A. Seegar and P. Haasen, Phil. Mag., 1958, *3*, p. 470.
46. O. D. Sherby and P. M. Burke, Prog. Mater. Sci., 1967, *13*, p. 324.
47. J. Weertman, in *Rate Process in Plastic Deformation of Materials*, edited by J. C. M. Li and A. K. Mukherjee, American Society for Metals, Metals Park, OH, 1975, p. 315.

48. J. P. Poirier, Acta Metall., 1978, *26*, p. 629.
49. A. K. Mukherjee, in *Plastic Deformation of Materials*, edited by R. J. Arsenault, Academic Press, New York, 1975, p. 164.
50. D. McLean, Rep. Prog. Phys., 1966, *29*, p. 1.
51. J. Weertman, in Proc. 2nd Int. Conf. on the Creep and Fracture of Engineering Materials and Structures, edited by B. Wilshire and D. R. J. Owen (Swansea, April 1-6, 1984), Pineridge, Swansea, 1984, p. 1.
52. D. Caillard and J. L. Martin, Acta Metall., 1982, *30*, p. 437.
53. D. Caillard and J. L. Martin, Acta Metall., 1982, *30*, p. 791.
54. D. Caillard and J. L. Martin, Acta Metall., 1983, *31*, p. 813.
55. J. Weertman, Trans. Am. Soc. Met., 1968, *61*, p. 681.

9

An Overview of the Finite-Element Method for the Analysis of Engineering Metals

E. THOMAS MOYER, JR.
School of Engineering and Applied Science
The George Washington University
Washington, DC 20052

Introduction

The finite-element method has become the most widely accepted numerical technique for the solution of partial differential equations. This is especially true in the field of solid mechanics, where application of the finite-element method has been extremely well developed. Research during the past 25 years has developed the finite-element method to the point where large, time-dependent, nonlinear problems can be solved routinely, provided sufficient computational resources are available.

This chapter will introduce the formulation of and concepts behind the finite-element method as applied to problems in the continuum deformation of engineering metals. While much of the material presented is relevant to other disciplines, emphasis in this work is placed on the implications and limitations imposed by the formulative assumptions on the accurate prediction of stress, strain, and displacement fields in the continuum deformation of engineering metals. Extensive reference is made to relevant publications in the open literature. Because of the vast number of publications in the field, no attempt is made to provide a complete bibliography of the discipline. The references are representative of the literature base and provide a useful starting point for further reading.

The finite-element method was first applied to the linear elastic deformation of engineering materials and structures. This is the most developed and accurate of all areas to which the finite element method is applied. In addition, the linear elastic solution is often employed as the initial guess to an iterative algorithm for the solution of a nonlinear problem. Linear elastic solutions provide a good model verification as a necessary precursor to a successful nonlinear analysis. The variational formulation of the finite-element method for linear elastic solids is presented in this chapter. Nonlinearities (due to either nonlinear constitutive response or geometrically nonlinear deformations) are introduced as extensions to the linear elastic formulation.

For the modeling of the deformation of engineering metals, it is usually necessary to address material nonlinearity. Material nonlinearity refers to the deformation response of a continuum element that is not linearly proportional to the applied load even when deformations are infinitesimal. Most engineering metals can be adequately modeled as either elastic-plastic continua or elastic-viscoplastic continua. For the purposes of this chapter, an elastic-plastic continuum is defined as a material which deforms elastically up to a predefined limit. Beyond the elastic limit, the deformation response is a nonlinear function of the applied load. In an elastic-plastic continuum, however, the response is independent of the loading rate and the total elapsed time. All hysteretic effects are contained in the loading history, independent of duration. The response may, however, be a function of temperature.

In contrast, an elastic-viscoplastic continuum is defined as a material whose response is proportional to the applied load and independent of load duration up to a predefined limit (this limit, or yield criteria, can be a function of the loading rate). Beyond that limit, the response is governed by the applied load level, loading rate, and test duration. The way in which these parameters affect the response is modeled by a set of evolutionary equations which often involve intermediate parameters usually referred to as internal variables. These variables allow for the description of various observed physical phenomena (such as Bauschinger effects, delayed reverse yielding, and compression resistance) which cannot be adequately modeled by the physical problem parameters (such as applied load or stress, loading rate, and duration of load). All parameters in these models may also be a function of temperature. Emphasis in this chapter is placed on the finite element implementation of these physical models and the resultant solutions. Choice of specific constitutive models and assumptions is problem-dependent. A complete discussion is beyond the scope of this chapter. When modeling the deformation of engineering metals, one is often faced with deformations which cannot be adequately modeled as infinitesimal.

The modeling of finite deformations and finite rotations is an area which is currently receiving much research attention. While still an open research topic, geometric nonlinearities are addressable to some degree with current theoretical methodologies. The most widely accepted approach to the modeling of larger than infinitesimal deformations is the adoption of the Green-Lagrange formulation and the use of a total Lagrangian coordinate system.

While this formulation is limited (as discussed subsequently), it is the most tested and understood technique available. The Green-Lagrange formulation is presented in this chapter. Reference is made to other alternatives undergoing current research and debate.

Having established a variational principle and constitutive model for the material and deformation range being addressed, the finite-element method requires the discretization of the domain of interest by subdividing the material into smaller subdomains called elements. The definition of the mesh is probably the most important component of a successful finite-element solution. In this chapter, discussion is limited to a class of elements called *isoparametric.* These elements are the most widely employed and successful for two- and three-dimensional problems. Formulation of these elements, assumptions involved, and proper application are discussed. While the use of isoparametric elements allows for the solution of a wide class of problems, certain applications (such as the bending of plates and shells) require special elements and formulation changes beyond the scope of this chapter.

After the choice of constitutive behavior and model definition, the finite-element program attempts to solve for the required deformations and stresses. How successful this is depends on the problem being solved, the solution algorithm employed, and several parameters controlling the iterative and evolutionary characteristics of the solution algorithm. A knowledge of how these parameters are used is often essential for the attainment of an accurate solution. The most popular solution algorithms employed for solving nonlinear finite element systems are discussed in this chapter, including the relevant parameters employed.

The primary results of a finite element analysis are the values of the displacement components at each of the nodes in the model. Often, however, other quantities are of more interest, such as, stresses, strains, and strain rates. These variables are not primary unknowns to the finite-element system. They can be derived from the nodal displacements; however, their accuracy is less than the displacement components from which they are derived. The calculation of gradient components (stresses, strains, strain rates) is presented and their reliability and convergence are discussed. The importance of graphical postprocessing and the use of contour maps is emphasized.

The final section introduces those formulation changes required for the uncoupled thermomechanical analysis of metals. Fully coupled analysis is a field undergoing constant research and changes. No fully coupled model has gained total acceptance in the field. For most problems of interest, uncoupled thermomechanical analysis will yield sufficient insight into the physical processes being modeled. This approach requires a knowledge of the temperature field at every node in the model for the entire loading and time history. The temperature field may be known a priori or obtained from a heat-transfer analysis using the same mesh for the thermal problem as for the mechanical analysis. The finite-element formulation for heat-transfer problems is beyond the scope of this chapter; however, the interested reader can find abundant reference in the open literature (see, for example, Ref 1–4).

This chapter presents (in concise form) those aspects of the finite-element method required for the solution of problems involving the deformation of engineering metals. It is written primarily for those engineers and scientists who may not be thoroughly familiar with finite-element analysis. Where open questions in theoretical foundations remain, the most established and widely employed methodologies have been discussed. This chapter should provide the starting point for practitioners wanting to employ the finite-element method in their work.

Variational Principles Governing the Deformation of Engineering Metals

The finite-element method is a numerical method for the solution of partial differential equations. A critical element in this process is the identification of a variational statement equivalent to the governing differential system. For arbitrary differential systems, several general-purpose variational approaches are available (such as the Ritz technique and Galerkin methods[5-7]). For many physical problems, however, the governing equations can often be written in variational form directly from the first principles of the governing physics. If this approach is possible, it is to be preferred as the error norm will have physical meaning indicative of the solution accuracy.

For a continuum solid undergoing noninertial deformations, the total work done by external forces can be written as

$$\mathrm{W} = \mathbf{F}\mathbf{u}_{\mathrm{B}} - \int_{\mathrm{V}} \mathbf{b}\mathbf{u}\,\mathrm{dV} - \int_{\mathrm{A}} \mathbf{T}\mathbf{u}\,\mathrm{dA}$$

where $\mathbf{F}$ are concentrated forces applied at discrete boundary points, $\mathbf{u}_{\mathrm{B}}$ are the corresponding displacements at those points, $\mathbf{b}$ is the body force density at any given point in the domain (for example, due to gravity), $\mathbf{T}$ are applied tractions to the surface of the domain, and $\mathbf{u}$ is the displacement field at any point in the body. For an elastic material whose stress ($\boldsymbol{\sigma}$) and strain ($\boldsymbol{\epsilon}$) fields are related linearly the total strain energy of the body can be written as

$$\mathrm{U} = \frac{1}{2}\int_{\mathrm{V}} \boldsymbol{\sigma\epsilon}\,\mathrm{dV} - \int_{\mathrm{V}} \boldsymbol{\sigma\epsilon}_0\,\mathrm{dV} - \int \boldsymbol{\epsilon\sigma}_0\,\mathrm{dV}$$

where $\boldsymbol{\sigma}_0$ and $\boldsymbol{\epsilon}_0$ are stress and strain states existing prior to loading. The total potential energy can be written as

$$\pi = \mathrm{U} - \mathrm{W}$$

For linear elastic problems, the total potential energy must be a minimum stationary value for variations of admissible displacements. This requirement for equilibrium can be written, therefore, as

$$\delta[\pi] = 0$$

For a solid undergoing nonlinear material deformation, the statement of minimum potential energy does not apply. Consider a general solid undergoing a quasistatic deformation process. Starting with the principle of energy balance, the solution which minimizes the functional given by

$$\Phi = \int_{V} \bar{\sigma}\dot{\bar{\epsilon}}\, dV - \int_{S_T} \mathbf{T}\dot{\mathbf{u}}\, dS$$

is the equilibrium solution. As shown above, $\bar{\sigma}$ is the effective stress, $\dot{\bar{\epsilon}}$ is the effective strain rate, **T** is the vector of applied surface tractions applied to a portion of the surface S_T, and $\dot{\mathbf{u}}$ is the displacement rate vector. This functional is valid for both elastic-plastic and elastic-viscoplastic materials. In the case of rate-independent elastic-plastic deformation, the time rate concept is replaced by the load history. Rates, therefore, are relative to load changes that are not explicitly time-dependent. Full discussion of the functionals applicable to rate-independent elastic-plastic problems can be found in Ref 8. Functional formulations for elastic-viscoplastic analysis are discussed fully in Ref 9.

For certain specific applications, the elastic deformation is negligible relative to the total deformation (for example, the elastic strain is small relative to the total strain). In these problems, it is computationally more efficient to assume that the total strain is incompressible and that the deformation is either entirely plastic or entirely viscoplastic. This assumption requires the introduction of the constraint that

$$\dot{\epsilon}_{ii} = 0$$

where $\dot{\epsilon}_{ii}$ is the volumetric strain rate. To incorporate this constraint into the governing variational principle, the Lagrange multiplier technique is employed. The governing functional is modified and given in the form

$$\Phi' = \int_{V} \bar{\sigma}\dot{\bar{\epsilon}}\, dV + \int_{V} \lambda\dot{\epsilon}_{ii}\, dV - \int_{S_T} \mathbf{T}\cdot\mathbf{u}\, dS$$

where λ is a Lagrange multiplier. From a computational viewpoint, strict Lagrange multiplier theory is difficult to implement, as the solution must be obtained in the limit as the multiplier increases without bound. It usually suffices in practice to assume a large, finite value for the Lagrange multiplier. The value is problem-dependent and needs to be verified as new problem classes are solved. The use of a large, finite value for the Lagrange multiplier is referred to as the Penalty Function Method. The formulation of the rigid-viscoplastic approach is completely discussed in Ref 10 and 11.

The above functionals are formulated assuming purely mechanical effects (for example, no thermal effects). If thermal effects can be modeled through the material constant behaviors only (for example, by considering material constants to be a quasistatic function of temperature), the above functionals are sufficient for analysis. In many problems, however, thermal gradients cause strain effects which influence the fundamental response (through thermal expansion) in addition to the material properties. Functionals which include thermal influences are discussed in a subsequent section.

Finite-Element Formulation

After a valid variational principle is identified, the finite-element method employs several simplifying assumptions and mathematical manipulations to reduce the problem to computational form. The fundamental assumption employed in the finite element method is that the entire domain can be subdivided into a finite number of discrete elements. In a given element, the displacement field can be written as

$$\mathbf{u}(x,y,z) = \sum_{i=1}^{n} N_i(x,y,z)\mathbf{u}_i \;\; ; \quad n = \text{No. of nodes in an element}$$

where N_i are functions defined in terms of the element geometry only, and $\mathbf{u}_i$ are displacements at predefined discrete points along the element borders (or, for some element formulations, points interior to the element may be included). These predefined boundary points are called *nodes*. Shape functions are normally chosen as polynomials in the geometric coordinates. This choice, however, is not unique. The functions can be chosen quite arbitrarily, but the variational calculus imposes some restrictions (see Ref 12 for a complete description of shape-function restrictions). Since the shape functions are only geometry-dependent, the velocity at any point in a given domain can be written as

$$\dot{\mathbf{u}}(x,y,z) = \sum_{i=1}^{n} N_i(x,y,z)\dot{\mathbf{u}}_i$$

where $\cdot$ denotes time differentiation (this assumes that the coordinate reference frame is fixed in time; thus, a Lagrangian description of the deformation is employed). Strains and strain rates at a point can be written as

$$\epsilon = \mathbf{B}\mathbf{u} \;\; ; \quad \dot{\epsilon} = \mathbf{B}'\dot{\mathbf{u}}$$

where $\mathbf{B}$ and $\mathbf{B}'$ are nonlinear functions of the nodal displacements. If only infinitesimal strains are considered, these matrices are functions of the coordinate geometry only. This has significant impact on the computational requirements for a given problem solution.

If the finite-element approximations are used in conjunction with the appropriate functional, the governing equations are reduced from an integral system in space and time to a system of discrete ordinary differential equations in time (or, for rate- and time-independent problems, in load space). The only unknowns are the nodal displacements, assuming the constitutive relations between stress, strain, strain rate, and other pertinent variables are known. Constitutive assumptions for plastic and viscoplastic problems are discussed subsequently.

Constitutive Relations

For nonlinear material problems, the total rate of strain tensor is usually subdivided into two components: the linear elastic strain rate and the nonlinear strain rate (either plastic or viscoplastic).

$$\dot{\epsilon} = \dot{\epsilon}_{\mathrm{L}} + \dot{\epsilon}_{\mathrm{NL}}$$

The elastic strain rate is related to the total stress rate as

$$\dot{\sigma} = \mathbf{D}\dot{\epsilon}_{\mathrm{L}}$$

where $\mathbf{D}$ is the elasticity matrix. It only remains, therefore, to relate the stress (or stress rate) to the strain, strain rate, and other relevant parameters.

For viscoplastic problems, the onset of viscoplastic behavior is governed by a scalar yield function involving the stress state, the strain state, any internal variables involved (such as hardening parameters) and, in some formulations, the strain rate. In general, the yield condition can be written in the form

$$\mathrm{F}(\sigma, \dot{\epsilon}_{\mathrm{NL}}) - \mathrm{F}_0 = 0$$

where F is denoted as the "flow function." The parameter F_0 is the current value of the "yield" function (the value of the flow function at which continued viscoplastic deformation occurs). In general, the flow function is related to the viscoplastic strain rate through the form

$$\dot{\epsilon}_{\mathrm{NL}} = \phi(\mathrm{F}) \frac{\partial \mathrm{F}}{\partial \sigma}$$

where ϕ is an increasing monotonic function of F and can also depend on other parameters or internal variables (note that this formulation is the standard "associated" flow law with normality assumptions included). This formulation is by no means unique. It is, however, the most widely employed for viscoplastic problems. The choice of a specific function, ϕ, is problematic (for example, it depends on the materials involved). Complete discussions of viscoplasticity can be found in Ref 13. The finite-element implementation is discussed in Ref 14. Alternative approaches are exemplified by Ref 15–17.

For problems involving most engineering metals, the von Mises yield criterion is the most widely employed. The most general form of the yield criterion can be written as

$$(J_2')^{1/2} = R(K_j)$$

where J_2' is the second stress tensor invariant in "Updated Stress Space" (a stress space in which the "origin" can convect to account for kinematic hardening). The term R is the yield radius which can depend on any number of "internal variables," K_j, to account for isotropic hardening. This relation can be modified to account for such phenomena as anisotropic hardening, strain softening, and cyclic effects. For a complete discussion of yield and hardening see Ref 18 and 19.

Having established an appropriate yield function, a relation between the viscoplastic strain rate and the yield variable, F, must be established. For many materials, simple laws work very well. The simplest approach is to assume that no internal parameters are involved, that F is the "effective stress" (standard von Mises stress), and that hardening is isotropic only. Simple power-law behavior

$$\phi(F) = \frac{(F - F_0)^N}{F_0}$$

or exponential behavior

$$F = (J_2')^{1/2}$$

$$\phi(F) = \exp\left[M \frac{(F - F_0)}{F_0}\right] - 1$$

model some problems in metal forming extremely well.[20] For other problems, more sophisticated models are necessary. The past 10 years have produced much research in the area of constitutive modeling of viscoplastic materials. Little widespread agreement, however, has been obtained. For a discussion of "state-of-the-art" constitutive modeling, the reader is referred to Ref 21.

For rate-independent elastic-plastic problems, the von Mises criterion is also the most common for problems involving engineering metals. It only remains, therefore, to relate the effective stress (or von Mises stress) to the plastic (or total) strain. A constitutive law in the form

$$\frac{d\bar{\sigma}}{d\bar{\epsilon}} = H'(\bar{\epsilon})$$

is therefore sought. Various forms of the function H′ have been used for different engineering materials. Power laws, exponentials functions, combinations of the two and many others have been suggested. From a computational viewpoint, a table of slopes for various stress and strain values often better approximates the behavior of real materials. The representation of the uniaxial response, therefore, is multilinear. Various approaches to modeling elastic-plastic behavior can be found in Ref 22–23.

This section has summarized the remaining theoretical modeling for elastic-plastic and viscoplastic behavior of engineering metals. Special cases of the above include rigid-plastic or rigid-viscoplastic behavior (in which all loading deformation is nonlinear, yield criteria are unnecessary, and computation times are reduced). Rigid nonlinear models, however, do not accurately model problems in which unloading or stress relaxation occur. This problem limits their utility. For well-chosen problems, however, rigid nonlinear models can be accurately employed.[24]

Another special case of the above formulation is the case of perfectly plastic or perfectly viscoplastic behavior. In these problems, the strain response continues, unrestricted, for any condition surpassing yield. These models can be accurately employed for materials which exhibit little or no hardening. The main disadvantage is that all boundary conditions must be kinematic (as opposed to stress or force type) for a unique solution to exist. This is the major limitation of perfectly plastic models.

Modeling Geometric Nonlinearities

For many engineering problems, especially in metal-forming applications, the usual assumption of infinitesimal strains is often an insufficient description of the deformation. The modeling of geometric nonlinearities is an area of much current research. Unfortunately, no "best" method has been established to date. Several of the more modern approaches being investigated can be found in Ref 25–27.

The most popular approach to the modeling of geometric nonlinearities involves the use of the Green strain tensor. For a Cartesian coordinate system (X,Y,Z) which remains stationary for all time and loading history, the position of any point in a body can be written as

$$x = X + u$$

$$y = Y + v$$

$$z = Z + w$$

where (u,v,w) are the displacements in (X,Y,Z) from the initial configuration and (x,y,z) are the coordinates after displacement. The standard infinitesimal strains are defined as

$$\epsilon_x = \frac{\partial u}{\partial X}$$

$$\gamma_{xy} = \frac{\partial u}{\partial Y} + \frac{\partial v}{\partial X}$$

with the other components ($\epsilon_y, \epsilon_z, \gamma_{xz}, \gamma_{yz}$) defined by suitable permutation of the coordinates. The Green strain tensor is defined as

$$\epsilon_x = \frac{\partial u}{\partial X} + \frac{1}{2}\left[\frac{(\partial u)^2}{\partial X} + \frac{(\partial v)^2}{\partial X} + \frac{(\partial w)^2}{\partial X}\right]$$

$$\gamma_{xy} = \frac{\partial u}{\partial Y} + \frac{\partial v}{\partial X} + \left[\frac{\partial y}{\partial X}\frac{\partial u}{\partial Y} + \frac{\partial v}{\partial X}\frac{\partial v}{\partial Y} + \frac{\partial w}{\partial X}\frac{\partial w}{\partial Y}\right]$$

with similar expressions for the other components. This strain tensor, therefore, is a second-order strain tensor in a Lagrangian coordinate reference.

The Green strain tensor can be used in the constitutive models discussed previously, completing the theoretical formulation for the finite-element method. While this is often sufficient, several open questions remain. A major issue is the "stress" measure this implies. From a practical viewpoint, it is more important to understand the stress measure implications. Since the relation of the mathematical stress to the laboratory stress is not clear, the best approach is to: (1) choose a constitutive model linking the Green strain to the "stress"; (2) extract parameters from a uniaxial test; (3) perform a finite-element analysis on the tensile specimen; and (4) compare all measurable geometric quantities to the test results. The material parameters can then be adjusted until the model matches the test results. This pragmatic approach has yielded good results for many problems.[(28)]

The major competing method for the above formulation is the use of an "Updated Lagrangian" coordinate reference with appropriate modifications and corrections to the strain and stress definitions.[(29)] Essentially, this approach moves the coordinate system with the deformation; however, the coordinate directions and length parameters are incrementally related back to the reference configuration. The main advantage to the Updated Lagrangian approach is computational efficiency. It is more limited, however, than the total Lagrangian description (with Green strains) with respect to the range of deformation that can be accurately modeled. While theoretically the formulations are of equivalent order, the iterative requirements of the Updated Lagrangian approach become prohibitive for many problems. Comparisons of the two approaches can be found in Ref 30.

If significant rotations are involved in a given application, corrections can be made to improve predictions. Without significant rotations, the basic constitutive relations presented are all in the form

$$\dot{\boldsymbol{\sigma}} = \mathbf{C}\dot{\boldsymbol{\epsilon}}$$

where **C** is the constitutive matrix (a general function of the deformation), and $\dot{\boldsymbol{\epsilon}}$ are the Green strain rates. A modified relation for finite strains rewrites the constitutive assumption in the form

$$\dot{\boldsymbol{\sigma}} + \boldsymbol{\sigma}\boldsymbol{\Omega} - \boldsymbol{\Omega}\boldsymbol{\sigma} = \mathbf{C}\dot{\boldsymbol{\epsilon}}$$

where $\boldsymbol{\Omega}$ is the rotation rate tensor defined as

$$\boldsymbol{\Omega} = \dot{\mathbf{R}}\mathbf{R}$$

The matrix, **R**, is found by applying the polar decomposition theorem to the deformation gradient matrix, **G**, and is given by

$$\mathbf{G} = \mathbf{R}\mathbf{U}$$

This formulation is the Green-Naghdi approach to finite rotations. While much debate exists about the validity and accuracy of this approach, it is the most thoroughly tested and understood theory available. Further discussions of this theory can be found in Ref 31 and 32.

Element Formulations

The theoretical basis for the finite-element approach has now been fully established in the previous sections. The only remaining task is the subdivision of the domain into elements and the definition of the shape functions for these elements. Previously, the fundamental assumption of the finite element method had been that the displacement field at any point inside of an element could be written as

$$\mathbf{u}(x,y,z) = \sum_{i=1}^{n} N_i(x,y,z)\mathbf{u}_i$$

where $N_i(x,y,z)$ are the shape functions and $\mathbf{u}_i$ are the displacements of discrete points in the element called nodes. If the further assumption is made that the coordinates of any point in the element can be related to the coordinates of the nodes by

$$x = \sum_{i=1}^{n} N_i x_i$$

$$y = \sum_{i=1}^{n} N_i y_i$$

$$z = \sum_{i=1}^{n} N_i z_i$$

the elements are called isoparametric (x_i, y_i, z_i are the coordinates of the ith node). This assumption has many computational and convergence advantages for accurate finite element solutions.[(1)] Only isoparametric elements will be discussed subsequently.

For two-dimensional problems, the two standard element shapes employed are the generalized triangle and quadrilateral shown in Fig. 1. If three-noded triangles are used (nodes A,B,C), the shape functions are linear in the coordinates (x,y). If four-noded rectangles are employed (D,E,F,G), the shape functions are generalized linear relations. The shape functions for these elements are in the form

$$N_i = \alpha_i + B_i x + \gamma_i y$$

for three-noded triangles and

$$N_i = \alpha_i + \beta_i x + \gamma_i y + K_i xy$$

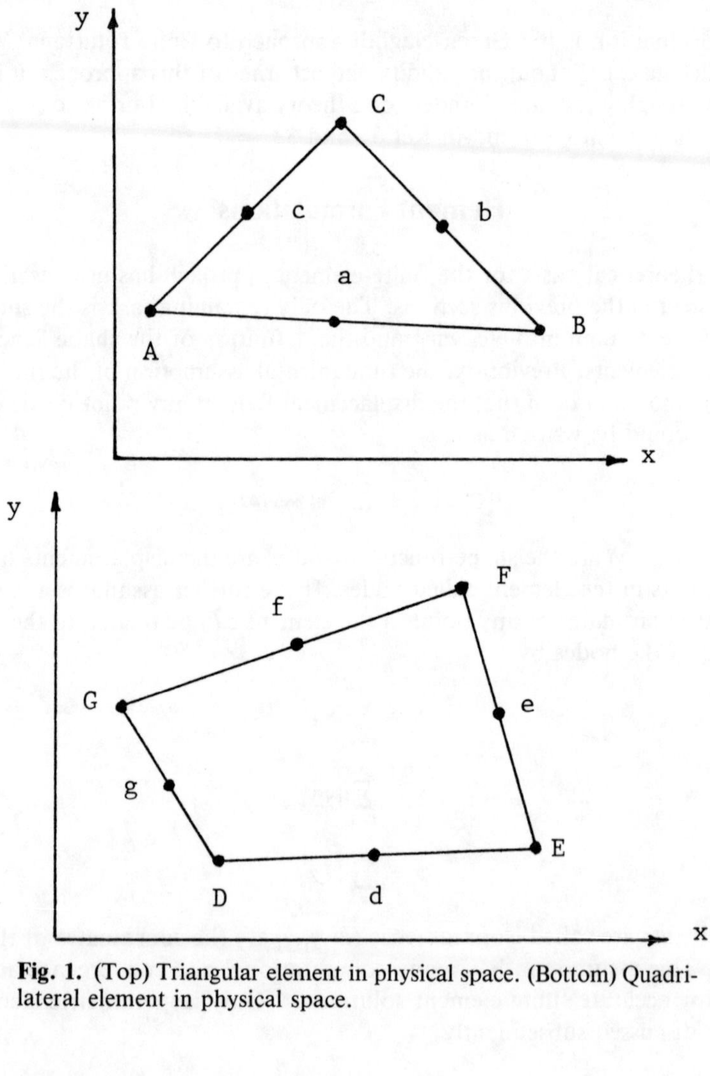

Fig. 1. (Top) Triangular element in physical space. (Bottom) Quadrilateral element in physical space.

for four-noded rectangles. If three nodes are used on an element edge, generalized quadratic functions can be defined as shape functions. In the formulation of isoparametric elements, it is convenient to define a reference element in a predefined "mapped" coordinate system (shown in Fig. 2). Here the ′ denotes the corresponding nodes in this system. Generalized quadratic functions in this "mapped" space are defined for the shape functions. The positioning of the nodes in physical space (x,y) define the shape-function behavior in (x,y). If the midside nodes (d,e,f,g) are at the midpoint of the parabolic arc between the corner nodes (D,E,F,G), the shape functions in physical space will also be parabolic. By placing midside nodes at other positions, however, it is possible to obtain many different shape functions which may be more appropriate for special problems.

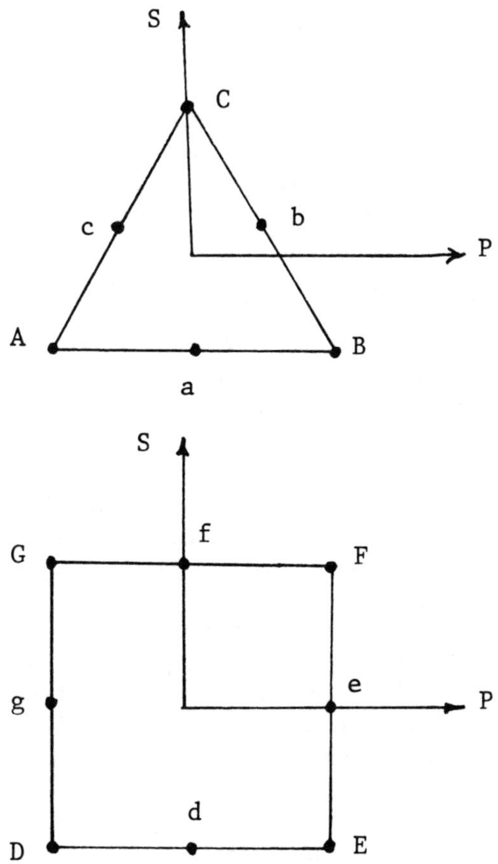

Fig. 2. (Top) Triangular element in mapped space. (Bottom) Quadrilateral element in mapped space.

For three-dimensional problems, the corresponding elements are the pentahedron and hexahedron shown in Fig. 3. The shape functions can be derived from their planar counterparts by extrusion. If linear shape functions are chosen, the pentahedron will have 6 nodes and the hexahedron will have 8 nodes. For quadratic shape functions, pentahedral elements will have 15 nodes and hexahedral elements will have 20 nodes.

It is obviously possible to define elements of arbitrary order. Much experience over the past 20 years has indicated that, in general, little accuracy is gained using elements above quadratic order and the computational overhead (caused by increased degrees of freedom and solution matrix bandwidth) is significant. Complete discussions on isoparametric element formulation can be found in Ref 33.

Choosing an element for a given application is not necessarily obvious. Accurate solutions can be obtained employing any of the elements described above. Several features of the various elements, however, will make the choice for a given problem more obvious. Both the linear triangle and linear rectangle have edges which are straight. Modeling curved boundaries with these elements is, therefore, difficult and requires large numbers of elements. For curved boundaries, therefore, quadratic elements are to be preferred.

Since the strain is a function of the shape-function gradients, linear triangles produce a constant strain value in each element. The solution using linear triangles, therefore, produces a patch-wise discontinuous strain field. If strain gradients are large, many elements may be required to achieve an accurate solution. The linear rectangle, however, does not have as severe a problem since the term K_ixy admits a linear strain variation. In general, linear rectangles are superior to linear triangles for finite-element analysis.

The choice between linear and quadratic elements is usually based on the degree of strain gradient in the problem being solved. For relatively uniform deformation problems, quadratic elements often introduce far more degrees of freedom than are required for accurate solutions. When gradient effects are appreciable, however, the quadratic elements are superior performers. In general, when geometry constraints do not necessitate the use of generalized triangles, quadrilateral elements are superior to generalized triangles.

Having chosen an element degree for a problem (either linear or quadratic), a mesh is developed which covers the domain of the problem. It is impossible to give general guidelines for the choice of element size. A good approach is often to do an analysis with a coarse mesh and look at the size of the strain gradients. If the strain variations within an element are large, increase the mesh density and observe the changes. This can be repeated until convergence is achieved. For nonlinear problems, no general methods for error estimation exist. Each problem, therefore, must be considered a numerical experiment.

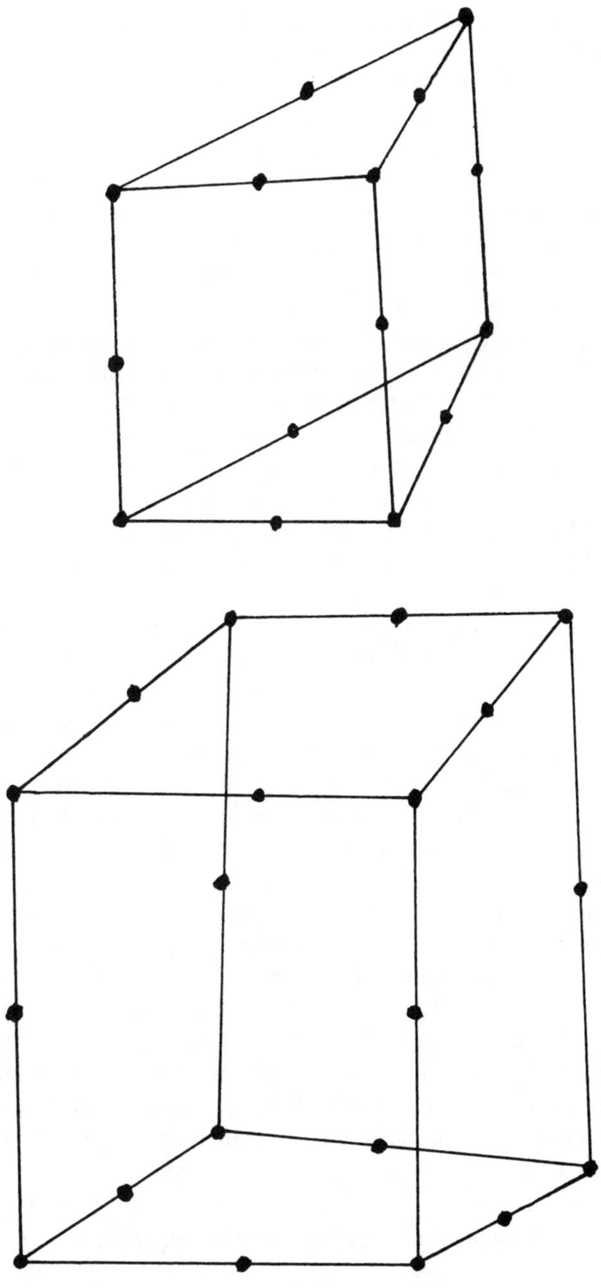

Fig. 3. (Top) 15-noded pentahedral element. (Bottom) 20-noded hexahedral element.

Two important considerations in mesh design are the element aspect ratio and internal angles. The element aspect ratio is defined as the ratio of the longest element edge to the shortest element edge. This ratio must be less than 10 for any of the elements presented to perform well. It is better if the ratio can be kept to less than 5. In addition, the angle between any two adjacent edges (or faces) should be between 35 degrees and 145 degrees to ensure confidence in the solution. Angles outside of this range may yield accurate solutions for a given problem; however, this cannot be guaranteed. These guidelines for element aspect ratio and internal angles should be followed unless the geometry being modeled cannot be meshed by any other means.

From the above discussion, the obvious use for triangular elements is for modeling geometries in which the use of quadrilaterals would violate the guidelines on aspect ratio and internal angles. Triangular elements can be extremely useful in conjunction with quadrilateral elements. Provided that all elements are either linear or quadratic, quadrilaterals and triangles can be employed in the same mesh without the introduction of modeling inconsistencies. When a free choice is available, however, quadrilaterals are superior to triangular elements. Comparisons of element performance can be found in Ref 34 and 35.

Mesh design and data preparation for finite element analysis has, historically, been a time-consuming and expensive task. Fortunately, this is changing rapidly with the introduction of Computer-Aided Design (CAD) software with finite-element mesh-design capabilities. Most CAD packages (which support finite-element design) allow the user to design a mesh manually or to use an automated mesh-generation algorithm which is part of the package. If severe geometric discontinuities are inherent in the problem being addressed, automated mesh-design algorithms will produce either an overly complex and dense mesh (increasing computing costs) or a mesh that will not produce a good solution. For these problems, manual meshing is required. If the geometry of a given problem is relatively regular, however, automated mesh generation is the technique of choice. In addition to allowing for the efficient design of finite-element meshes, most CAD packages will also check the aspect ratios and internal angles of the mesh prior to analysis. This will alert the user to potential problems in the model. As mesh-design packages are becoming available for virtually all computers (including personal computers), the need for hand designing of finite-element meshes has been virtually eliminated.

Solution of Finite-Element Equations

The formulation presented in the previous section will yield a system of ordinary differential equations in time (for the case of viscoplastic behavior) or a system of nonlinear algebraic equations (for the case of rate-independent plastic behavior). For the viscoplastic formulation, the governing system is in the general form

$$\mathbf{C}\dot{\boldsymbol{\delta}} + \mathbf{K}\boldsymbol{\delta} = \mathbf{F}$$

where $\boldsymbol{\delta}$ is a vector of the nodal displacements. The matrices **C** and **K** are known functions of the displacement vector and **F** is a known, time-dependent, load vector. Hundreds of techniques exist for the numerical solution of first-order differential initial value systems. For viscoplastic problems the most widely employed is the "θ" method described here.

Consider the ordinary differential equation given by

$$\dot{\phi} = \mathrm{G}(\phi,\mathrm{t})$$

where

$$\phi(\mathrm{t} = 0) \Rightarrow \mathrm{KNOWN}$$

If ϕ_n is the solution at any time, t_n, the solution at the next time step, t_{n+1}, can be written as

$$\phi_{n+1} = \phi_n + \Delta t[(1 - \theta)G(\phi_n,t_n) + \theta G(\phi_{n+1},t_{n+1})]$$

The parameter θ can take the values from 0 to 1. For $\theta = 0$, the method is fully explicit (that is, the value at t_{n+1}, can be directly calculated from the values at t_n). For all other values, the method is implicit. The employment of this algorithm reduces the problem to be a solution of an algebraic system of equations at each discrete time (the system is linear if the explicit algorithm is chosen and nonlinear if the implicit value is chosen).

The only difficulty with the numerical integration of the temporal ordinary differential equations is the choice of an appropriate time-step size. In general, there exists no estimator for time-step choice. The decision is problem-dependent. Most finite-element codes require the user to choose an initial time-step size and a minimum time-step size. The initial size is used to start the solution. The time step is reduced until an error tolerance is met. This size is then used to advance the time to the user-set step size. Thus, the method is based on numerical experiment with the choice of an appropriate convergence criterion. The minimum step-size choice ensures that the program will not run indefinitely without reporting trouble to the user. In addition, some programs contain algorithms for estimating new time-step sizes after the solution has been started. A full discussion of the theoretical bases for these methods can be found in Ref 1 and 36. Since no method is universally valid, the supplied error tolerance is the limiting quantity and must be continually monitored to ensure that the solution remains valid.

Having reduced the finite-element problem to the solution of a system of algebraic equations, the final task is the choice of an appropriate algorithm for their solution. If the system is linear, the solution may be computationally time-consuming; however, the theory is trivial and the accuracy can be to an arbitrary precision. The most popular algorithms involve a modification of Gauss elimination. All involve taking the banded structure of the system into account. The most widely employed are the frontal solution technique[12] and the skyline technique[1] (also referred to as the *profile storage method*). The method does not affect the accuracy of the solution but does affect computational requirements. The choice of a "best" algorithm depends on both the problem being solved and the computer hardware being employed. The algorithms both depend on an efficient mesh-numbering scheme. Most CAD packages have optimizing routines built into them, relieving the user of concern. For the interested reader, full discussions can be found in Ref 33 and 12.

The solution of nonlinear algebraic equations is more difficult than the solution of linear equations. The algorithms involved, however, are well founded and have been established for many years. The field of finite-element methodology has employed several methods throughout its history for the solution of nonlinear algebraic systems. Full discussions of several techniques can be found in Ref 1 and 14.

The most popular technique for the solution of nonlinear equations is the Newton-Raphson technique (in finite-element literature, this is often referred to as the Tangent-Stiffness technique). Consider the nonlinear algebraic equation given as

$$F(u) = 0$$

Assume the initial guess, u_0, is known. The first iterative solution given is

$$u_1 = u_0 - \frac{F(u_0)}{F'(u_0)}$$

The solution is continued, employing the recursion relation

$$u_{K+1} = u_K - \frac{F(u_K)}{F'(u_K)}$$

until a predetermined convergence criterion is met.

Having formulated the differential and algebraic equations involved in the finite-element method and presenting the solution algorithms, the full formalism has been outlined. No discussion, however, would be complete without discussing convergence (or accuracy) of the solutions. To measure the convergence of a solution, an appropriate norm must be introduced. Essentially, an error norm can be considered as a measure of how far from equilibrium the numerical solution is. This can be measured using a force equilibrium approach, an energy balance, an inertial balance, etc. The "best" choice is highly problem-dependent. Some general comments, however, may help guide their use in general.

For most applications, the error norm (whichever is employed) will be a relative measure of the error. That is, it will measure how far (usually in percent) from the "theoretical" solution the numerical solution is. Since the "theoretical" (or true) solution is unknown, this measure is approximate. It is wise, therefore, to choose a value between 5 to 10 times more restrictive than the largest error tolerable (for example, if displacement solutions are required to be ±1%, a norm of 0.5% to 0.1% should be employed). The best approach to monitoring error is to employ a "local" norm at each time step (or iteration) and, simultaneously, a global norm. The local norm is a measure of how far a given step deviates from equilibrium while the global norm indicates how accurate the total solution is. Local criteria, in general, need to be much smaller than global criteria for obtaining accurate solutions. The exact amount is highly problem-dependent.

In addition to accuracy, another numerical problem is the stability of a problem. Often (especially if solution parameters are inappropriate), a solution will increase without bounds yielding physically absurd results. This is due to a numerical "divergence" or instability. While mathematically quite different, the results are similar (solutions which are impossible or highly suspect). Assuming the finite-element code employed has reasonable integrity, the solution is simply to choose smaller time or load steps in the solution algorithm and try again. Several mathematical approaches to avoiding instabilities have been proposed in the past 10 years and are available in some finite-element codes. Ongoing research is abundant in this field.

The questions of convergence, error norms, stability, and accuracy of solutions are very involved theoretical questions. It is beyond the scope of this chapter to address these issues from more than a pragmatic and superficial viewpoint. Introductory discussions can be found in Ref 1 and 12. For more-advanced treatments, the technical journals referenced should be consulted.

Secondary Results—Stresses and Strains

The primary quantities obtained from finite-element analyses are the nodal displacements through the history of the problem. From an engineering viewpoint, stresses and strains are often the required unknowns. The most-accurate quantities predicted by the finite-element method, however, are the displacements. Nevertheless, stress and strain predictions can be made.

To obtain strains, the basic strain-displacement relationships are employed (these relations have been presented previously). For time-evolutionary systems, the strains are calculated incrementally from the displacement rates, utilizing the same time integrator as for the governing system. For time-independent problems, the strain-displacement equations are employed for each increment of displacement accumulated through the loading history.

Extensive experience has demonstrated that the strains evaluated on (or near) element boundaries are far less accurate than those predicted in the interior of the element. In addition, strains evaluated at the Gauss-Legendre integration points (employed in the evaluation of the functional volume and area integrals) are the most accurate strains in the finite-element mesh. Strains, therefore, are calculated to these Gauss points in most finite-element codes. If strains at other points are required, they are either interpolated or extrapolated from the Gauss point strains.

Often, nodal strains (or strains at other points on an element edge) are required. To calculate boundary strains, a technique called "Strain Smoothing" is employed.[37,38] This approach, however, involves an averaging technique which involves all elements containing a given boundary. If large strain gradients exist, the boundary values are unreliable. If at all possible, Gauss point strains should be employed in analyzing finite-element results.

Stresses are calculated in a manner like that used to calculate strains. The constitutive relations are employed incrementally whenever the strains are updated. The Gauss-point stresses, therefore, are the most accurate in the domain.

For enhanced interpretation of finite-element results, the emphasis in modern computational environments is in graphical postprocessing. Stress and strain contours, deformed geometries, local distortions, etc., can be displayed graphically over the entire domain and in desired blowups of critical subdomains. This reduces the manual effort in interpreting finite-element results and in their understanding. The use of graphical postprocessors is highly encouraged. It is important, however, to realize that graphical postprocessing does not change the accuracy of the results. Smooth contour values are no more accurate than the numbers in the output. In addition, if nodal values are used for contouring, the predicted contours can only be as accurate as the nodal results. If possible, choose a postprocessing package which employs Gauss-point stresses and strains as primary input.

Another major concern at postprocessing time is the final distortion of the mesh. For problems in which substantial deformation takes place, elements (which originally had valid aspect ratios and internal angles) may become extremely distorted, rendering the predicted results erroneous. It is necessary, therefore, to periodically examine the deformed grid and, when appropriate, remesh. Many of the finite-element codes in use today have the ability to stop the solution when a given criterion is met (such as the violation of aspect ratio or internal angle rules), save the results, admit a new geometry, and continue the solution procedure with a new mesh. All variables from the initial solution (such as, stresses, strains, displacements) are extrapolated to the new mesh.

Temperature-Dependent Problems

The final topic to be addressed in this chapter is the issue of thermal effects. In the formulation presented previously, it was assumed that the temperature was either constant, or that the temperature only affected the deformation through the same mechanical constitutive relations as in the isothermal problem (for example, the material constants could be temperature-dependent). It is well known, however, that when deformation takes place in a nonisothermal environment, thermal expansion takes place. This is easily accounted for in the finite-element formulation by introducing the "pre-strain" term in the functional as

$$\dot{\boldsymbol{\epsilon}}_0 = \boldsymbol{\alpha}\dot{\mathrm{T}} \qquad \boldsymbol{\alpha} = \begin{bmatrix} \alpha_x \\ \alpha_y \\ \alpha_z \\ 0 \\ 0 \\ 0 \end{bmatrix}$$

The total functional (for the viscoplastic example), therefore, can be written as

$$\Phi' = \int_V \bar{\sigma}\dot{\bar{\epsilon}}\,dV - \int_V \bar{\sigma}\dot{\bar{\epsilon}}_0\,dV - \int_{S_T} \mathbf{T}\dot{\mathbf{u}}\,dS$$

This requires a knowledge of the temperature field at every node in the domain for the entire history.

If the thermal effects are due to the energetics of the environment only (for example, uninfluenced by the mechanical state of the body), the temperature distribution can be obtained from the solution of the standard heat-conduction equations. Most finite-element programs can perform this analysis and employ the same mesh as that used in the mechanical analysis. The formulation for the finite-element analysis of heat-conduction problems can be found in Ref 1 and 2.

If the mechanical deformations influence the temperature distribution (so-called stress heating effects), the problem is much more complicated. In this case the thermal analysis and the mechanical analysis are fully coupled and must be performed simultaneously. Several formulations for this problem exist in the open literature, and the interested reader is referred to Ref 39 and 40. The theoretical basis for fully coupled analysis, however, is not well established. It is a problem in irreversible thermodynamics which is a discussion of much theoretical debate currently.

Several of the general-purpose finite-element codes are introducing fully coupled thermomechanical analysis capabilities. Unfortunately, with the lack of established theory and known benchmark solutions, the predicted results remain suspect. Active research is proceeding in this area, however, and a better understanding of analysis capabilities should be forthcoming.

Concluding Remarks

This chapter has presented the formulation, implementation, and basic background necessary to perform finite-element analysis of engineering metals. Emphasis has been placed on presenting theory which is implemented in currently available finite-element codes and on the basic understanding of this theory. While many open issues remain, the use of the theoretical approach in this chapter (and implemented commercially by many finite-element codes) yields excellent solutions to many engineering problems.

Ongoing research is addressing the issues of coupled thermomechanical effects, more-accurate representations of finite deformations, and alternative constitutive models. The reader is cautioned against the blind acceptance of every new approach that may become available. For the engineer not actively working in the field of finite element methods for large strain, nonlinear, thermo-viscoplastic problems, it is wise to wait until methods are time-tested and compared with existing technology. With existing technology, however, accurate solution to many problems can be obtained.

References

1. O. C. Zienkiewicz, *The Finite Element Method*, McGraw-Hill, London, 1977.
2. E. L. Wilson and R. E. Nickell, Nuclear Engineering Design, Vol. 4, p. 1, 1966.
3. O. C. Zienkiewicz, E. Onate, and J. C. Heinrich, in *Application of Numerical Methods to Forming Processes*, ASME-AMD, Vol. 18, p. 107, 1978.
4. M. Zlaml, in *The Mathematics of Finite Elements and Applications*, Vol. II, Academic Press, New York, p. 85, 1977.
5. K. Washizu, *Variational Methods in Elasticity and Plasticity*, Pergamon Press, London, 1975.
6. B. A. Finlayson, *The Method of Weighted Residuals and Variational Principles*, Academic Press, New York, 1972.

7. J. T. Oden, International Journal for Numerical Methods in Engineering, Vol. 1, pp. 205 and 247, 1969.
8. A. Samuelsson and M. Froier, in *The Mathematics of Finite Elements and Applications*, Vol. III, Academic Press, New York, p. 241, 1979.
9. M. B. Kanchi, O. C. Zienkiewicz, and D. R. J. Owen, International Journal for Numerical Methods in Engineering, Vol. 12, p. 169, 1978.
10. C. H. Lee and S. Kobayashi, ASME Journal of Engineering for Industry, Vol. 95, p. 865, 1973.
11. S. I. Oh, N. Rebelo, and S. Kobayashi, in *Metal Forming Plasticity*, IUTAM Symposium Proceedings, Tutzing, Germany, p. 273, 1978.
12. B. M. Irons and S. Ahmad, *Techniques of Finite Elements*, Ellis Horwood, Chichester, 1980.
13. P. Perzyna, in *Recent Advances in Applied Mechanics*, p. 243, Academic Press, New York, 1966.
14. D. R. J. Owen and E. Hinton, *Finite Elements in Plasticity: Theory and Practice*, Pineridge Press, Swansea, 1980.
15. B. D. Coleman and W. Noll, Archives of Rational Mechanics, Vol. 4, p. 97, 1959.
16. F. Sidoroff, Archives of Rational Mechanics, Vol. 27, p. 807, 1975.
17. S. R. Bodner and Y. Partom, Journal of Applied Mechanics, Vol. 4, p. 385, 1975.
18. J. L. Chaboche and G. Rousselier, Journal of Pressure Vessel Technology, Vol. 105, p. 159, 1983.
19. J. J. Gilman, *Micromechanics of Flow in Solids*, McGraw-Hill, New York, 1969.
20. S. Kobayashi, Journal of Applied Metalworking, Vol. 2, p. 163, 1982.
21. J. Eftis and D. L. Jones, Final Technical Report, AFOSR, 1983.
22. A. Mendelson, *Plasticity-Theory and Application*, Macmillan, New York, 1968.
23. G. C. Nayak and O. C. Zienkiewicz, International Journal for Numerical Methods in Engineering, Vol. 5, p. 113, 1972.
24. C. H. Lee and S. Kobayashi, Journal of Engineering for Industry, Vol. 95, p. 865, 1973.
25. S. N. Atluri, J. Structural Mechanics, Vol. 8, p. 61, 1980.
26. E. H. Lee, International Journal of Solids and Structures, Vol. 17, p. 859, 1981.
27. S. J. Kim and J. T. Oden, International Journal of Engineering Science, Vol. 22, p. 1235, 1984.
28. N. A. Cyr and R. D. Teter, International Journal of Computers and Structures, Vol. 3, p. 849, 1973.
29. K. A. Derbalian, E. H. Lee, R. L. Mallet, and R. M. McMeeking, in *Applications of Numerical Methods to Forming Processes*, ASME-AMD, Vol. 28, 1978.
30. W. Key, R. D. Kreig, and K. J. Bathe, Computer Methods in Applied Mechanics and Engineering, Vol. 17, p. 597, 1979.
31. R. Hill, Journal of the Mechanics and Physics of Solids, Vol. 7, p. 29, 1959.
32. L. E. Malvern, *Introduction to the Mechanics of a Continuous Medium*, Prentice-Hall, Englewood Cliffs, 1969.
33. K. J. Bathe and M. L. Wilson, *Numerical Methods in Finite Element Analysis*, Prentice-Hall, Englewood Cliffs, 1977.
34. D. R. J. Owen, G. C. Nayak, A. P. Kfouri, and J. R. Griffiths, International Journal for Numerical Methods in Engineering, Vol. 6, p. 63, 1973.
35. A. M. Winslow, Journal of Computational Physics, Vol. 1, p. 149, 1967.

36. O. C. Zienkiewicz and I. C. Cormeau, International Journal for Numerical Methods in Engineering, Vol. 8, p. 821, 1974.
37. E. Hinton and D. R. J. Owen, *Finite Element Programming*, Academic Press, London, 1977.
38. E. T. Moyer, Jr., International Journal of Computers and Structures, Vol. 18, No. 1, p. 185, 1984.
39. N. Rebelo and S. Kobayashi, International Journal of Mechanical Science, Vol. 22, p. 699, 1980.
40. N. Robelo and S. Kobayashi, International Journal of Mechanical Science, Vol. 22, p. 707, 1980.

10

Friction Modeling in Forging

William R. D. Wilson

Director, Center for Manufacturing Engineering
Professor, Mechanical Engineering Department
Northwestern University
Evanston, IL 60201

Abstract

Conditions of friction and lubrication are important to the development of processes for quality forged parts in a timely and cost-effective way. In order to create friction models which accurately account for the influence of processing variables, it is necessary to approach the problem on a fundamental basis. This requires the recognition that any of a number of different regimes can occur in a forging operation and that different types of friction models are appropriate to each regime. In order to determine the active regime or regimes, the processes by which lubricant films are formed and transported and changes in surface roughness that occur during forming must be understood. Once the pertinent regime has been determined, applicable friction models can be developed using fundamental knowledge about the regime. In general, this will involve tradeoffs between accuracy and complexity characteristic of any modeling activity.

Importance of Friction and Lubrication

Friction and lubrication are of great importance in forging operations. In most cases reducing friction is beneficial since it reduces the force and energy required for a given operation. This will reduce the stresses imposed on dies and may allow the use of smaller hammers or presses for a given part. Alternatively, larger changes of shape can be achieved with a given level of force or energy. In some operations a controlled amount of friction is necessary to control material flow in order to promote die filling or reduce workpiece spreading. In such cases too little friction is as bad as too much, and the lubrication system must be carefully specified and controlled to achieve optimum friction levels.

Low friction levels tend to promote more-homogeneous deformation

NOMENCLATURE

A : fractional contact area
C : thermal correction factor
D : $\alpha(T_b - T_d)$, nondimensional temperature difference
D : $4v/\pi d_1^2$, nondimensional displacement
G : $\gamma\sigma$, nondimensional pressure coefficient
K : $\alpha\sigma/\rho c$, plastic heating parameter
L : $\mu_0\alpha U_1/4k$, viscous heating parameter (wedge)
L^* : $\dfrac{\mu_0\alpha V^2}{k}\left(\dfrac{x_1\sigma}{3\mu_0 V}\right)^{2/3}$, viscous heating parameter (squeeze)
Q : q/σ, nondimensional extrusion pressure
R : r/r_n, nondimensional roughness
T_b : billet (workpiece) temperature
T_d : die (tooling) temperature
U : workpiece surface velocity
U_1 : workpiece inlet velocity
U_t : lubricant transport velocity
V : tooling (forming) velocity
c : workpiece specific heat
d_1 : inlet billet diameter
h_a : applied lubricant film thickness
h_0 : initial or inlet film thickness
k : lubricant thermal conductivity
ℓ : asperity spacing
m : friction factor
p : interface pressure
p_a : pressure to flatten asperity
q : extrusion pressure
r : workpiece roughness
r_n : workpiece natural roughness
v : nose volume
v_f : asperity flattening velocity
x_1 : initial half-width
x_2 : half-width of lubricated region
x_3 : final half-width
α : lubricant temperature coefficient of viscosity
γ : lubricant pressure coefficient of viscosity
ϵ : workpiece strain
$\dot{\epsilon}$: workpiece strain rate
θ : inlet angle
μ : coefficient of friction
μ : lubricant viscosity
μ_0 : lubricant viscosity at reference condition
ρ : workpiece density
σ : workpiece flow strength
τ_b : boundary friction stress
τ_f : friction stress
τ_ℓ : lubricant shear strength
τ_t : thick-film friction stress
τ_y : workpiece shear strength

which may increase the effective workability of a material and improve product quality by reducing the incidence of harmful residual stresses and internal defects.

Effective lubrication can also improve product quality by reducing surface defects such as scoring which are associated with metal-to-metal contact. Die life may also be extended by the presence of a lubricant film which will reduce wear and act as an insulating shield between the die and a hot workpiece.

In addition to guiding the improvement of lubrication systems for forging, a fundamental understanding of friction and lubrication phenomena is critical to the development of computer-based simulations of forging. These have already proved useful in process and tooling development, and as their accuracy and range of applicability improve they will be essential to producing high-quality forgings in a timely and cost-effective way. The purpose of this chapter is to present an overview of the factors which contribute to friction in forging and to suggest ways of providing more-realistic friction models than those which have been used to date in computer simulations of forging. It is hoped that this will be useful to those involved in developing more-sophisticated simulations.

Lubrication Regimes

In forging, as in most other metal forming operations, friction modeling is complicated by the fact that any of several different regimes of lubrication can exist at the billet-die interface. These regimes can be characterized by the thickness of the lubricant film relative to the surface roughness and by the fraction of the interface load that is carried by the contact of roughness peaks or asperities. Wilson[1] has described four main regimes as shown in Fig. 1.

In the thick-film regime shown in Fig. 1(a), the surfaces are separated by a continuous film of lubricant which is much thicker than the roughness of either of the surfaces. In forging this will also generally imply that the film is much thicker than the size of the lubricant molecules. Thus, surface roughness has little influence on lubrication and the lubricant can be modeled as

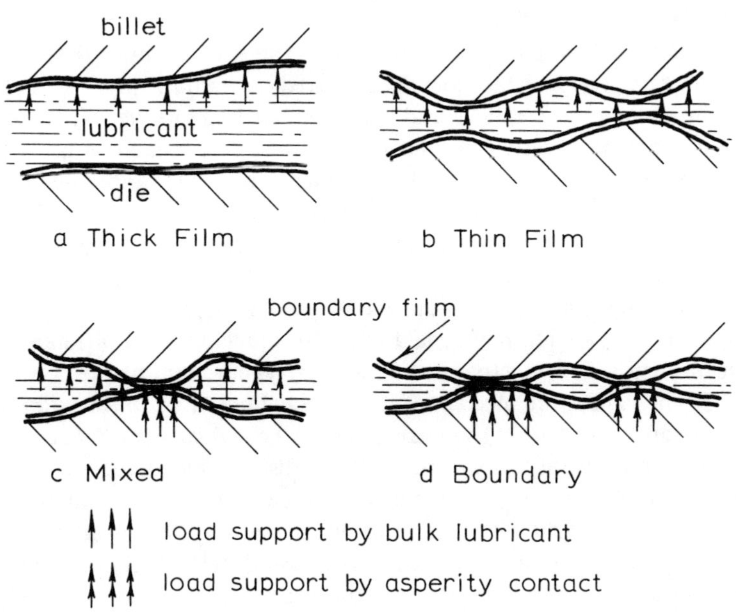

Fig. 1. Regimes of lubrication.

a continuum between smooth surfaces. Lubrication is essentially a mechanical process, and friction is decided by the shear resistance of the lubricant film under the conditions at the interface. Friction is usually quite small in the thick-film regime with measured coefficients of friction generally less than 0.1 and often less than 0.05. Since the surfaces are not in contact, wear is unlikely in the thick-film regime. However it can occur as a result of corrosion, or erosion by abrasive particles in the lubricant or surface fatigue. The thick lubricant film is relatively compliant; thus the billet surface roughness will increase during forming to the level that would occur on a free surface under the same strain conditions.

In the thin-film regime shown in Fig. 1(b), the mean lubricant film thickness is between approximately three and ten times the RMS composite roughness of the surfaces. In this regime, roughness influences lubricant flow, but the film thickness is still generally much larger than the lubricant molecular size and "contacts" between individual asperities carry a negligible fraction of the interface load. Thus the lubricant can be modeled as a continuum between rough surfaces. Friction and wear levels in the thin-film regime are comparable to those in the thick-film regime. Roughening of the workpiece surface may occur in the thin-film regime, but if the film thickness is small the roughness may stay essentially constant or even decrease during forming. Roughness changes are especially important in this regime because they can influence the lubrication process.

In the mixed lubrication regime shown in Fig. 1(c), the mean lubricant film thickness is less than about three times the RMS composite roughness of the surfaces. The interface loading is shared between the pressurized lubricant film in roughness valleys and "contacts" at asperity peaks. In reality, if the lubricant is properly formulated, actual metal-to-metal contact will not occur at asperity peaks because of the presence of tightly adhering "boundary" films formed as a result of chemical reaction or physical adsorbtion of active lubricant species at the surfaces. Even though these films are very thin (on the order of the lubricant molecular size), they prevent metal-to-metal contact, adhesion, and pickup during asperity collisions.

The mixed lubrication regime is difficult to model because different lubrication mechanisms are active in the asperity valleys and at the contacting peaks. In the valley, the mechanics-based techniques used in thin-film lubrication are appropriate. At the peaks, chemistry and surface physics as well as the mechanics of asperity deformation must be considered.

Friction coefficients in the mixed regime can have values ranging from 0.05 or less, associated with thick-film lubrication, to 0.4 or more, associated with poor boundary lubrication (discussed below). Significant wear levels can occur, particularly if conditions are severe enough to cause intermittent disruption of the boundary films. The friction level in a given system is controlled by the degree of load sharing between the pressurized lubricant film in the asperity valleys and the contacting peaks. Load sharing is often sensitive to process geometry and conditions. Thus variations in these can result in large changes in friction in the mixed regime. Workpiece surface roughness tends to decrease during forming in the mixed regime. However, if boundary films are disrupted, local adhesion and pickup can lead to billet roughening or scoring.

In the boundary regime shown in Fig. 1(d), all of the interface load is carried by the thin films at the contacting asperity peaks. The mechanics of asperity deformation, the surface physics and chemistry of the billet and die, and the active components of the lubricant control the lubrication process. The Amontons-Coulomb[(2)] model of friction (constant coefficient of friction) is applicable to the boundary lubrication of lightly loaded elastic contacts. The conditions of high loading and plastic deformation of the workpiece seem to make this simple model invalid in forging. However, as a rough guide, coefficients of friction in the boundary-lubrication regime are generally greater than 0.1 and may be as high as 0.4, particularly if partial breakdown of the boundary film occurs. Under good boundary-lubrication conditions, workpiece surface smoothing will occur but film breakdown may lead to roughening through surface damage.

An example of the importance of lubrication regime on system performance is given in Fig. 2. This figure shows the variation of measured coefficient of friction and product surface roughness with die semiangle in a strip drawing operation with wedge-shaped dies. Aluminum alloy strips (1100-H14) with a paraffin wax lubricant coating were used in the experiments.[(3)] As the die angle is increased, less lubricant is drawn into the die/strip interface and the system goes progressively through the thick-film, thin-film, mixed, and boundary regimes. Figure 2(a) shows that as the regimes change, the coefficient of friction increases from the very small values associated with good thick-film lubrication to values characteristic of good boundary lubrication. The influence on workpiece surface roughness is shown in Fig. 2(b). Roughening due to unconstrained deformation in the thick-film regime gives way to smoothing in the mixed regime, which in turn gives way to roughening due to scoring in the boundary regime.

It is evident from Fig. 2(a) that even with a particular combination of workpiece, die, and lubricant materials, a wide variation in frictional conditions is possible depending on the lubrication regime that prevails. This is largely decided by the amount of lubricant which is present between the workpiece and die, which is in turn decided by the mechanics by which lubricant films are formed and transported at the interface. Thus in order to develop realistic frictional models for forging, we must understand more about the mechanics of lubrication in forging operations.

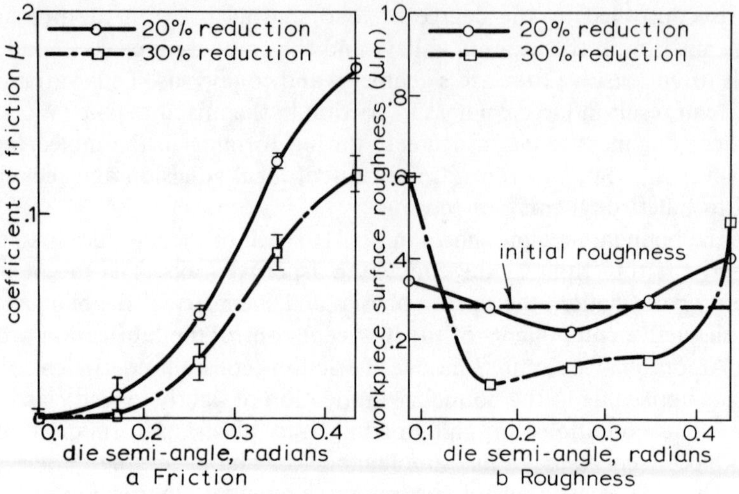

Fig. 2. Strip drawing of aluminum with wax lubricant.[3]

Film Formation

For a forging lubricant to be effective, it must find its way into the high-pressure region between the workpiece and die surfaces. This film formation process generally occurs by either squeeze or wedge action or some combination of these.

Squeeze action is illustrated in Fig. 3. As the tooling and workpiece approach each other, the intervening lubricant film is squeezed out. As the film thickness decreases, the pressure gradients required to squeeze out the film become so high that the film pressure reaches the level required to initiate plastic flow of the workpiece, and the remaining film is trapped between the tooling and plastically deforming workpiece.

The only metal forming operation in which squeeze action has been extensively studied is simple upsetting with homogeneous deformation and flat dies, shown in Fig. 4. In the initial approach phase, both the workpiece and dies are rigid as shown in Fig. 4(a). As the process continues, the system enters the yielding phase shown in Fig. 4(b), where the pressure in the lubricant film is high enough to cause plastic flow near the center of the workpiece. In this plastic region the pressure gradients are too small to cause significant flow of the lubricant film, which is consequently essentially trapped. Eventually the plastic region spreads until the workpiece is involved and the system enters the deformation phase shown in Fig. 4(c). At this point the film thickness at the edge of the conjunction has been reduced to zero.

In modeling the squeeze lubrication process with a liquid lubricant, it is

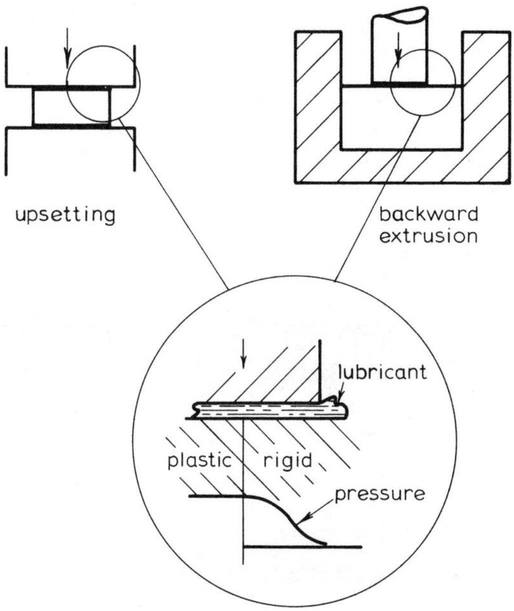

Fig. 3. Squeeze lubrication.

usual to assume Newtonian behavior with a lubricant viscosity μ which varies with pressure p and temperature T (both measured above ambient) according to

$$\mu = \mu_0 e^{\gamma p - \alpha T} \tag{1}$$

For the isoviscous case ($\gamma = 0$, $\alpha = 0$) it has been shown using thick-film theory[4,5] that the central film thickness h_0 in both the axisymmetric and plane strain cases is given by

$$h_0 = \left(\frac{3\mu_0 V x_1^2}{\sigma} \right)^{1/3} \tag{2}$$

where V is the approach velocity of the dies, x_1 is the workpiece half-width, and σ is the workpiece flow strength. For the isothermal case ($\gamma \neq 0$, $\alpha = 0$) the equivalent equation is

$$h_0 = \left(\frac{3\mu_0 \gamma V x_1^2}{1 - e^{-\gamma\sigma}} \right)^{1/3} \tag{3}$$

It is evident that both Eq 2 and 3 predict that the film thickness increases with the lubricant viscosity μ_0, forming speed V, and workpiece size x_1 and decreases with increasing workpiece strength σ. However, in the case of pressure-dependent viscosity (Eq 3), the pressure coefficient of viscosity γ

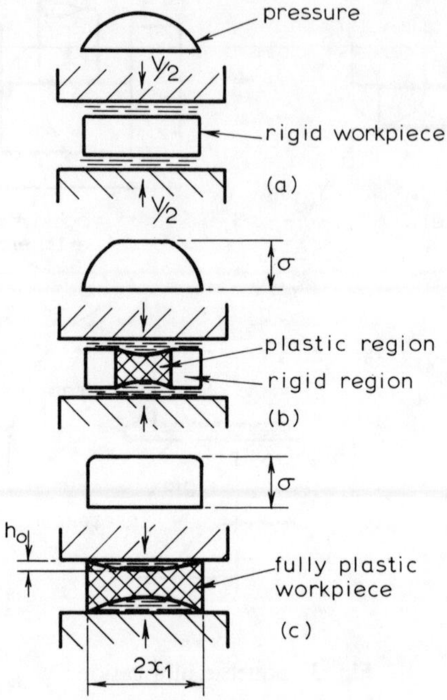

Fig. 4. Stages in upsetting.

is just as important as the base viscosity μ_0. This explains why lubricants with high pressure coefficients such as mineral oils have much better film-forming capabilities than simple liquids such as alcohol or water, which have low pressure coefficients. Also, in the pressure-dependent case the film thickness is relatively insensitive to workpiece strength σ, since the only term in which this appears ($e^{-\gamma\sigma}$ in the denominator) is generally small compared with unity.

In the case where viscosity varies with both pressure and temperature ($\gamma \neq 0$, $\alpha \neq 0$), the central film thickness h_0 can be written[6] as

$$h_0 = C\left(\frac{3\mu_0 \gamma V x_1^2}{1 - e^{-\gamma\sigma}}\right)^{1/3} \tag{4}$$

where C is a correction factor for viscous heating of the lubricant given approximately by

$$C = (1 + 0.16L^{*0.56})^{-1} \tag{5}$$

L^* is the so-called thermal loading parameter, a dimensionless group which controls viscous heating in squeeze lubrication and is defined by

$$L^* = \frac{\mu_0 \alpha V^2}{k}\left(\frac{x_1 \sigma}{3\mu_0 V}\right)^{2/3} \tag{6}$$

where k is the thermal conductivity of the lubricant. For small values of L^*, C approaches unity and Eq 4 reduces to Eq 3. At large values of L^* the film thickness is greatly reduced by viscous heating. In particular, under high-speed conditions the film thickness may actually decrease with increasing forming speed rather than increase as predicted by isothermal theory.

In general, it is difficult to test the isothermal theory under realistic conditions. In cases where forming speed and lubricant viscosity are large enough to generate a measurable film thickness, they are also large enough to result in major thermal effects. However, in this case the thermal theory seems to be in good agreement with experimental film thickness measurements as shown in Fig. 5. This compares the variation of the thermal correction factor C with thermal loading parameter L^* predicted by theory (Eq 5) with that determined in a series of experiments with 6063-T52 aluminum alloy workpieces lubricated with a variety of liquid lubricants.[7] The workpieces were upset using a drop hammer, and the trapped film thickness was estimated using a profilometer to measure the depression left in the workpiece faces after forming. It is evident that viscous heating results in a substantial reduction in film thickness and that the theory does a good job in accounting for this effect.

Solid lubricants such as polymers, soft metals, and conversion coatings are commonly used in forging. Such materials behave more like plastic solids than viscous liquids. They generally have superior film-forming capabilities and are less sensitive to speed and thermal effects than liquid lubricants. Their film-forming capability often means that the trapped film thickness is limited by the thickness of the lubricant coating prior to forging rather than by the mechanics of squeeze lubrication. In lubrication parlance the system is operating in a "starved" state. For a simple upsetting operation[8] with a

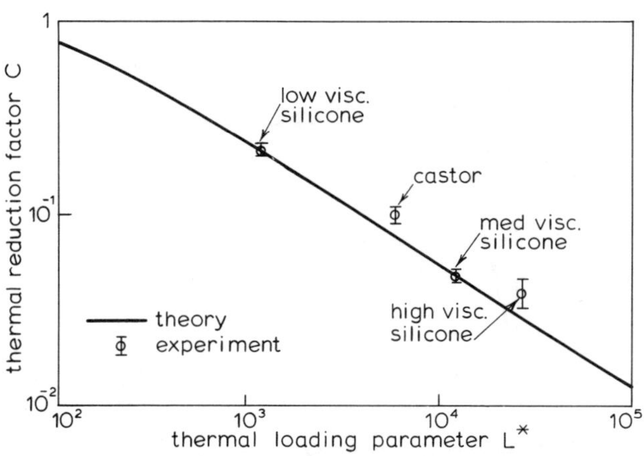

Fig. 5. Thermal effect on film thickness in upsetting.[7]

perfectly plastic lubricant of shear strength τ_ℓ, the central film thickness h_0 is given by the smaller of

$$h_0 = h_a \tag{7}$$

where h_a is the applied film thickness, or

$$h_0 = \frac{2\tau_\ell x_1}{\phi\sigma} \tag{8}$$

where ϕ is a redundant work factor. Equation 7 is for the starved state while Eq 8 is where excess lubricant is available, the so-called flooded state.

Figure 6 compares the starved and flooded predictions with experimentally measured trapped central-film thickness for a variety of applied film thicknesses. Lead was used as the lubricant in upsetting aluminum alloy (6063-T52) workpieces. At low values of the applied film thickness h_a, all of the film is trapped according to the starved model. At large values of h_a the trapped film thickness is close to that given by the flooded theory. However, in an intermediate range all of the film is trapped even when this leads to film thicknesses greater than those predicted by flooded theory. The reasons for this are not currently understood. Furthermore, lead has relatively simple rheology and additional work is needed to improve our understanding of realistic materials which exhibit more complicated behavior.

The second main way of forming lubricant films in forging, by wedge action, is illustrated in Fig. 7. A coordinate system which moves with the edge of the region of plastic deformation or work is assumed in the case of unsteady processes such as forging. In lubrication by wedge action, the motion of the workpiece and/or die into the work zone drags the lubricant

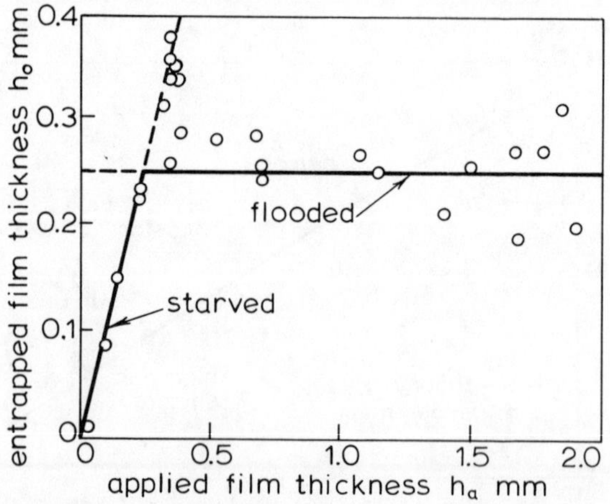

Fig. 6. Film thickness in upsetting aluminum with lead lubricant.[8]

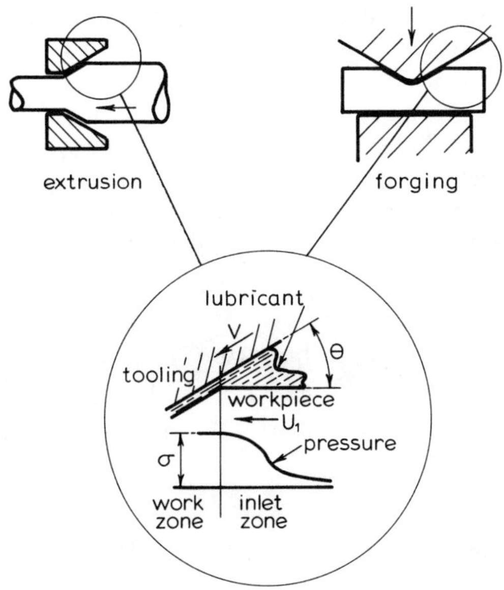

Fig. 7. Wedge lubrication.

against the steep pressure gradient, which is necessary to allow the pressure to build up to the high value required for plastic flow. The dynamic balance between the dragging action assisting lubricant entrainment and the pressure gradient resisting lubricant entrainment determines how much lubricant passes into the work zone.

Investigations of wedge action in metal forming processes have generally been limited to steady processes such as hydrostatic extrusion[9] or rolling.[10] For a Newtonian liquid with viscosity given by Eq 1, if the surfaces are rigid in the inlet zone, the inlet film thickness h_0 can readily be determined for the isoviscous case ($\gamma = 0$, $\alpha = 0$) as

$$h_0 = \frac{3\mu_0(U_1 + V)}{\sigma \tan\theta} \tag{9}$$

and for the isothermal case ($\gamma \neq 0$, $\alpha = 0$) as

$$h_0 = \frac{3\mu_0\gamma(U_1 + V)}{\tan\theta(1 - e^{-\gamma\sigma})} \tag{10}$$

where μ_0 is the lubricant base viscosity, γ the pressure coefficient of viscosity, U_1 and V the inlet workpiece and die velocities relative to the boundary between the work and inlet zones, σ the workpiece flow stress, and θ the inlet angle. These equations must be modified to allow for the influence of upstream pressurization as would occur, for example, in extrusion.[11] As in the squeeze case, wedge theory predicts that the film thickness increases with lubricant viscosity μ_0, pressure coefficient γ (if this is significant), and forming speed as typified by U_1 and V. However, these are now related linearly rather than through the cube-root relationship of Eq 2 and 3. The film thickness is also independent of the size of the system but is strongly affected by the inlet angle θ, smaller values of θ leading to thicker films. The low sensitivity of the film thickness to the workpiece flow strength in the case of pressure-dependent viscosity is also apparent.

In the case where thermal effects are important, the film thickness h_0 can be expressed as

$$h_0 = C\,\frac{3\mu_0\gamma(U_1 + V)}{\tan\theta(1 - e^{-\gamma\sigma})} \tag{11}$$

where C is a thermal correction factor. For the case of simple sliding (V = 0), C is given approximately by[12]

$$C = (1 + 0.56G^{0.38}L^{0.8})^{-1} \tag{12}$$

where G is the nondimensional pressure coefficient defined by

$$G = \gamma\sigma \tag{13}$$

and L is the parameter which controls viscous heating in wedge lubrication, defined by

$$L = \frac{\mu_0\alpha U_1^2}{4k} \tag{14}$$

As in the squeeze case, for small values of L, C approaches unity and the system behaves essentially isothermally. At large values of L, the film thickness is substantially reduced by viscous heating so that it may actually decrease with increasing speed. A comparison of the values of L (Eq 14) and L^* (Eq 6) shows that the main difference is the presence of the term $(x_1\sigma/3\mu_0V)^{2/3}$ in the latter. This is associated with the "squish" effect present in squeeze lubrication whereby the lubricant velocities can be much larger than the forming velocity V. Thus, for a given forming speed, viscous heating effects tend to be much larger in squeeze lubrication than in wedge lubrication, and it is possible to form films under essentially isothermal conditions in wedge lubrication of forming operations, whereas this is very difficult with squeeze lubrication.

As with the theoretical work, there has been relatively little experimental research on wedge lubrication with liquid lubricants in forging. However, experimental research on hydrostatic extrusion[12] and rolling[13] has indicated the soundness of the theoretical approach described above.

Research on wedge lubrication by solid lubricants is still in its infancy. Early work attempted to adapt viscous theory by using artificial viscosities which varied with speed or temperature. However, this approach fails to capture the essentially plastic behavior of solid lubricants under low-speed conditions.

Approaches have also been made using the upper-bound method of plasticity to model lubricant flow in drawing and hydrostatic extrusion and to confirm the results experimentally.[14-17] In general this research has shown that the inlet film thickness h_0 is sensitive to the applied film thickness h_a (or that "starved" conditions often exist) and to the inlet geometry (including the inlet angle θ and the presence of "rounding in" if this occurs). Increasing h_a or decreasing θ increases h_0. The relative strength of the coating compared to the workpiece is also important. Increasing the lubricant strength τ_ℓ or decreasing the workpiece strength σ tends to increase h_0. The models and experiments also show that the entrained film thickness h_0 is generally insensitive to speed over a wide range.

Lubricant Transport

After the lubricant enters the high-pressure work zone between the die and the workpiece, it is further transported by the motion of the workpiece and/or die surfaces. The smearing and stretching of the lubricant film modifies its thickness so that the final thickness and consequently the lubrication regime can be quite different from that inferred from studies of film formation. In extreme cases, even though a particular area of the workpiece surface may be adequately supplied with lubricant during the initial stages of a process, at later stages the lubricant film may be so depleted that lubrication breakdown and metal-to-metal contact will occur. It is thus of great practical importance to understand the mechanics of lubricant transport during the forming process.

In analyzing lubricant transport it is convenient to use the idea of a mean lubricant transport velocity V_t, which is the average lubricant velocity across the film thickness. While local pressure gradients associated with roughness may influence lubricant flow, the average pressure gradients in the work zone associated with the deformation process are relatively small and have a negligible influence on lubricant transport. If surfaces are smooth and thermal effects are neglected, the transport velocity U_t is simply the mean of the workpiece surface velocity U and die velocity V parallel to the film:

$$U_t = (U + V)/2 \tag{15}$$

If the die is stationary (or moving normal to the film), U_t is simply half the workpiece speed:

$$U_t = U/2 \tag{16}$$

This behavior has important consequences in a number of processes and in particular tends to cause a breakdown in full film lubrication in simple upsetting with overhanging dies, as shown in Fig. 8, and during the start of extrusion, as shown in Fig. 9.

Figure 8 shows two stages. The situation at the end of the yielding phase with a trapped film of lubricant generated by squeeze action over each end face is shown in Fig. 8(a). As the process proceeds and bulk workpiece deformation takes place, the workpiece surface slides outward against the die. However, the lubricant film is transported at only half the workpiece speed and thus cannot keep up with the outward motion of the edge of the specimen. As a result there is a region near the specimen edge which is unlubricated except for boundary films, as shown in Fig. 8(b). A similar effect can occur in upsetting of rings at both the inner and outer edges.[5] However, in upsetting with an overhanging workpiece, as in plane-strain compression testing, the effect is not present and average friction levels will generally be lower than in upsetting with overhanging dies.

Figure 9 shows a similar phenomenon which occurs during the start of an extrusion operation. Figure 9(a) shows the initial condition with the billet pressed lightly against the die. As extrusion starts, the billet is nosed down by the die, as shown in Fig. 9(b). During this stage the lubricant film being generated by wedge action fails to keep up with the motion of the billet into the die, and a region which is unlubricated except by boundary films is formed near the front of the billet. The unlubricated region reaches a maximum size when the billet just fills the die, as shown in Fig. 9(c). At this point the higher friction associated with the unlubricated region results in a peak in extrusion pressure (called the breakthrough peak). As extrusion proceeds, the unlubricated surface passes through the die and the unlubricated contact decreases in size, as shown in Fig. 9(d). This results in a drop in fric-

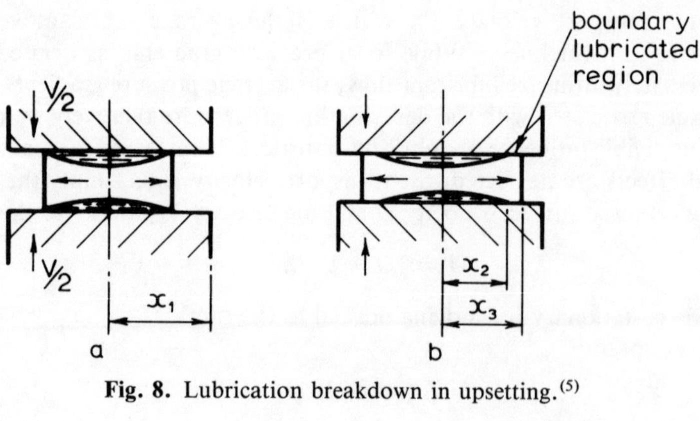

Fig. 8. Lubrication breakdown in upsetting.[5]

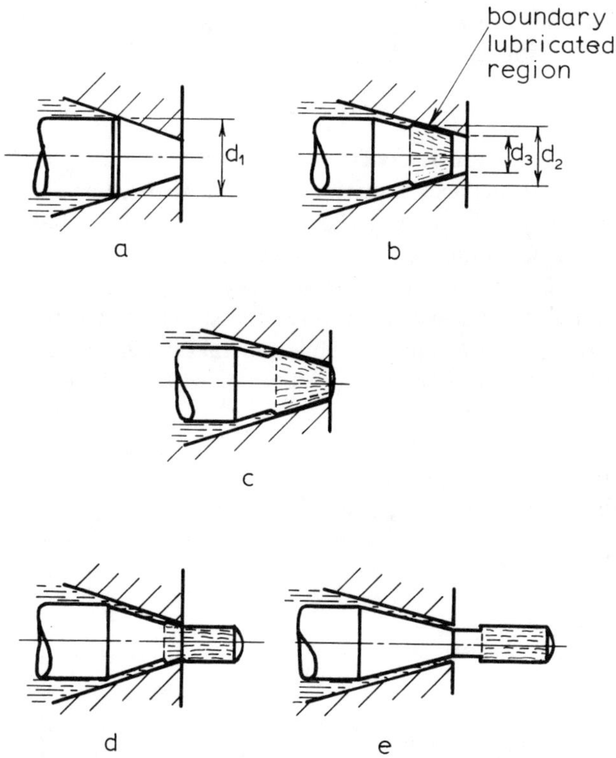

Fig. 9. Lubrication breakdown in extrusion.[18]

tion and extrusion pressure unless excessive surface roughening due to pickup or scoring has occurred. Eventually the lubricant film reaches the die throat and a steady-state condition is achieved with effective lubrication, as shown in Fig. 9(e).

The lubricant transport phenomena in upsetting and extrusion are not only of direct practical interest but also provide a means of experimentally investigating the factors which influence lubricant transport. In particular, upsetting has been used to investigate thermal and surface roughness effects in lubricant transport.[19-22]

The transport of liquid lubricants is strongly influenced by temperature gradients across the film. For the viscosity model of Eq 1, with smooth surfaces, negligible viscous heating, and a stationary die (V = 0), the lubricant transport velocity U_t is given by[13]

$$U_t = \frac{U(e^D - D - 1)}{D(e^D - 1)} \tag{17}$$

where U is the surface velocity. D is the nondimensional temperature difference across the film defined by

$$D = \alpha(T_w - T_d) \tag{18}$$

where α is the temperature coefficient of viscosity and T_w and T_d are the workpiece and die surface temperatures, respectively. Figure 10 shows the relationship between U_t/U and D from Eq 7. At low values of D, U_t/U approaches 1/2 according to the half-speed isoviscous model discussed earlier. Increasing D decreases the transport velocity and thus worsens problems with lubrication breakdown. Even relatively small temperature differences can be significant. For example, with a mineral oil a 25 °C temperature difference corresponds to a D value of about 1, which will result in almost a 20% reduction in transport speed.

In an actual forming operation the calculation of the nondimensional temperature difference D is quite difficult. However, for simple upsetting with relatively thick lubricant films, D can be approximated by assuming that the temperature difference is that which would be achieved if the workpiece were

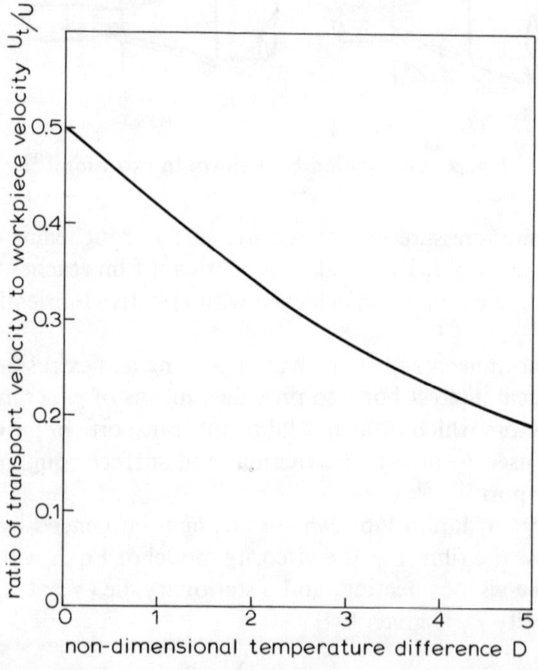

Fig. 10. Variation of lubricant transport velocity with temperature difference.[19]

adiabatic and all of the energy of plastic deformation were transformed into heat. This yields

$$D = K\epsilon \tag{19}$$

where ϵ is the workpiece strain. K is the nondimensional parameter controlling plastic heating effects on lubrication defined by

$$K = \alpha\sigma/\rho c \tag{20}$$

where σ is the workpiece flow strength, ρ the workpiece density, and c its specific heat. This estimate of D can then be used in Eq 17 to find the lubricant transport velocity and, by integration, the extent of the unlubricated region determined theoretically.[19]

It is evident from Eq 20 that K and hence the lubricant transport velocity can be influenced by changing either lubricant or workpiece properties. This has been demonstrated and the theory validated quantitatively by a series of upsetting experiments using aluminum alloy workpieces with a variety of silicone oils as lubricants. Figures 11 and 12 compare the results of these experiments with theory. In each case the nondimensional width $(x_3 - x_2)/x_1$ of the unlubricated region is plotted against the ratio x_3/x_1 of specimen final half-width to initial half-width. The dimensions x_1, x_2, and x_3 are defined in Fig. 8. Figure 11 shows the effect of changing lubricant properties. The silicone oils of 50 cs and 1000 cs nominal viscosities have similar values of the temperature coefficient α, while the 10,000 cs oil has a somewhat larger value of α. The higher α results in a higher value of D with a resulting decrease in U_t and an increase in the unlubricated width. The theoretical curves for the low-viscosity oils ($K = 3.6$), the high-viscosity oil ($K = 20$), and the ideal cases of half-speed transport ($K = 0$) and no transport ($K = \infty$) are also shown. The theoretical curves are in excellent agreement with the experiments. Early in the process, with a cool workpiece, the curves for realistic values of K are close to the half-speed curve. As deformation proceeds, the workpiece heats up, the lubricant transport velocity decreases, and the curves become parallel to the no transport case. Figure 12 shows the effect of changing the workpiece material. Aluminum alloy 6063 has about three times the flow strength of aluminum alloy 1100. This not only increases K directly by increasing σ but the higher interface pressure also increases α substantially. The result is that lubricant transport rates are much lower and the unlubricated region much larger with the higher-strength workpiece.

Lubricant transport is also influenced by the relative roughness of the workpiece and die surfaces. In general, increasing the roughness of a surface tends to cause the lubricant to be transported at the same rate as the surface. Thus, increasing the roughness of the die in an upsetting operation tends to

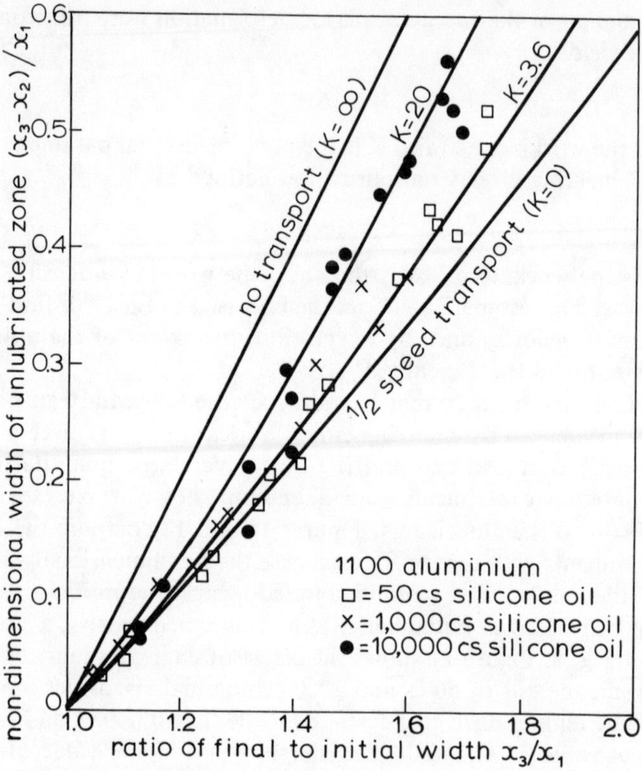

Fig. 11. Influence of lubricant on unlubricated width.[18]

reduce outward lubricant transport, and increasing workpiece roughness tends to increase outward lubricant transport. The latter effect is shown in Fig. 13 for 6063 aluminum alloy workpieces of different roughnesses lubricated with a 50 cs silicone oil. It is evident that increasing workpiece surface roughness reduces the unlubricated width by enhancing outward lubricant transport. The scatter on the experimental results also increases with increasing roughness, reflecting the more stochastic nature of the breakdown process and the resulting lack of definition of the unlubricated boundary.

The theoretical analysis of the influence of surface roughness on the transport of liquid lubricants presents a number of difficulties. Figure 14 shows the results of a simple model[20] which treats the case of an isoviscous lubricant with a smooth die and periodic unidirectional workpiece roughness with its lay normal to the direction of motion. For low values of roughness the transport velocity approaches the half-speed behavior of the isoviscous case. At higher workpiece roughnesses, the transport velocity is substantially increased. While this simple model shows qualitatively the influence of

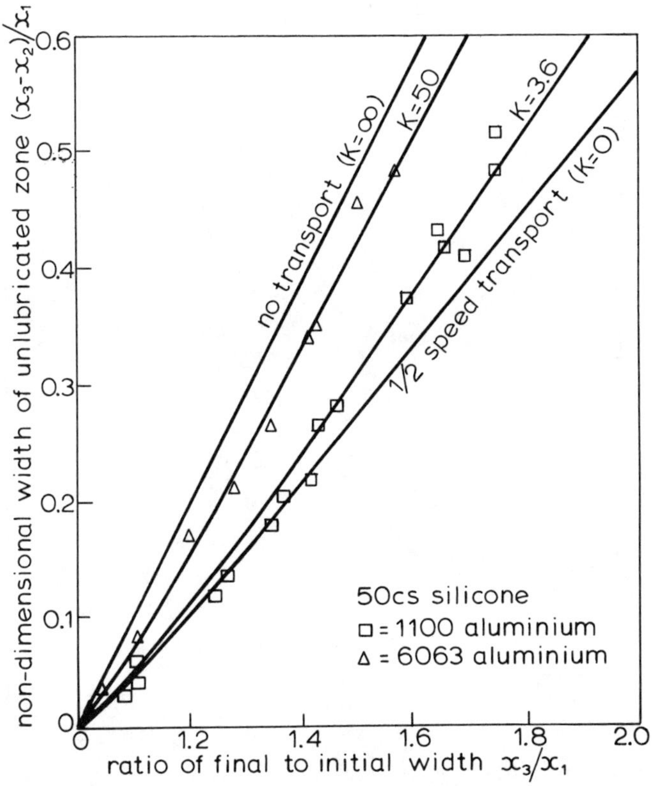

Fig. 12. Influence of workpiece on unlubricated width.[18,21]

roughness on lubricant transport, in order to make useful quantitative predictions it must be combined with the thermal model discussed above. Moreover, actual surfaces are more complicated than the simple isotropic characterization adopted in the model. The lubrication of such surfaces has been an important area of research in lubrication.[24]

Since the transport phenomena described above are basically governed by mechanics, they are unaffected by addition of additives which promote boundary lubrication.[21] However, such additives reduce friction in the region which must rely on boundary lubrication and help prevent irreversible surface damage. Solid lubricants also behave in a similar way, although roughness effects seem to be more important than thermal effects.[21,22,23]

Surface Roughness Changes

If the forging surface roughness is specified, then changes in the roughness during forming are important from a part-quality standpoint. Even if

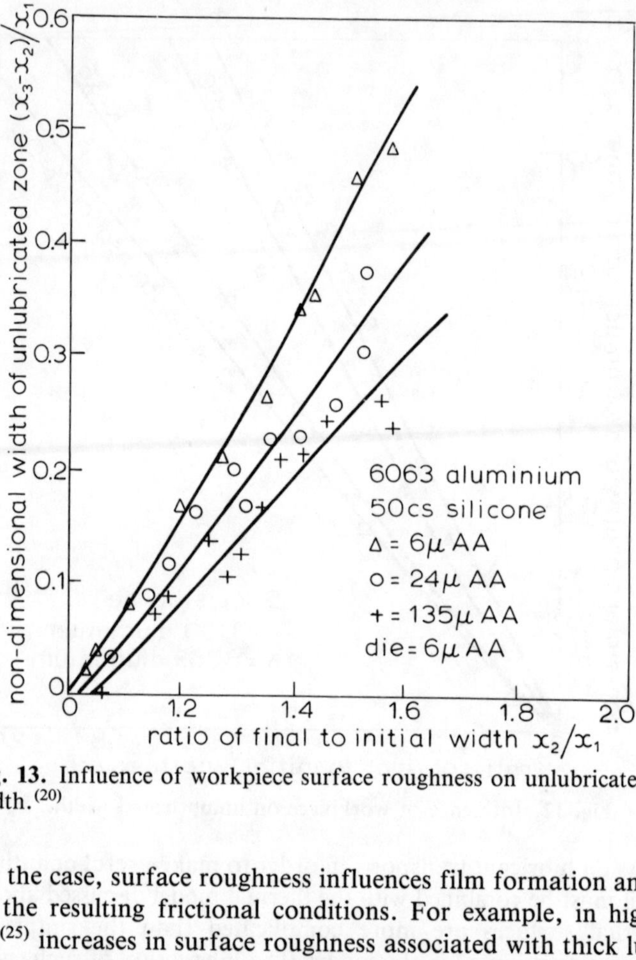

Fig. 13. Influence of workpiece surface roughness on unlubricated width.[20]

this is not the case, surface roughness influences film formation and transport and the resulting frictional conditions. For example, in high-speed upsetting,[25] increases in surface roughness associated with thick lubricant films can lead to enhanced lubricant transport, reduced unlubricated regions, and lowered forming forces. There is thus ample reason to study roughness changes.

Roughness changes can be classified into three types: roughening in the full film regions due to unconstrained surface deformation, smoothing in the mixed and boundary regimes due to asperity flattening, and roughening in the boundary regime due to metal-to-metal adhesion. The first and second of these are of most interest from the point of view of modeling and will be discussed in greater detail. The third condition is generally indicative of an unsatisfactory lubrication system which can be improved by changes in the factors controlling film formation and transport or by improving the boundary-lubrication capabilities of the lubricant.

Roughening under thick-film conditions seems to be associated with the

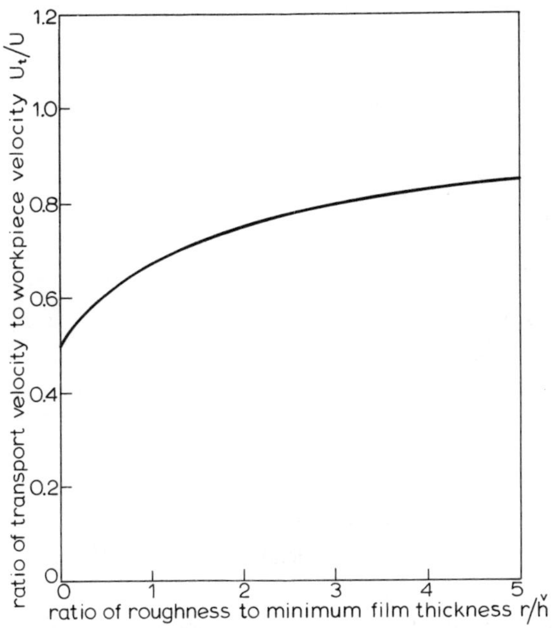

Fig. 14. Variation of lubricant transport velocity with roughness.[20]

unconstrained differential deformation of individual grains. Figure 15 shows some experimental results for 6063 aluminum alloy workpieces lubricated with silicone oils over a wide range of reduction and film thickness conditions. The workpieces were initially lapped smooth and, after forming the roughness near the center of a contact face, were measured using a profilometer. The data are plotted as a nondimensional roughness R versus nondimensional film thickness H. These are defined by

$$R = r/r_n \tag{21}$$

and

$$H = h_0/r_n \tag{22}$$

respectively, where r is the roughness after forming, r_n is the natural roughness or roughness achieved at a free surface under the same strain, and h_0 is the central lubricant film thickness. It is evident that with small nondimensional film thicknesses the data lie close to the line

$$R = H/4 \tag{23}$$

or that the roughness is limited to about one-quarter the film thickness. This implies that workpiece roughness peaks are close to the die surface. At larger nondimensional film thicknesses the data are close to the line

$$R = 1 \tag{24}$$

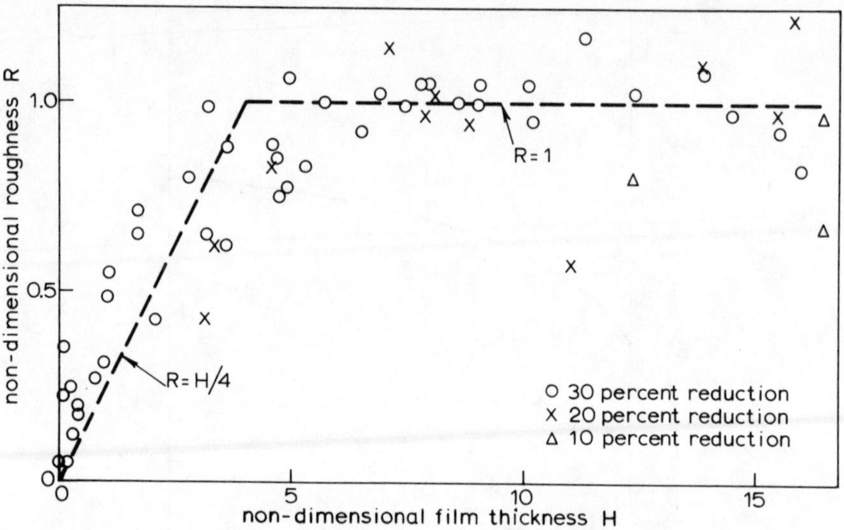

Fig. 15. Roughening in upsetting.[28]

implying that the roughening process is essentially unconstrained and the roughness approaches the natural roughness.

Under conditions of mixed or boundary lubrication with smooth tooling where a substantial fraction of the loading between the die and workpiece is borne on the summits of workpiece asperities, flattening of the asperities leads to a progressive reduction in workpiece roughness. In modeling the deformation of asperities on an elastic substrate such as is found in machine elements, it is usual to assume that the pressure p_a on the asperity is approximately six times the yield strength τ_y of the surface material. Thus, in an operation such as rolling or upsetting where the average interface pressure is about twice the workpiece shear strength, the conventional model would imply that asperity flattening would occur immediately on entering the high-pressure region and would proceed to the point where about one-third of the surface area was in contact. In fact in a real deformation process[29] the degree of asperity flattening increases monotonically with surface strain until the asperities are almost completely eliminated by flattening and the workpiece conforms to the tooling surface. This behavior can be explained by an analytical model[30] of the flattening of an asperity on a plastically deforming substrate. The model uses two-dimensional upper-bound plasticity theory to estimate the pressure required to indent or flatten asperities on the surface of a material undergoing bulk plastic deformation in a state of plane strain. The results are plotted in Fig. 16 as nondimensional hardness p_a/τ_y on a base of nondimensional strain rate $\dot{\epsilon}\ell/v_f$, for various values of the fractional contact area A, where $\dot{\epsilon}$ is the bulk strain rate, ℓ is the asperity

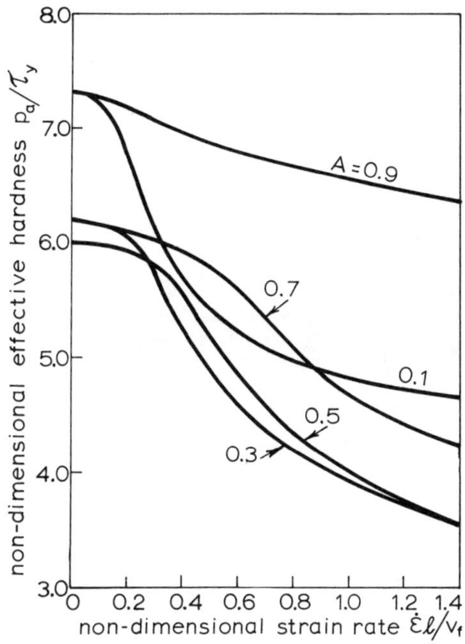

Fig. 16. Influence of underlying plastic flow on asperity hardness.[30]

spacing, and v_f is flattening velocity. The variable $\dot{\epsilon}\ell/v_f$ can be interpreted as the ratio of the bulk strain rate to that typical of the local flattening process. It is evident that increasing the nondimensional strain rate tends to greatly reduce the pressure required to flatten an asperity. A more sophisticated analysis[31] shows that at very large values of nondimensional strain rate the nondimensional hardness is approximately inversely proportional to nondimensional strain rate and eventually approaches zero.

If it is assumed that the average interface pressure is approximately twice the shear yield stress τ_y, then the data in Fig. 16 can be used to calculate the development of asperity flattening with workpiece strain in a forming operation.[30] The resulting theoretical variation of A with workpiece strain ϵ is compared with experimental data from rolling in Fig. 17. While the theory shows the same trend as the experiments, it tends to underestimate the actual level of flattening. This is because the theory is based on the upper-bound approach and thus tends to overestimate the resistance of the asperities to flattening.

Simple Friction Models

The most commonly used method of modeling friction between contacting surfaces, whether lubricated or not, is the Amontons-Coulomb law.[2]

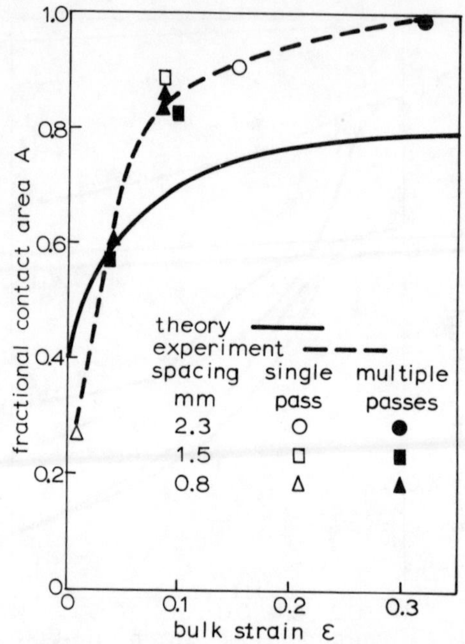

Fig. 17. Increase in contact area in rolling.[30]

In the present situation this implies that the friction stress τ_f opposes relative surface motion and is proportional to the interface pressure p:

$$\tau_f = \mu p \tag{25}$$

where the constant of proportionality μ is called the coefficient of friction. The coefficient of friction model arises from the study of lightly loaded boundary-lubricated contacts. In this situation, the real area of contact increases in proportion to normal load, resulting in a similar increase in frictional resistance to tangential motion due to adhesion. In boundary-lubricated metal forming operations where the loading is high enough to cause plastic flow of the bulk material and fractional contact areas are large enough so that asperity flattening does not occur in isolation,[31] there is little fundamental support for the use of the model. The connection with reality is even more tenuous when operating in other regimes where contacting asperities do not contribute to friction or where they bear only part of the interface load.

An alternative method of characterizing friction which is also widely used for metal forming involves the use of a friction factor m such that the friction stress τ_f is given by

$$\tau_f = m\tau_y \tag{26}$$

where τ_y is the shear strength of the workpiece material. While this model has a fundamental basis where the surfaces are completely separated by a film of a perfectly plastic lubricant, the main reason for its use is pragmatic empiricism since it uncouples the friction stress from the interface pressure. This facilitates analysis particularly in upper-bound analyses where the interface pressure is not explicit, or in numerical solutions where many iterations are necessary to match pressure and friction conditions.

In using either the friction coefficient μ or friction factor m in metal forming analyses, it is generally assumed that these quantities are properties of the lubricant and workpiece materials and independent of other factors such as process geometry, global forming speed, local surface sliding speed, and surface topography. Given the complexity of the processes involved, it is unlikely that such an approach will be useful in predicting system performance over a wide range of conditions. This is particularly true when transitions between lubrication regimes are involved or when more than one regime is present concurrently at the tooling/workpiece interface.

The inability of the simple friction models to represent the influence of process variables on friction does not mean that the models are useless. Their simplicity makes them valuable in studying the influences of the general level of friction on processes. They are also useful as a design tool to simulate processes which are relatively insensitive to friction or where general friction levels can be established by experiment under conditions similar to those used in the process. Alternatively, if a variety of similar processes are involved, suitable values for the friction factor or coefficient can be "backed" out of the analysis by comparison with measured process behavior.

Realistic Friction Models

In cases where simple friction models prove unsatisfactory, it is necessary to adopt more-realistic models based on the fundamental mechanisms of friction. Such models will generally be required where the process is very sensitive to friction and friction is strongly influenced by changes in the process variables under study. The latter condition often occurs when the system is operating in the mixed regime where small changes in film thickness or surface roughness can produce large changes in friction. Since the mixed regime combines low friction with potential for controlled surface roughness, this will often represent the optimum lubrication condition. More-realistic models are also necessary when information on factors such as tooling wear or interface transfer, which are sensitive to lubrication conditions, are required.

The key to developing better friction models is to take account of the local, current lubrication regime. This requires a knowledge of the local, current film thickness and surface roughnesses. Film thickness can be estimated from experimental evidence or by using the mechanics of lubricant film formation and transport described earlier. Surface roughness must take account of roughness changes occurring during the process as described earlier. The ratio of film thickness to composite roughness* is generally characteristic of the regime. Values of ten or more indicate the thick-film regime. Values of three to ten indicate the thin-film regime. Values of one to three indicate the mixed regime, and values less than one generally imply a mainly boundary regime. While this regime analysis process will use theory based on thick-film mechanics, it will generally yield useful results. Once the local current regime is determined, an appropriate model can be used for friction.

For the thick-film regime the friction model depends on the rheology. In the simplest case for a liquid Newtonian lubricant of viscosity μ, the friction stress τ_f is given by

$$\tau_f = \frac{\mu(U - V)}{h} \tag{27}$$

where U and V are the workpiece and tooling speeds and h is the local film thickness. In practice, allowance must be made for pressure-induced viscosity changes using a constitutive relation similar to Eq 1. Corrections must also be made for thermal effects[33] and for non-Newtonian effects,[34] both of which tend to reduce friction. For a perfectly plastic lubricant of shear strength τ_ℓ, the friction stress τ_f is simply given by

$$\tau_f = \tau_\ell \tag{28}$$

which is essentially equivalent to the use of a constant friction factor. As with liquid lubricants, such a model should probably be modified to account for thermal effects and more complex rheology, particularly with organic lubricants such as waxes, soaps, and polymers. However, there has been much less research in this area on solid lubricants than on liquid lubricants.

In the thin-film regime, the basic approaches used in the thick-film regime must be modified to allow for surface roughness. In the case of liquid lubricants this will generally result in a small increase in friction.[24] Little is known about the direct influence of roughness on friction with solid lubricants in the thin-film regime. It is worthy of note that friction in both the thick- and thin-film regimes is generally small compared with that in other regimes which may exist concurrently. This means that even large errors in friction estimation may not contribute significantly to overall errors in the process model.

Mixed lubrication is an area of active research in tribology. It appears that

*Composite roughness is the square root of the sum of the square of the individual surface roughness.

the mixed lubrication of metal forming processes is particularly complicated and that sophisticated models[31] will be necessary to capture the load-sharing process critical to a realistic friction model. Existing models assume that the friction stress τ_f is a weighted average of the friction stress τ_b associated with boundary lubrication and the friction stress τ_t associated with thick-film lubrication:

$$\tau_f = A\tau_b + (1 - A)\tau_t \tag{29}$$

where A is the fractional contact area. Two main approaches have been used to estimate A: the constant-pressure model and the pressure-sharing model. In the constant-pressure model[35] it is assumed that the pressure in the interface is that calculated using bulk plasticity theory. This is then used in an analysis of film formation to estimate the mean film thickness which is combined with surface topography statistics to estimate A. The pressure-sharing model[36] uses the inverse approach. It is assumed that the mean film thickness is approximately equal to the roughness. A film formation analysis is used to estimate the pressure generated by the lubricant film. Since the remainder of the interface pressure required by bulk plasticity must be provided by asperity contact, the fractional contact area A may be estimated.

Figure 18 compares the predictions of the constant-pressure and pressure-sharing methods with the experimental data for strip drawing at 20% reduc-

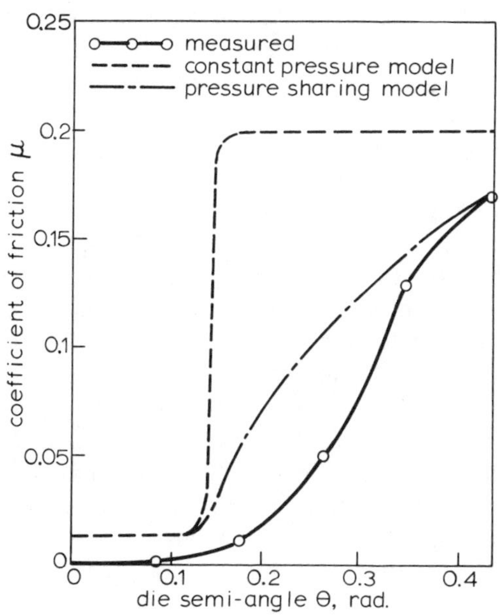

Fig. 18. Comparison of mixed friction models with experiment.[36]

tion shown in Fig. 1(a). The constant-pressure model predicts a more rapid transition from thick film to boundary conditions than actually occurs. The pressure-sharing model yields more accurate predictions but tends to overestimate friction in the thin-film and mixed regimes. This may be due to workpiece surface roughness increasing the lubricant film thickness in these regimes. Both models tend to overestimate friction in the thick-film regime at small die angles, probably because thermal effects are neglected.

In boundary lubrication of forging the conditions of large fractional contact area and a plastically deforming substrate render the Amontons-Coulomb model of questionable value. A better understanding of the mechanics of asperity flattening or surface indentation on a plastically deforming surface are necessary before a better model can be developed. Based on preliminary work in this area[30,31] which shows that the hardness of workpiece asperities is greatly reduced by underlying plastic flow, it would appear that the plowing of tooling asperities through the workpiece surface may be the major contribution to friction. This would result in a real area of contact and friction stress which is proportional to interface pressure and local strain rate and inversely proportional to sliding speed. Flattening of workpiece asperities could also be involved in friction where dies are relatively smooth. This would result in a frictional contribution which would be independent of pressure and sliding velocity and which would increase with workpiece strain. Additional theoretical work and careful experiments will be necessary to understand the fundamentals and to develop a useful frictional model for the boundary regime.

A case of great practical importance is where multiple lubrication regimes are involved in a single process. The breakdown of full film lubrication due to lubricant transport phenomena in upsetting with overhanging dies and in the nosing of an extrusion billet are cases where this may occur. If conditions are appropriate in either process, part of the tooling/workpiece interface will operate in the full film regimes with the remainder in the boundary regime. In modeling such systems it is important to use different frictional characterizations for the two portions of the interface. In such circumstances even the simple friction models such as the constant-friction-factor model can yield useful results with different factors for each regime. Figure 19 compares theoretical predictions of pressure made using this approach with experimental results using three different lubrication conditions.[37] In the normal lubrication case a single coating of a commercial solid lubricant was applied prior to forming. In the ideal lubrication case the coating was reapplied periodically. Unlubricated tests were also run. The results are plotted as nondimensional pressure q on a base of nondimensional displacement D. These are defined by

$$Q = q/\sigma \tag{30}$$

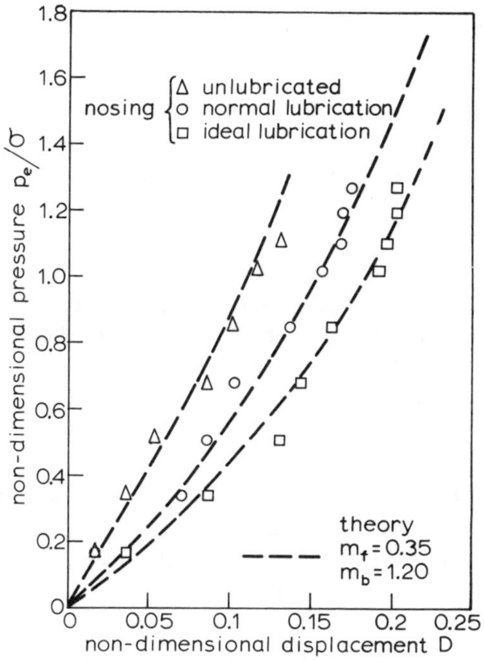

Fig. 19. Influence of lubrication condition on nosing pressure.[37]

and

$$D = 4v/\pi d_1^2 \tag{31}$$

where q is the nosing pressure, σ the workpiece flow strength, v the nose volume, and d_1 the inlet diameter. The results for the normal lubrication case (part boundary, part full film) are between the ideal (full film) and unlubricated (boundary) cases. The results can be fitted by assuming friction factors $m_b = 1.2$ and $m_b = 0.35$ for the full film and boundary regimes, respectively. In the normal lubrication case the extent of the two regimes was determined experimentally.

Conclusion

Four different lubrication regimes can occur in forging: the thick-film regime, the thin-film regime, the mixed regime, and the boundary regime. The regimes are distinguished by the lubricant film thickness compared with the roughness of the surfaces involved and the degree of load sharing between the bulk lubricant in surface valleys and boundary films at roughness peaks. Different mechanisms determine friction in each regime, and this must be taken into account in developing friction models.

Simple friction models, such as the use of a constant coefficient of friction or friction factor, are useful in investigating the influence of friction stress on a particular process or in process-design simulations where friction variability is small. However, they fail to represent the influence of processing conditions on friction, particularly when the system undergoes a transition from one regime to another or where more than one regime exists concurrently.

More-realistic models require a knowledge of the local current lubrication regime. This can be achieved through an understanding of the processes by which lubricant films are formed and transported in the tooling/workpiece interface. Once the regime is determined, appropriate frictional models based on the fundamental friction mechanisms involved can be employed.

Additional research is necessary before a relatively complete understanding of friction and lubrication in forging is obtained. In particular, more work is needed on the peculiar character of mixed and boundary lubrication in metal forming because of the large fractional contact area and underlying plastic deformation which occur. More data are also required on the rheological properties of practical forging lubricants under the high pressures and temperatures which occur in forging.

Acknowledgments

The author wishes to thank the National Science Foundation for its support through grant MEA-8307152. Any opinions, findings or conclusions, or recommendations expressed in this publication are those of the author and do not necessarily reflect the views of the N.S.F.

References

1. W. R. D. Wilson, *J. Applied Metalworking*, 1979, Vol. 1, No. 1, pp. 1-19.
2. F. P. Bowden and D. Tabor, *The Friction and Lubrication of Solids, Vol. I*, pp. 98-101, Oxford Clarendon Press, 1953.
3. W. R. D. Wilson and P. Cazeault, *Proc. NAMRC IV*, 1976, pp. 165-170.
4. M. Oyane and K. Osakada, *Bull. J.S.M.E.*, 1969, Vol. 12, pp. 149-155.
5. W. R. D. Wilson, *J. Lub. Technol.*, 1974, Vol. 95, pp. 539–546.
6. W. R. D. Wilson and C. J. Wong, *J. Lub. Technol.*, 1974, Vol. 95, pp. 605-610.
7. W. R. D. Wilson and C. J. Wong, *Proc. NAMRC III*, 1975, pp. 168-178.
8. W. R. D. Wilson, *Proc. First Int. Conf. on Lubrication Challenges*, IITRI, Chicago, 1978.
9. W. R. D. Wilson and J. A. Walowit, *J. Lub. Technol.*, 1971, Vol. 93, pp. 69-74.
10. W. R. D. Wilson and J. A. Walowit, *Tribology Convention–1971*, pp. 169-172, Inst. Mech. Engrs., London, 1972.
11. W. R. D. Wilson, *J. Lub. Technol.*, 1971, Vol. 93, pp. 75-78.
12. W. R. D. Wilson and S. M. Mahdavian, *J. Lub. Technol.*, 1976, Vol. 98, pp. 27-31.

13. P. Tsao and W. R. D. Wilson, *Proc. Int. Conf. on Steel Rolling*, The Iron and Steel Institute of Japan, 1981.
14. W. R. D. Wilson and K. Halliday, *Wear*, 1977, Vol. 42, pp. 135-148.
15. K. Halliday and W. R. D. Wilson, *Proc. NAMRC V*, 1977, pp. 165-170.
16. D. R. White and W. R. D. Wilson, *Trans. A.S.L.E.*, 1980, Vol. 23, pp. 305-314.
17. W. R. D. Wilson and D. R. Johnson, *Trans. A.S.L.E.*, 1981, Vol. 24, pp. 307-318.
18. W. R. D. Wilson, *Int. J. Mech. Sci.*, 1971, Vol. 13, pp. 17-28.
19. W. R. D. Wilson and W. B. Carpenter, *Wear*, 1973, Vol. 24, pp. 351-360.
20. W. R. D. Wilson and W. P. Delmolino, *Wear*, 1974, Vol. 25, pp. 1-10.
21. W. R. D. Wilson and E. A. Kokoška, *Wear*, 1975, Vol. 26, pp. 25-32.
22. W. R. D. Wilson and S. Lak, *J. Lub. Technol.*, 1977, Vol. 99, pp. 230-235.
23. S. Lak and W. R. D. Wilson, *Proc. 1977 Leeds-Lyon Symposium*, pp. 301-307, Inst. Mech. Engrs., London, 1978.
24. N. Patir and H. S. Cheng, *J. Lub. Technol.*, 1978, Vol. 100, pp. 12-17.
25. W. R. D. Wilson and W. J. Farrel, *Proc. NAMRC IV*, 1976, pp. 130-137.
26. D. D. Ratnagar, H. S. Cheng, and J. A. Schey, *J. Lub. Technol.*, 1974, Vol. 96, pp. 591-594.
27. W. R. D. Wilson, *J. Lub. Technol.*, 1977, Vol. 99, pp. 10-14.
28. W. R. D. Wilson and J. G. Siletto, *Metalworking Lubrication*, pp. 87-94, ASME, New York, 1977.
29. R. A. Atala and G. W. Rowe, *Wear*, 1975, Vol. 32, pp. 249-268.
30. S. Sheu and W. R. D. Wilson, *Proc. NAMRC XI*, 1983, pp. 172-178.
31. S. Sheu, Ph.D. Thesis, Mechanical Engineering, Northwestern University, 1985.
32. T. Wanheim and N. Bay, *Annals C.I.R.P.*, 1978, Vol. 27, pp. 189-194.
33. S. M. Mahdavian and W. R. D. Wilson, *J. Lub. Technol.*, 1976, Vol. 98, pp. 22-26.
34. S. Barr and W. O. Winer, *J. Lub. Technol.*, 1982, Vol. 104, pp. 357-364.
35. Y. H. Tsao and L. B. Sargent, *Trans. A.S.L.E.*, 1980, Vol. 23, pp. 70-76.
36. W. R. D. Wilson, *Proc. G. M. Symposium on Sheet Metal Forming*, pp. 156-177, Plenum, New York, 1978.
37. W. R. D. Wilson, B. B. Aggarwal, A. B. Norelius and H. Quist, *Proc. NAMRC VII*, 1979, pp. 129-136.

11

Constitutive Equations for High-Temperature Deformation

J. F. THOMAS, JR. AND R. SRINIVASAN

Materials Science and Engineering
Wright State University
Dayton, OH 45435

Abstract

Significant advances in the finite-element modeling of high-temperature deformation processes such as forging have led to increased capabilities and wider application of computer simulation. Such models often couple deformation and thermal analyses, and their implementation requires realistic constitutive equations for material flow. These constitutive equations describe the effect of strain rate, temperature, and plastic strain increment on the flow stress and take account of effects of stress state and material anisotropy when necessary and achievable. To determine constitutive equations, it is necessary to measure the flow stress over a range of temperatures and strain rates which correspond to deformation process variables. This is most often done by compression tests, and a computer-based test facility for such measurements and the requisite data analysis is described. A primary goal of computer process simulation is to describe accurately the local strains within the workpiece and the dependence of strain and strain rate distribution on deformation process variables. Thus, process variables can be optimized and product integrity can be assessed.

Introduction

Analytical modeling of metal deformation processes requires the characterization of the deforming material under processing conditions. In general for such a description, this characterization is expressed as a set of constitutive equations which relate the flow and effort variables[(1)] in the current state of the material. In particular, for the modeling of deformation, the effort variable is the flow stress, the flow variable can be either strain or strain rate, and the state of the material is determined by the temperature and thermomechanical history. While stress, strain, and strain rate can often be treated as scalar variables, in some cases the state of stress is important, and the full tensor nature of these variables must be taken into account. In these cases, the set of constitutive equations will include a yield function and some assumption in regard to the directionality of strain hardening.

The types of equations that have been developed depend on the nature of the tests performed to collect data, and, more importantly, the application for which they were developed. It is, in general, fruitless to attempt to develop one type of constitutive equation for all materials covering all magnitudes of deformation and all ranges of external variables.[2] As an example, when considering large plastic deformation, it is possible to get reasonably accurate results using a constitutive equation involving only the plastic strains or strain rates, while ignoring elastic strains entirely.[3] On the other hand, when elastic deformation is of the same order of magnitude as the plastic strains, such as when predicting residual stresses, constitutive equations should include elastic as well as plastic deformation behavior.

One can consider two general approaches to the development of constitutive equations for material flow. In one approach, specific deformation mechanisms are assumed, and stress-strain or stress–strain rate equations are worked out in terms of relevant material parameters. While this approach has a strong appeal, it does have limitations even beyond our ability to accurately describe specific mechanisms. For example, deformation mechanisms in metals and alloys frequently change with both temperature and strain rate. Any one constitutive form should, therefore, be applied only within the boundaries of the domain for which it was intended.

A second approach is based on careful measurement of material flow stress over a range of temperatures and strain rates which are characteristic of a deformation process of interest. The results are then described analytically to produce equations which can be used in computer-based models such as a finite-element process model. Admittedly, this approach is empirical, but it is likely to produce a result which is valid over a relatively broad range of process variables. Hopefully, a careful empirical description will lend insight which will lead eventually to equations with some foundation in deformation mechanisms. It is particularly hoped that evolutionary state variables which describe the current thermomechanical state of the material can be identified.

Finite Element Modeling of Deformation

Since the advent of high-speed digital computers, modeling of metal deformation has been conducted at several levels of material behavior and over a wide geometrical scale. Models range from the micromechanical level for which the approach relates stresses and strains due to dislocation movement, to the macroscopic continuum material behavior level, in which the average deformation of all the grains in a polycrystalline metal is used to give relations between stresses and strains, and lattice scale effects are ignored. The finite-element method has been applied across this broad spectrum with interesting results in all cases.

Inelastic deformation in metals occurs by several mechanisms including slip and climb of dislocations, diffusional mass transport accompanied by grain-boundary sliding, martensitic shear transformations, and twinning.[4] Most metal deformation occurs by dislocation movement. Early work by Taylor and co-workers[5,6] and by Schmid[7] established the relation between deformation of single crystals and the magnitude of resolved shear stress along certain crystallographic directions on certain crystallographic planes. After the discovery of dislocations, relations between stress and strain rate due to dislocation movement were also established.[8,9] Asaro and others[10,11,12] used the finite-element method to model deformation of single crystals and were able to predict the strains due to slip on two slip systems operating simultaneously and the corresponding lattice rotations. The constitutive equations used related shear stress on a slip plane to shear strain and accounted for hardening due to dislocation interactions. This work was then extended to polycrystalline materials,[13] and the results clearly showed the development of deformation texture resulting from lattice rotations due to slip. This texture led to a geometrical softening effect and the development of strain localization in shear bands.

At a larger scale, Margolin and Ankem[14] used the FEM to model plastic yield and small strain plasticity in two-phase alloys. Using linear elastic, strain hardening flow behavior for each of the two phases, they were able to show the effect of second-phase grain size and volume fraction on the yield behavior of the two-phase alloy. This is essentially a continuum scale calculation but does account for the effects of deformation at the grain-size level including the effects of microstructure.

In the past several years, newly developed finite-element codes, such as ALPID[3] and MARC,[15] have been used in metal forming analyses to obtain the simultaneous variations of field quantities such as stress, strain, and temperature during the deformations associated with bulk forming. Constitutive equations used in these computer programs assume the deforming material to be a continuum, ignoring the behavior of individual metallic grains or phases. As an example of the capabilities of these techniques,[3] the rigid-plastic finite-element code ALPID was used to compare the deformation under uniaxial compression of the titanium alloy Ti-6Al-2Sn-4Zr-2Mo (Ti-6242) with two different microstructures. For one microstructure, termed transformed β, the alloy flow softens dramatically, while the other microstructure, $\alpha + \beta$, does not. Using constitutive equations that show this difference in behavior, the computer simulation quite accurately predicted the occurrence of shear bands in the flow softening microstructure, since flow softening operates to induce strain localization. This prediction is illustrated in Fig. 1, which compares a macrograph showing shear bands in a side-pressed Ti-6242 cylinder with a contour plot of local strain rates from the ALPID simulation.[16] Note that the strain rates in the section vary by up to a factor of 10.

The capability of using the finite-element method with realistic constitutive equations has provided the opportunity for flow-based analysis[17] of metals deformation processing. Earlier analyses addressed only the geometrical effects of shape change and average or integrated quantities such as press load. By partitioning a workpiece into a grid of finite elements, it is now possible to calculate strain, strain rate, and the resulting stresses for each element at each step of the deformation. Hence, it is possible to determine the deformation (and temperature) history of each material element. This allows determination of the distribution of microstructure and properties in a formed component and, hence, an assessment of overall product integrity.

Constitutive Equations

A constitutive equation for the flow stress dependence in high-temperature bulk forming can be expressed as

$$\sigma = \sigma(\dot{\epsilon}, T, S) \tag{1}$$

where $\dot{\epsilon}$ is the plastic strain rate, T is the absolute temperature, and S is a parameter representing those features of the microstructure which determine

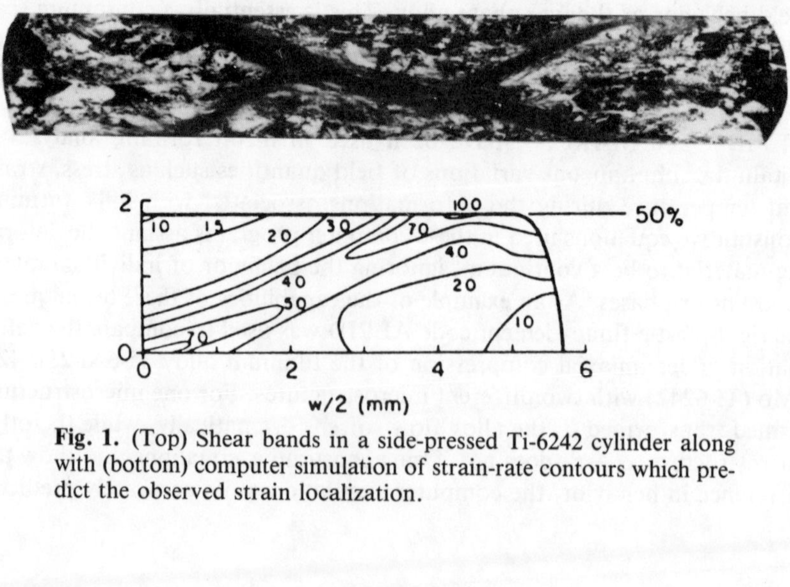

Fig. 1. (Top) Shear bands in a side-pressed Ti-6242 cylinder along with (bottom) computer simulation of strain-rate contours which predict the observed strain localization.

the current state of the material and, hence, its current flow behavior. The parameter S is evolutionary in that it evolves with the thermomechanical history of the material. For example, if we think of S as a characteristic stress, it would increase with strain at low temperature (hardening), and decrease with thermal or flow softening histories. If we know S for the current state and how it would evolve, the previous history is no longer of concern.

A common approximation for Eq 1 is to replace S with the value of accumulated plastic strain, ϵ, for a specific process,

$$\sigma = \sigma(\dot{\epsilon}, T, \epsilon) \tag{2}$$

which is a constitutive equation originally proposed by Zener and Hollomon.[18] This is an approximation which must be carefully considered, since the plastic strain is certainly not a state variable. It is, however, nearly an unavoidable path variable in deformation processes, and a constitutive equation which includes plastic strain explicitly can be extremely useful as long as the approximation is kept firmly in view.

Equation 2 is usually considered in differential form,

$$d \ln \sigma = \gamma \, d\epsilon + \nu d \ln \dot{\epsilon} + \phi \, dT \tag{3}$$

where the coefficients of the strain, strain rate, and temperature increments are given by the respective partial derivatives,

$$\gamma = \left. \frac{\partial \ln \sigma}{\partial \epsilon} \right|_{\dot{\epsilon},T} \tag{4}$$

$$\nu = \left. \frac{\partial \ln \sigma}{\partial \ln \dot{\epsilon}} \right|_{\epsilon,T} \tag{5}$$

$$\phi = \left. \frac{\partial \ln \sigma}{\partial T} \right|_{\epsilon,\dot{\epsilon}} \tag{6}$$

In the analysis of large strain deformation, these derivatives represent material properties which are as important as the flow stress for characterizing material behavior, particularly in terms of workability during high-temperature processing and identification of optimum process variables.

Strain Rate and Temperature Effects

The strain rate and temperature dependencies of the flow stress are of crucial importance for high-temperature deformation processing. When strain hardening and strain or thermal softening are in balance to produce a strain independent flow stress, the constant-temperature flow stress equation is usually written as

$$\sigma = K\dot{\epsilon}^{m} \tag{7}$$

where K is a temperature-dependent strength parameter. The strain rate sensitivity parameter, ν, defined by Eq 5, is then $\nu = m =$ constant. More generally, of course, the strain rate sensitivity is a function of strain rate and temperature. For high-temperature deformation, values of m can be sufficiently large to produce significant flow stress variations over strain rate variations which could be encountered in a single process. For example, for $m = 0.3$, a factor of ten change in strain rate will produce a factor of two change in flow stress.

For the temperature dependence of the flow stress, it is usually assumed that high-temperature deformation proceeds by thermally assisted mechanisms such that the strain rate at a given stress depends on temperature through an Arrhenius factor,

$$\dot{\epsilon} = f(\sigma)\exp(-Q/RT) \tag{8}$$

where Q is an apparent activation energy and R is the gas constant. This relation is most useful when Q is a constant, independent of stress and temperature. Rearranging Eq 8, we obtain

$$f(\sigma) = \dot{\epsilon}\exp(Q/RT) = Z \tag{9}$$

where Z is known as the Zener-Hollomon parameter. This form shows the equivalence of strain rate and temperature in determining the material flow stress. If the flow stress at constant structure depends only on Z, then the flow stress temperature coefficient ϕ, defined by Eq 6, can be written as

$$\phi = -\nu Q/RT^2 \tag{10}$$

Under this assumption, the variation in flow stress at constant strain and strain rate can be written as

$$d \ln \sigma = \nu Q d(1/RT) \tag{11}$$

This suggests a form of data analysis for flow stress temperature dependence which will be pursued below.

In order to develop these phenomenological relationships, tests that show the effect of varying one independent variable on the dependent flow stress variable while holding all other variables constant have to be conducted. For forging and other large strain process modeling, the test we will choose to emphasize is the constant true strain rate compression test. This will be discussed in the next section.

Flow Stress Measurements

For the constant true strain rate compression test, consider a specimen of length l, which is being compressed between two flat platens (Fig. 2). In an

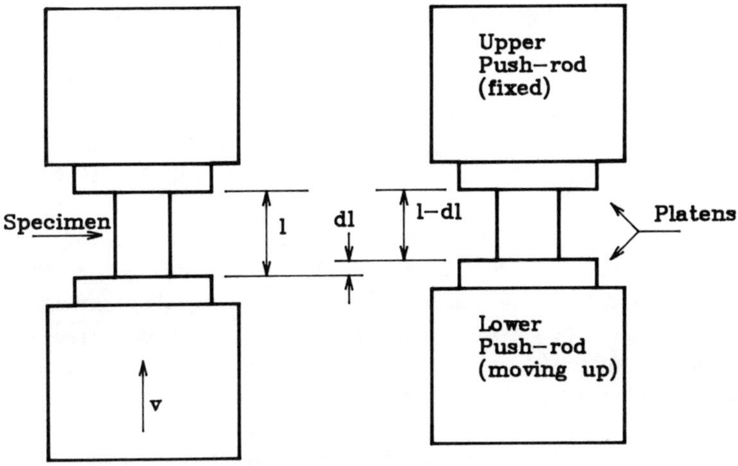

Fig. 2. Experimental configuration for the compression test.

increment of time dt, the length of the specimen changes by dl, and the corresponding increment of true strain is defined as

$$d\epsilon = dl/l \tag{12}$$

The instantaneous true strain rate is defined as

$$\dot{\epsilon} = \frac{d\epsilon}{dt} = \frac{1}{l}\frac{dl}{dt} = -\frac{v}{l} \tag{13}$$

where v is taken as the upward velocity of the lower platen. The negative sign accounts for the fact that dl is negative for compression since the specimen length is decreasing. Thus, to conduct a constant true strain rate compression test, the velocity of the moving platen should decrease continuously, while maintaining a constant (v/l) ratio. Equation 13 can be integrated to give an equation for the relative platen position versus time,

$$\Delta y = l_0(1 - \exp(\dot{\epsilon}t)) \tag{14}$$

where $\dot{\epsilon}$ is taken to be negative. This is the relevant equation for an actual test since it is the position of the platen, Δy, which is actually controlled. This position control can be achieved by either mechanical means, such as in a cam plastometer, analog electrical means, with a signal from an RC circuit, or digital electronic means. An example of the last method will be discussed here.

High-Temperature Test Facility

The requirements for compression testing for characterizing a material for forging are the ability to maintain constant true strain rate and accurate control of temperature. Figure 3 shows a schematic of such a facility at Wright State University. It consists of a 200 Kip MTS servohydraulic load frame with a Brew controlled-atmosphere (vacuum or inert gas) radiant heat furnace. The displacement of the actuator of the MTS machine can be controlled by a function generator, or by the signal from a microcomputer (IBM PC) equipped with a Data Translation DT2801A digital-to-analog (D/A) converter.

Specimens to be tested are compressed between TiC platens supported by TZM molybdenum alloy tooling. Equation 14 represents the displacement of the lower platen as a function of time for a given initial specimen length l_0 and strain rate $\dot{\epsilon}$. The IBM PC generates such a displacement-versus-time file and sends an analog signal to the MTS servovalve controller via the D/A converter. The actuator then moves under closed loop control in response to this command signal. A very wide range of constant strain rates can be obtained using this digital electronic control, from less than 10^{-4}/s to 20/s, which is a limit imposed by the hydraulic system. With specimens of about 20 mm initial length, the testing machine is able to achieve strain rates up to 1/s without any appreciable delay. To achieve higher strain rates, however, the actuator needs an acceleration distance before it can reach the necessary speed. To conduct tests at strain rates of 10/s and greater, the test is started with a gap of about 3 mm between the specimen and the upper platen. The actuator accelerates to the test speed and impacts the specimen against the

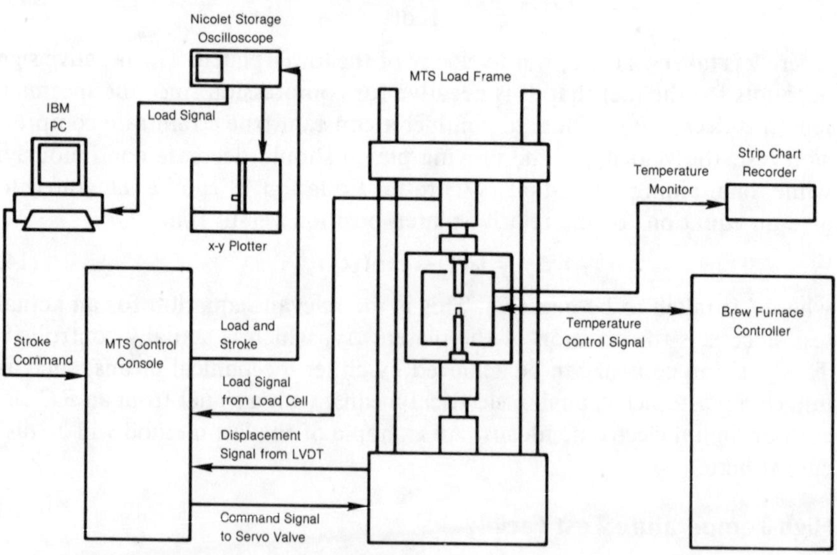

Fig. 3. High-temperature testing facility for constant strain rate compression tests.

upper platen. An illustrative set of flow curves for the titanium alloy Ti-6Al-4V for strain rates of 0.05/s to 20/s are shown in Fig. 4.

For this system, data acquisition is accomplished by a Nicolet 3091 digital storage oscilloscope capable of collecting 4000 points each in two channels simultaneously at sampling rates from 1 point per microsecond to 1 point every 200 seconds. The two channels are used for load and stroke, respectively. The data that is collected is transferred after the completion of the test to the microcomputer for conversion to true stress vs. true strain flow curves.

Currently, the temperature capacity of the Brew furnace, which uses tungsten heating elements, is approximately 1650 °C due to the use of molybdenum shields. The use of TZM tooling further restricts the maximum temperature to 1350 °C, whereas graphite or SiC tooling would allow the higher temperature. In conducting constant true strain rate compression tests, it is desired to maintain isothermal conditions as closely as possible. The use of a radiant furnace which heats the tooling as well as the specimen is one factor in accomplishing this. The entire system design is aimed at thermal stability, and it is generally possible to maintain temperatures constant to better than 1°C. However, it is not possible to avoid deformation heating of the specimen, and this can lead to an effective temperature change of several tens of degrees during a test. Since isothermal flow curves are desired for analysis, it is necessary to correct the flow curves to isothermal conditions.

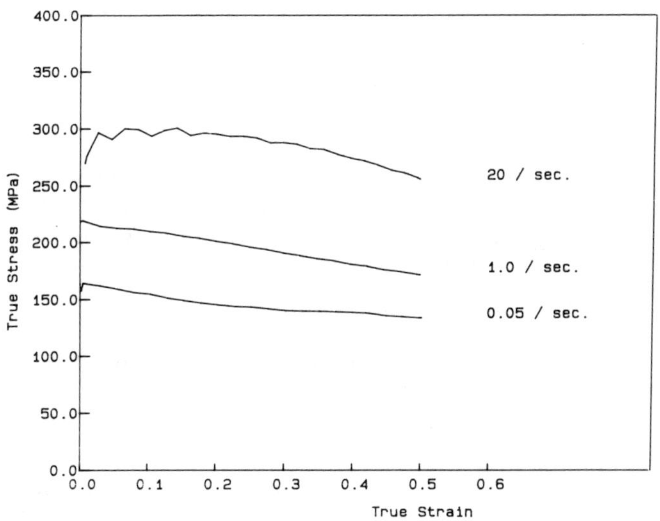

Fig. 4. True stress vs. true strain curves for Ti-6Al-4V at 870 °C.

Analysis of Compression Test Data

The analysis of compression test data proceeds by several well-defined steps. To carry these through, it is necessary to have load-stroke data for a matrix of compression tests representing at least three temperatures and four strain rates. A large number of tests will improve the accuracy of the results, and a matrix including from 25 to 50 tests is usually considered to be optimum.

The various stages in the analysis of load-time data proceed in terms of a series of computer routines and illustrative plots:

1. Convert load-time data to load-stroke data using Eq 14.
2. Convert load-stroke data to true stress vs. true plastic strain. This requires correction for the specimen and machine elastic deflection.
3. Plot log σ vs. $(1/T + \Delta T)$ at each strain rate and at various plastic strains (as suggested by Eq 11) where ΔT is the temperature rise at that strain.
4. Correct the true stress values at the strains used in (3) to those which would occur at the nominal test temperature. Plot the isothermal true stress vs. true strain curves and fit these to empirical equations such as a piecewise quadratic.
5. Determine log σ vs. log $\dot{\epsilon}$ at each test temperature and for a selection of plastic strains. Fit these data sets with a quadratic or cubic equation. This allows computation of the strain rate sensitivity (Eq 5).
6. Use a graphics package such as ISSCO DISSPLA to produce both three-dimensional and contour plots of true stress and strain rate sensitivity vs. temperature and strain rate for selected values of true plastic strain.

Isothermal Flow Curves

The temperature correction to obtain isothermal stress-strain plots requires some further discussion. To calculate this correction, it is necessary to compute the temperature rise as a function of strain during a test. The relation is

$$\Delta T = \frac{\eta}{\rho C} \int_0^{\epsilon} \sigma \, d\epsilon \tag{15}$$

In this equation, C is the heat capacity per unit mass, and ρ is the density. The parameter η accounts for the fraction of deformation energy which appears as a temperature rise. One factor in this is the fraction of stored energy, but a more important factor, at least for slower strain rate tests, is the fraction of heat which flows out of the specimen into the tooling or surroundings. Determination of η has been analyzed in some detail by Charpentier et al.[19] in relation to the measured temperature rise during a test.

In general, η is primarily a function of strain rate, and results show that tests above a strain rate of 1/s are adiabatic ($\eta > 0.95$) while tests at or below a strain rate of 10^{-3}/s are isothermal ($\eta = 0$). In between these rates, a reasonable interpolation is assumed, say η depending linearly on $\log \dot{\epsilon}$. Once the temperature increments are calculated, the plots of $\log \sigma$ versus inverse temperature ($1/T + \Delta T$) can be obtained.

In several cases studied to date, it is found that these plots are piecewise linear. An example, for 2024 Al, at a strain of 0.3, is shown in Fig. 5. As expected, the flow stress values for the lowest temperature (highest stress) show the greatest temperature rise with strain rate. The effect of these corrections for the compressive stress-strain curves for 2024 Al at 145 °C is shown in Fig. 6. The apparent saturation of flow stress in Fig. 6(a) is removed by the correction to isothermal flow curves as shown in Fig. 6(b), showing that the flow behavior is still in a strain hardening regime. Isothermal flow curves obtained in this manner are those which should be used to derive empirical constitutive equations for input to finite-element simulation programs.

Processing Maps

Once the temperature and strain rate dependence of the flow stress have been determined, and the strain dependence has been corrected to isothermal conditions, it is possible to derive empirical constitutive equations to be

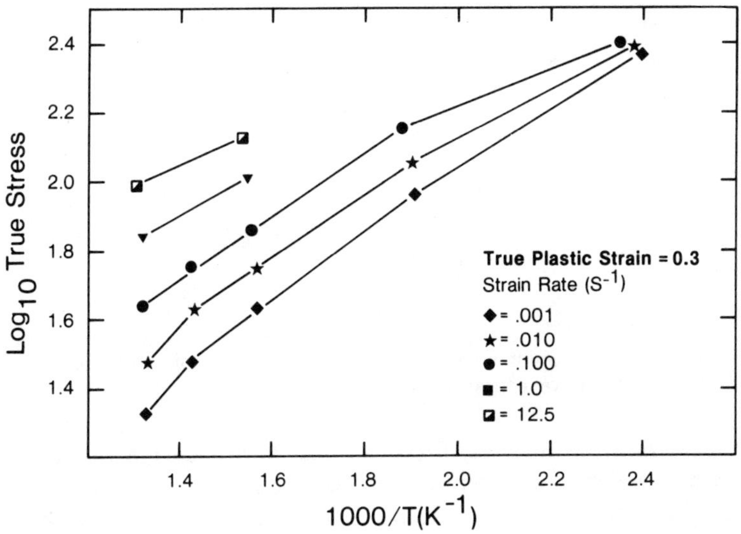

Fig. 5. Plots of log true stress vs. inverse temperature for 2024-0 aluminum to be used for correction to isothermal flow curves.

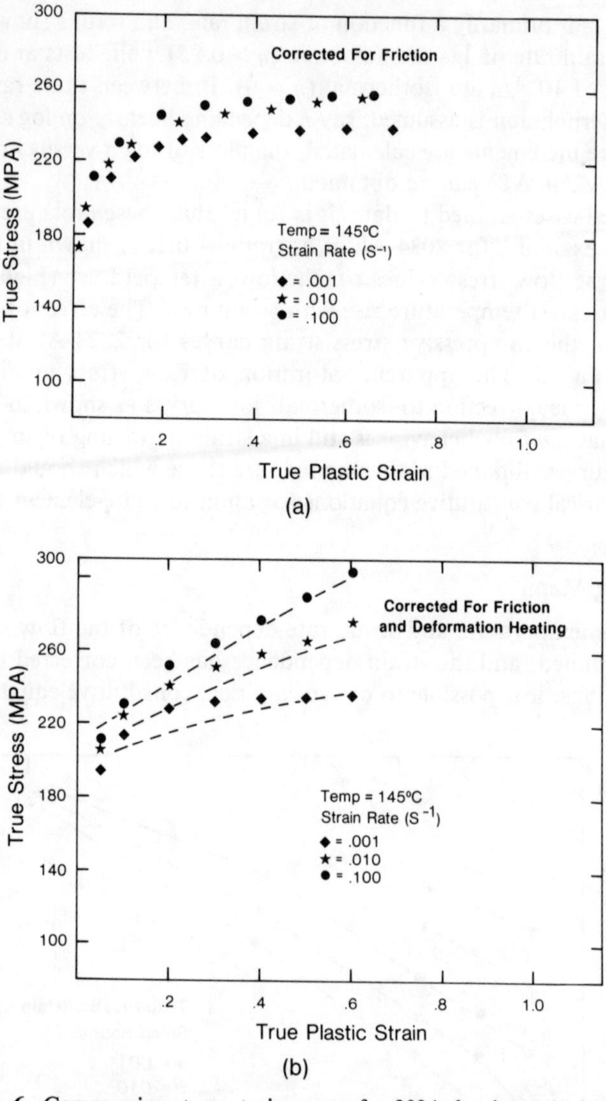

Fig. 6. Compression stress strain curves for 2024 aluminum (a) before and (b) after correction to isothermal conditions.

used for subsequent analyses. These can then be displayed in various graphical formats which visually assist in understanding the material flow behavior. To illustrate this, the generation of three-dimensional and contour plots of stress and strain rate sensitivity will be described for the transformed beta microstructure of the titanium alloy Ti-6242.

Constitutive equations for the flow behavior of Ti-6242 have been determined by Dadras and Thomas.[20] This material is processed in one of two starting microstructures. A microstructure of equiaxed alpha phase in a matrix of transformed beta, termed $\alpha + \beta$, is obtained by hot working below the beta transus temperature (the temperature above which the alloy is single-phase beta) and has good fatigue properties. The microstructure obtained by cooling from temperatures above the beta transus temperature is called transformed β and has high strength and good creep properties. For both microstructures, the flow stress rises to a peak at a plastic strain of approximately 0.05 and then decreases. For the $\alpha + \beta$ structure, this decrease is primarily due to deformation heating, and the isothermal flow curve is assumed independent of strain. For the transformed β structure, there is real flow softening as the initial structure begins transformation toward the more stable $\alpha + \beta$ microstructure. The constitutive equations are set up to describe the temperature and strain rate dependence of the peak stress for both microstructures, with the flow softening behavior of the transformed β structure treated separately. The temperature and strain rate dependence of the peak stress for both the $\alpha + \beta$ and the transformed β microstructures are given by the following equations:

$$\begin{aligned}
\log \sigma &= 9850/T - A \text{ for } (\alpha + \beta) \quad 1283\,K \geq T \geq 1144\,K \\
\log \sigma &= 13100/T - B \text{ for } (\beta) \quad 1283\,K \geq T \geq T' \\
\log \sigma &= 5230/T - C \text{ for } (\beta) \quad T' \geq T \geq 1144\,K \\
A &= 0.0350X^2 - 0.1090X + 5.992 \\
B &= 0.0180X^2 - 0.1710X + 8.398 \\
C &= 0.0345X^2 - 0.0765X + 1.956
\end{aligned} \tag{16}$$

where the temperature T is in kelvins and $X = \log \dot{\epsilon}$.

Once analytical equations have been obtained, the strain rate sensitivity, m, can be calculated as the partial derivative of $\log \sigma$ with respect to $\log \dot{\epsilon}$. The variations of the stress and strain rate sensitivity with strain rate and temperature can then be presented graphically in the form of either three-dimensional plots or two-dimensional contour maps. In order to give an explicit prescription, two short computer programs, written in FORTRAN 77 and designed to be linked with the device-independent graphics package ISSCO DISSPLA, are listed in the Appendix at the end of this chapter. The program TI6242B creates tables of stress and strain rate sensitivity as functions of temperature and strain rate and stores these in files with extensions .STS and .EMM. These are then passed to the graphical processing program CONTOUR which uses the DISSPLA commands to produce the 3-D and contour plots. The program in the form listed in the Appendix can produce figures which can either be viewed on a Tektronix device or output to a plotter driven by the Hewlett-Packard graphics language. Figures 7–10 show the resulting 3-D and contour plots for stress and strain rate sensitivity. For 3-D plots, the angle of view can be adjusted.

Concluding Remarks

With the motivation of implementing the finite-element method to understand complicated mechanical behavior, constitutive equations have been developed to represent both microplastic deformation and the rate-dependent deformation of bulk material from a continuum point of view. The emphasis in this chapter was to discuss the computer-assisted experimental test facilities and computer-based analysis of mechanical test results which can lead to empirical descriptions of high-temperature flow under temperature and strain rate conditions representative of bulk deformation processes such as forging. There is a need for additional testing of this type in order to develop a material database that is required to take advantage of the greatly enhanced capability for process simulation which results from the newer finite-element codes and cost-effective computer systems. In parallel with implementation of this technology, there is the hope that a greater understanding of high-temperature material behavior will evolve, particularly in terms of specifying evolutionary state parameters which capture the relevant

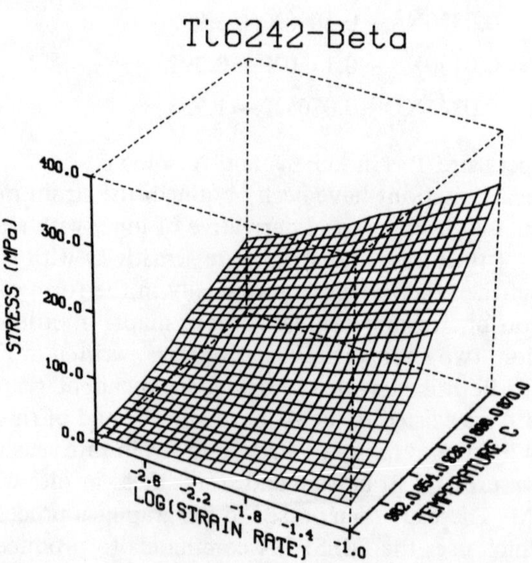

Fig. 7. Three-dimensional plot of the flow stress of Ti-6242 (Beta) vs. strain rate and temperature.

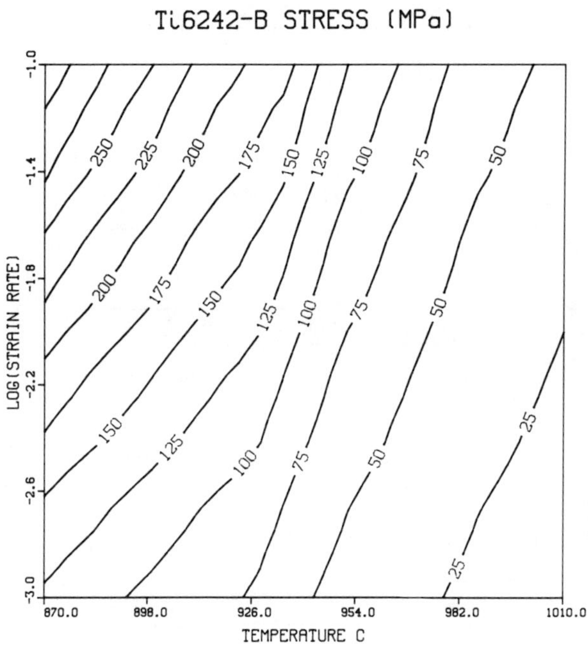

Fig. 8. Contour plot of the flow stress of Ti-6242 (Beta) vs. strain rate and temperature.

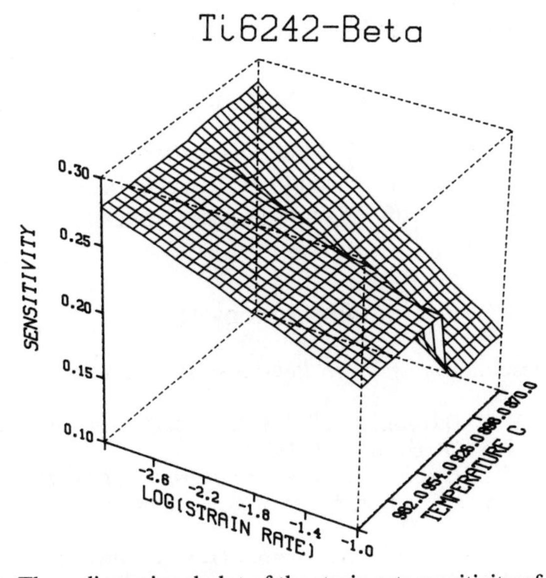

Fig. 9. Three-dimensional plot of the strain rate sensitivity of Ti-6242 (Beta) vs. strain rate and temperature.

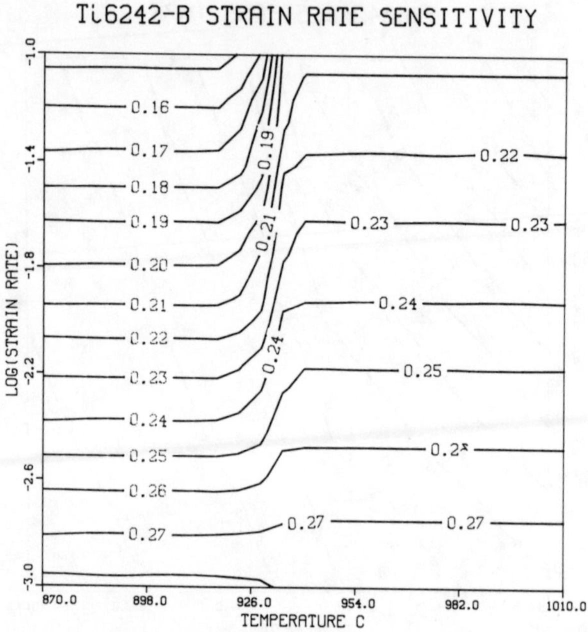

Fig. 10. Contour plot of the strain rate sensitivity of Ti-6242 (Beta) vs. strain rate and temperature.

factors of the material thermomechanical history and allow prediction of current material behavior under processing conditions.

Acknowledgments

The support of the AFWAL Materials Laboratory, of the Air Force Office of Scientific Research through grant AFOSR-85-0078, and of the General Electric Co. Aircraft Engine Business Group is gratefully acknowledged.

References

1. P. E. Wellstead, *Introduction to Physical System Modeling*, Academic Press (1979).
2. U. F. Kocks, P. R. Dawson, and P.S. Follansbee, *Los Alamos National Laboratory Report No. LA-UR-86-2672* (1986).
3. G. D. Lahoti, S. I. Oh, and T. Altan, in *Experimental Verification of Process Models*, edited by C. C. Chen, American Society for Metals, Metals Park, OH, p. 315 (1983).
4. A. S. Argon, in *Constitutive Equations in Plasticity*, edited by A. S. Argon, MIT Press, Cambridge, MA, p. 1 (1975).

5. G. I. Taylor and C. F. Elam, *Proc. Roy. Soc.*, *Vol. A102*, p. 643 (1923).
6. G. I. Taylor and C. F. Elam, *Proc. Roy. Soc.*, *Vol. A108*, p. 28 (1925).
7. E. Schmid, *Proc. Int. Cong. Appl. Mech.*, Delft, p. 342 (1924).
8. R. Hill, *J. Mech. Phys. Solids*, *Vol. 14*, p. 95 (1966).
9. R. Hill and J. R. Rice, *J. Mech. Phys. Solids*, *Vol. 20*, p. 401 (1972).
10. R. J. Asaro, *J. of Appl. Mech.*, *Vol. 50*, p. 921 (1983).
11. R. J. Asaro and A. Needleman, *Scripta Met.*, *Vol. 18*, p. 429 (1984).
12. D. Peirce, R. J. Asaro and A. Needleman, *Acta Met.*, *Vol. 31*, p. 1951 (1983).
13. R. J. Asaro and A. Needleman, *Acta Met.*, *Vol. 33*, p. 923 (1985).
14. S. Ankem and H. Margolin, *Met. Trans.*, *Vol. 13A*, p. 595 (1982).
15. MARC program of MARC Analysis Research Corp., Palo Alto, CA.
16. S. L. Semiatin and G. D. Lahoti, *Met. Trans.*, *Vol. 13A*, p. 275 (1982).
17. J. F. Thomas, Jr., in *Process Modeling Applied to Metalforming and Thermomechanical Processing*, *AGARD Lecture Series #137*, edited by J. F. Thomas, Jr., NATO/AGARD, Paris, p. 1 (1984).
18. C. Zener and J. H. Hollomon, *J. Appl. Physics*, *Vol. 15*, p. 22 (1944).
19. P. L. Charpentier, B. C. Stone, S. C. Ernst and J. F. Thomas, Jr., *Met. Trans.*, *Vol. 17A* (1986).
20. P. Dadras and J. F. Thomas, Jr., *Met. Trans.*, *Vol. 12A*, p. 1867 (1981).

APPENDIX

Computer Programs for Graphical Data Presentation

1. PROGRAM TI6242B. Generates tables of stress and strain rate sensitivity as a function of log(strain rate) and temperature.
2. PROGRAM CONTOUR. Produces three-dimensional and contour plots of a variable as a function of log(strain rate) and temperature using DISSPLA commands.

```
      PROGRAM TI6242B
C
C     THIS PROGRAM CALCULATES THE STRESS AND STRAIN RATE
C     SENSITIVITY OF TI-6-2-4-2 BETA MICROSTRUCTURE BASED ON THE
C     DATA OF DADRAS AND THOMAS
C
C     VARIABLE DECLARATIONS
C
      IMPLICIT REAL*8 (A-H,O-Z)
C
      CHARACTER*12 FILIN,FOUT
      CHARACTER*8 FILOUT
      COMMON/LIMITS/ NINTT,NINTS,FORTEMP,STRMIN,TSTEP,STSTEP
      DIMENSION SIGMA(60,60),EMM(60,60)
C
C     SIGMA --- PEAK STRESS
C     EMM   --- STRAIN RATE SENSITIVITY
```

PROGRAM TI6242B (*continued*)

```
C
C      INITIALIZE ARRAYS
C
       DO 1 I = 1,60
       DO 1 J = 1,60
       SIGMA(I,J) = 0.0
    1  EMM(I,J)   = 0.0
C
C      BACKGROUND AND FOREGROUND TEMPERATURES AND NO. OF STEPS
C
       BACTEMP = 1010.0
       FORTEMP = 870.0
       NINTT = 20
C
C      MIN. AND MAX. LOG(STRAIN RATE) AND NO. OF STEPS
C
       STRMIN = -3.0
       STRMAX = -1.0
       NINTS = 20
C
       TSTEP = (BACTEMP-FORTEMP)/NINTT
       DO 20 I = 1,NINTT+1
C
C      INTERPOLATE TEMPERATURE
C
       TINT = FORTEMP + (I-1)*TSTEP
C
           STSTEP = (STRMAX - STRMIN)/NINTS
           DO 50 K = 1,NINTS+1
C
C          INTERPOLATE LOG(STRAIN RATE)
C
           STRINT = STRMIN + (K-1)*STSTEP
C
C            INTERPOLATE DATA
C
             X = STRINT
C
             IF(TINT.GE.930.0.AND.TINT.LE.1010.0) THEN
             B = 0.0180*X*X - 0.171 * X + 8.398
             SIGMA(I,K) = 13100./(TINT+273.) - B
             EMM(I,K) = -0.036*X + 0.171
             ENDIF
C
             IF(TINT.GE.870.0.AND.TINT.LT.930.0) THEN
             CC = 0.0345 * X*X - 0.0765 * X + 1.956
             SIGMA(I,K) = 5230./(TINT+273.) - CC
             EMM(I,K) = -0.069*X + 0.0765
             ENDIF
C
C      CONVERT FROM LOG10(SIGMA) TO SIGMA
           SIGMA(I,K) = 10.0**SIGMA(I,K)
```

```
C
   50          CONTINUE
C
   30   CONTINUE
   20   CONTINUE
C
C       ALL INTERPOLATION IS COMPLETE. NOW SAVE DATA TO FILE
C
        WRITE(5,1000)
        WRITE(5,*) 'INTERPOLATION IS COMPLETE. ENTER SAVE-FILE NAME'
        WRITE(5,*) 'UPTO 8 CHARACTERS LONG. THE INTERPOLATED DATA'
        WRITE(5,*) 'WILL BE SAVED UNDER THIS NAME WITH DIFFERENT'
        WRITE(5,*) 'EXTENSIONS, AS FOLLOWS: '
        WRITE(5,*) '      ________.STS  ---> STRESS'
        WRITE(5,*) '      ________.EMM  ---> STRAIN RATE SENSITIVITY'
        WRITE(5,1000)
        WRITE(5,*) 'FILES ARE IN THE FORMAT FOR CONTOUR AND 3-D'
        WRITE(5,*) 'PLOTTING'
 C
        WRITE(5,1000)
        WRITE(5,*) 'ENTER SAVE-FILE NAME >>>> '
        READ(5,1100) FILOUT
C
        FOUT = FILOUT//'.STS'
        OPEN(7,FILE=FOUT,STATUS='NEW')
        CALL SAVER(SIGMA)
        CLOSE(7)
C
        FOUT = FILOUT//'.EMM'
        OPEN(7,FILE=FOUT,STATUS='NEW')
        CALL SAVER(EMM)
        CLOSE(7)
C
 1000   FORMAT(//)
 1100   FORMAT(A)
        STOP
        END

        SUBROUTINE SAVER(ZOUT)
C
C       SAVES IN A FILE FOR CONTOUR AND 3-D PLOTTING
C
        IMPLICIT REAL*8 (A-H,O-Z)
        COMMON/LIMITS/ NINTT,NINTS,FORTEMP,STRMIN,TSTEP,STSTEP
        DIMENSION ZOUT(60,60)
C
        NPTS = (NINTT+1)*(NINTS+1)
        WRITE(7,*) NPTS
        DO 10 I = 1,NINTT+1
        DO 10 J = 1,NINTS+1
        TEMPR = FORTEMP + (I-1)*TSTEP
        STRRT = STRMIN + (J-1)*STSTEP
        WRITE(7,*) TEMPR,STRRT,ZOUT(I,J)
```

PROGRAM TI6242B (*continued*)

```
   10  CONTINUE
C
       RETURN
       END


       PROGRAM CONTOUR
C
       DIMENSION X(2500),Y(2500),Z(2500),ZM(50,50)
       COMMON WORK(4000)
       CHARACTER*1 INFV(3)
       CHARACTER*20 INPFIL
       DATA INFV/'X','Y','Z'/
C
C      CHOOSE TERMINAL OR HP PLOTTER
C
   10  WRITE(5,*) 'CHOOSE OUTPUT MODE'
       WRITE(5,*) '     1. TERMINAL'
       WRITE(5,*) '     2. PLOTTER'
       WRITE(5,*) '     CHOOSE 1 OR 2'
       READ(5,*) IOUT
C
       IF(IOUT.NE.1.AND.IOUT.NE.2) GOTO 10
       IF (IOUT.EQ.1)  CALL TEKALL(4014,120,0,1,0)
C
       IF (IOUT.EQ.2) THEN
            CALL HPSETUP
            CALL HP7475(1)
            CALL HWSPEC(50.0,'PENS')
       ENDIF
C
C      CHOOSE VARIABLE TO BE PLOTTED
C
       WRITE(5,*) 'CHOOSE VARIABLE'
       WRITE(5,*) '     1.     STRESS'
       WRITE(5,*) '     2.     RATE SENSITIVITY'  WRITE(5,*) ' '
       READ(5,*) ICH
C
       WRITE(5,*) ' '
       WRITE(5,*) 'ENTER FILE NAME'
       READ(5,1100) INPFIL
 1100  FORMAT(A)
       OPEN(7,FILE=INPFIL,STATUS = 'OLD')
C
C      READ INPUT DATA FROM FILE
```

```
C
      REWIND(7)
      READ(7,*) NPTS
      DO 100 I = 1,NPTS
      READ(7,*) X(I),Y(I),Z(I)
  100 CONTINUE
C
C     DETERMINE LIMITS OF INPUT DATA
C
      CALL ORDER (X,XMIN,XINC,XMAX,NPTS)
      CALL ORDER (Y,YMIN,YINC,YMAX,NPTS)
      CALL ORDER (Z,ZMIN,ZINC,ZMAX,NPTS)
      WRITE(5,2000)
      WRITE(5,1000) INFV(1),XMIN,XMAX,XINC
      WRITE(5,1000) INFV(2),YMIN,YMAX,YINC
      WRITE(5,1000) INFV(3),ZMIN,ZMAX,ZINC
C
 1000 FORMAT(1X,A1,':  MIN = ',G12.5,' MAX = ',G12.5,' LENGTH = ',
     +         G12.5)
      XINC = XINC/5.
      YINC = YINC/5.
      ZINC = ZINC/25.
      INC = (ZINC + 2.5)*0.2
      ZINC = 5.0*INC
      CALL RESET('ALL')
C
      CALL NOBRDR
      CALL PAGE (11.0,8.5)
      CALL HEIGHT(0.2)
      CALL ZAXANG(90.0)
C
      WRITE(5,*) 'ENTER VIEW ANGLE FOR 3-D PLOT >>>'
      WRITE(5,*) 'XANG,YANG,ZANG'
      READ(5,*) XANG,YANG,ZANG
C
      CALL TITL3D('Ti-6242 Beta',12,8.5,6.5)
      CALL VUANGL(XANG,YANG,ZANG)
C
      IF (ICH.EQ.1)
     + CALL AXES3D('TEMPERATURE C',13,'LOG(STRAIN RATE)',16,
     +              'STRESS (MPa)',12,5.0,5.0,5.0)
      IF (ICH.EQ.2)
     + CALL AXES3D('TEMPERATURE C',13,'LOG(STRAIN RATE)',16,
     +              'SENSITIVITY',11,5.0,5.0,5.0)
      CALL GRAF3D(XMIN,XINC,XMAX,YMIN,YINC,YMAX,ZMIN,'SCALE',ZMAX)
C
      CALL BGNMAT(25,25)
      CALL GETMAT(X,Y,Z,NPTS,ZWORK)
      CALL ENDMAT(ZM,WORK)
      CALL SURMAT(ZM,1,25,1,25,WORK)
      CALL DASH
```

PROGRAM CONTOUR (*continued*)

```
      CALL BOX3D
      CALL ENDPL(0)
      CALL RESET('ALL')
C
   40 CALL NOBRDR
      CALL PAGE(11.0,8.5)
      IF(ICH.EQ.1)
     + CALL TITLE('Ti-6242 Beta STRESS (MPa)',25,'TEMPERATURE C'
     +           ,13, 'LOG(STRAIN RATE)',16,6.5,6.5)
      IF(ICH.EQ.2)
     + CALL TITLE('Ti-6242 Beta STRAIN RATE SENSITIVITY',36
     +           ,'TEMPERATURE C',13,'LOG(STRAIN RATE)'
     +            ,16,6.5,6.5) C
      CALL GRAF(XMIN,XINC,XMAX,YMIN,YINC,YMAX)
      CALL BCOMON(4000)

      CALL CONMAK(ZM,25,25,'SCALE')
      CALL CONLIN(0,'SOLID','LABELS',2,10)
C
      CALL CONMIN(2.0)
      CALL CONANG(30.)
      CALL CONTUR(1,'LABELS','DRAW')
      CALL MARKER(3)
      CALL GRID(0,0)
      CALL ENDPL(0)
      CALL DONEPL
      WRITE(5,2000)
 2000 FORMAT(//////)
      WRITE(5,*) 'DO YOU WANT TO RE-PLOT ?'  WRITE(5,*) ' '
      WRITE(5,*) '      0 - NO'
      WRITE(5,*) '      1 - YES'
      READ (5,*) IQUES
      IF(IQUES.EQ.1) GOTO 10
   20 STOP
      END
C
      SUBROUTINE ORDER(Z,XMIN,XINC,XMAX,NPTS)
      DIMENSION Z(1000)
      XMIN = 12000.
      XINC = 0.
      XMAX = -12000.
      DO 10 I = 1,NPTS
      IF(Z(I).LT.XMIN) XMIN = Z(I)
      IF(Z(I).GT.XMAX) XMAX = Z(I)
   10  CONTINUE
C
      XINC = XMAX-XMIN
      RETURN
      END
```

12

Synthesis of Atomistics and Continuum Modeling to Describe Microstructure

HAROLD L. GEGEL
Air Force Wright Aeronautical Laboratories
Materials Laboratory
Wright-Patterson Air Force Base, OH 45433

1. Introduction

In designing material-working processes for components made of complex materials with high information content, the most important task is the selection of the controlling process parameters which will ensure required part quality along with specific mechanical and physical characteristics. Controlling process parameters are the number and sequence of material flow operations, the heat-treating conditions, and the associated quality-assurance tests. When designing unit manufacturing processes for complex alloys and ceramic and polymeric materials, special features such as nonlinear, irreversible finite-deformation flow must be considered. Simultaneously, the complex interdependence of the unit process parameters and their effect on finished part quality, reliability, and inspectability must be considered.

An important goal in manufacturing is to determine the optimum means for producing defect-free parts on a repeatable basis. The optimization criteria depend on manufacturing goals and product specifications; the establishment of appropriate criteria requires indepth views—both global and local—of manufacturing processes and material behavior. From an optimization point of view, manufacturing processes require determination of material-flow mechanics to achieve proper process design and develop a rational strategy for process control. However, success in determining this rational strategy depends strongly on understanding the material behavior under processing conditions—conditions generally governed by irreversible thermodynamic processes and inhomogeneous flow behavior.

Modeling of the forging process involves both mechanics and thermodynamics. The process, as a rule, is inhomogeneous and transient over a large volume of workpiece material, and the material-flow process can be characterized as highly irreversible and stochastic in nature. The mechanics of the forging process are well established, and finite-element programs such as ALPID developed by Oh[1] are now being used by industry to analyze complex forging operations. Material behavior under processing conditions—commonly referred to as workability[2]—is the least understood aspect of forging process analysis. In light of its conventional definition, the term workability is ill-defined in relation to process simulation.

This chapter will address the subject of workability, that is, how the workpiece material responds to the demands of the process. Since the theme of this seminar is forging, the chapter is weighted toward the characteristics of the forging process; the fundamentals, however, apply to any unit process.

Deformation and fracture maps constructed from knowledge concerning atomic mechanisms will be used to provide the foundation for understanding the basic processes governing flow and fracture; processing maps based on mechanistic models for defect nucleation in terms of temperature and strain rate will be used to introduce the concept of workability. The deformation, fracture, and processing maps are complementary and contribute significantly to the understanding of the influence of "structure" on the basic workability of crystalline material. Microstructure evolution and defect generation will be related to energy-dispersal processes through the Second Law of Thermodynamics.

The mechanistic approach to understanding the fundamentals of flow, fracture, and workability has numerous practical limitations. For example, the approach is limited to processes which can be described by steady-state equations and to pure crystalline and simple alloys. In order to create a workability model which would be applicable to complex engineering materials, an approach was developed based on combining the science of continuum mechanics and thermodynamics; this approach will be described. Stability analysis, based on the Liapunov function, will be introduced to define the loci of all bifurcation processes which create either a continuum or structural defect. Processing maps constructed for assisting in defining a safe "processing window" in terms of temperature and effective-strain-rate space for complex, crystalline engineering alloys will be interpreted. In the final analysis the mechanistic and continuum approaches were merged for the purpose of understanding microstructure evolution and defect avoidance.

2. Fracture and Deformation Maps—Atomistic Processes

Ashby and Frost[3] have developed an approach to understanding an important aspect of workability which is related to the deformation dynamics of crystalline materials. The deformation and fracture maps which they developed for pure metals and simple alloys are constructed based on the assumption that the net steady-state shear strain rate ($\dot{\gamma}_{net}$) of a plastically deforming crystalline material depends on a number of basic *atomic processes* such as dislocation motion, diffusion, grain-boundary sliding, twinning, and α-phase transformation.

Deformation maps for any polycrystalline material can be constructed in normalized stress-temperature space, showing the area of dominance of each flow mechanism. Each flow mechanism must have an equation relating shear strain rate ($\dot{\gamma}$) to shear stress (τ), temperature T (kelvins), and structure. The term "structure" includes all parameters describing the atomic structure such as bonding, crystal class, defect structure, grain size, dislocation density and arrangement, solute concentration, and volume fraction of precipitate. Examples of Ashby-Frost deformation and fracture maps are given in Fig. 1(a) and (b), respectively. The equations for the various mechanisms are shown for the case of pure nickel.

The equations relate the normalized shear stress (τ/μ) (μ being the shear modulus of elasticity) to the homologous temperature [T/T(m)] [T(m) being the absolute melting temperature]. The ideal shear strength, defined by Kelly[4] to be no greater than $\mu/2$, is shown as a line marked "ideal strength." Polycrystalline materials normally deform at stresses much lower than this theoretical value by dislocation glide and/or climb caused by the diffusion of vacancies. At applied stresses near the yield stress, deformation generally occurs by dislocation glide. Deformation can occur at lower stresses by special dislocation-movement processes and by stress-directed vacancy flow. In these cases preferential diffusion paths such as grain boundaries and dislocation lines become very important, and it is by such short-circuiting diffusion paths that low activation energies can be observed.

The net shear strain rate ($\dot{\gamma}_{net}$) at temperature T and shear stress (τ) is the result of the superpositioning of several of the above-described mechanisms. Dislocation creep can occur simultaneously with diffusional flow but not with glide. Therefore, it is customary to choose the faster rate. The shear strain rates $\dot{\gamma}_i$ (i = 1,4) are of the form

$$\dot{\gamma}_1 = \dot{\gamma}_0 \exp\left[-\frac{\Delta F}{kT}(1 - \tau\kappa_0)\right] \tag{1a}$$

$$\dot{\gamma}_2 = \dot{\gamma}_p \exp\left[-\frac{\Delta F}{kT}\right] \tag{1b}$$

$$\dot{\gamma}_3 = \frac{A' D_{eff} \mu b}{kT}\left[\frac{\tau}{\mu}\right]^{m'} \tag{1c}$$

$$\dot{\gamma}_4 = 42\frac{\tau\Omega}{kTd^2} D_v + \frac{\pi\sigma}{\delta} D_b \tag{1d}$$

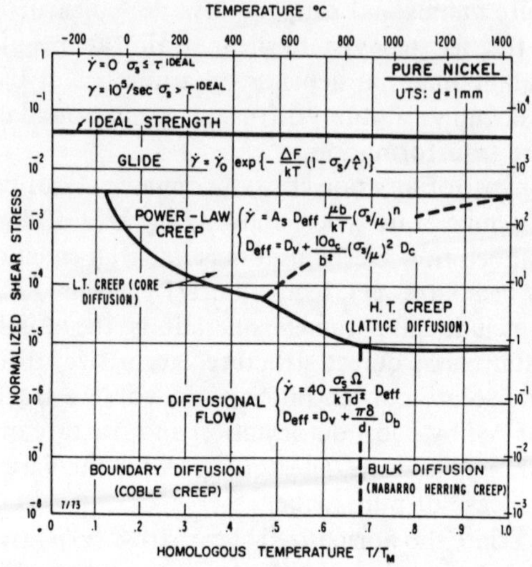

(a)

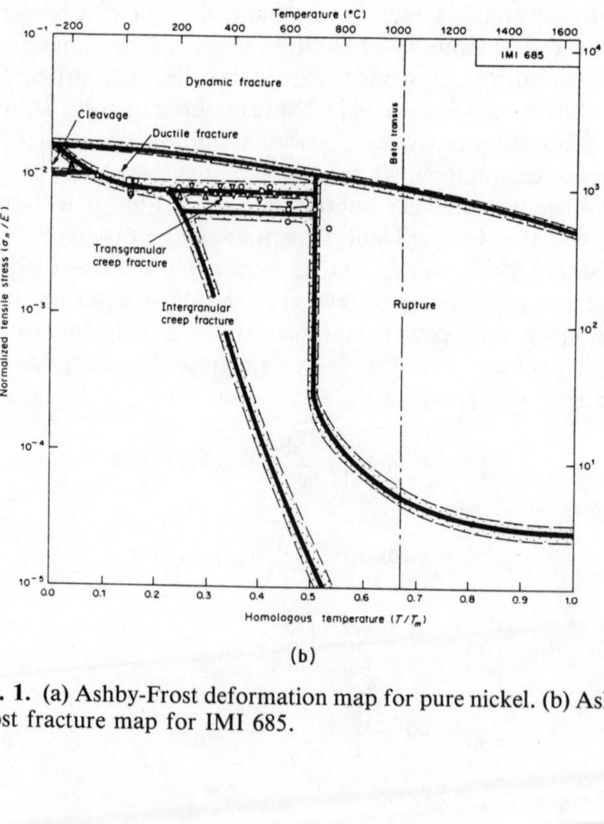

(b)

Fig. 1. (a) Ashby-Frost deformation map for pure nickel. (b) Ashby-Frost fracture map for IMI 685.

where $\dot{\gamma}_0$ and $\dot{\gamma}_p$ represent a number of mobile dislocation line segments; ΔF is the total free energy required to overcome the obstacle of a dislocation; τ_0 is the flow stress at 0 K = $\mu b/l$, where μ, b, and l are the elastic shear modulus, Burgers vector, and obstacle spacing, respectively. $\dot{\gamma}_0$ and $\dot{\gamma}_p$ are typically limiting strain rates on the order of 10^6 to 10^8 s^{-1}. A′ and m′ are material constants; D_{eff}, D_v, and D_b are diffusion constants; D_{eff} is an effective diffusivity defined as $D_{eff} = D_v f_v + D_c f_c$, where D_c is a diffusion coefficient associated with each type of diffusing species; Ω, d, and δ are the atomic volume, grain size, and effective thickness, respectively, of a boundary for diffusion transport.

Thus, the $\dot{\gamma}_{net}$ of a polycrystalline material subjected to a shear stress τ at a temperature T is approximately

$$\dot{\gamma}_{net} = \dot{\gamma}_4(\text{diffusional flow}) + \text{greatest of} \begin{bmatrix} \dot{\gamma}_3(\text{dislocation creep}) \\ \text{least of } \dot{\gamma}_1 \text{ and } \dot{\gamma}_2(\text{glide}) \end{bmatrix}$$

This information constitutes the deformation mechanism map. The Ashby-Frost deformation and fracture maps make an important contribution to the understanding of these phenomena in terms of atomic mechanisms. The atomic mechanisms described by the $\dot{\gamma}_i(i = 1,4)$ equations can combine to produce at least 12 distinct deformation mechanisms such as yielding, power-law creep, Nabarro-Herring creep, Coble creep, dynamic recrystallization, and superplastic flow.

This network of atomistic processes is significant since it provides the degrees of freedom required for shaping difficult-to-process materials such as high-temperature alloys and intermetallic compounds. These processes may increase the entropy of the universe while creating new material structures. Such structures may evolve during deformation as an outcome of a series of energy-dispersal processes. Microstructures which evolve as a result of these processes are called *dissipative structures.*(5,6) Thus, the atomistic approach provides a fundamental means for understanding the concepts of workability theory and, perhaps, the creation of microstructures as a result of energy-dispersal mechanisms.

3. Processing Maps

The deformation and fracture maps in their present form are not suitable for the design of a forging process, as represented in normalized stress and temperature space. The kinematic variables associated with this process are temperature and effective strain rate. Wray(7) suggested that fracture processes might be categorized by representing them in temperature-strain rate space; and Raj,(8) taking note of this possibility, quantified the mechanisms and constructed the first processing map.

A map for the nucleation of cavities at hard second-phase particles in an aluminum matrix, as shown in Fig. 2, illustrates the procedure. Palmer and Smith[9] demonstrated that ductile fracture can initiate at hard particles if tensile stresses are produced by inhomogeneous flow around the particle. A micrograph (150,000×) illustrating this phenomenon is presented in Fig. 3(a). Figure 3(b) is a schematic diagram of the state of stress producing the cavity-nucleation condition, and Fig. 3(c) shows the nucleated cavity. If diffusional stability had existed at the particular temperature T and applied effective strain rate $\dot{\bar{\epsilon}}$, the tensile stresses would have been relaxed and the cavity shown in Fig. 3(c) would not have formed. The critical strain rate for this cavitation mechanism is given by[9]

$$\dot{\bar{\epsilon}}_d = \frac{118(1-\nu)(1-2\nu+2/\pi)}{\left(\frac{5}{6}-\nu\right)^2} \frac{\mu\Omega}{kT} \frac{f_v \delta D_B}{p^3} \qquad (2)$$

The cavitation process in the alloy should not be initiated until the applied effective strain rate $\dot{\bar{\epsilon}}$ becomes greater than the critical strain rate $\dot{\bar{\epsilon}}_d$. The new parameters in Eq 2 are ρ (particle size) and f_v (volume fraction of hard particles). The cavity-initiation map in Fig. 2 shows changes in the calculated boundary with particle size. The map represents a plot of $\dot{\epsilon}_d$ as a function

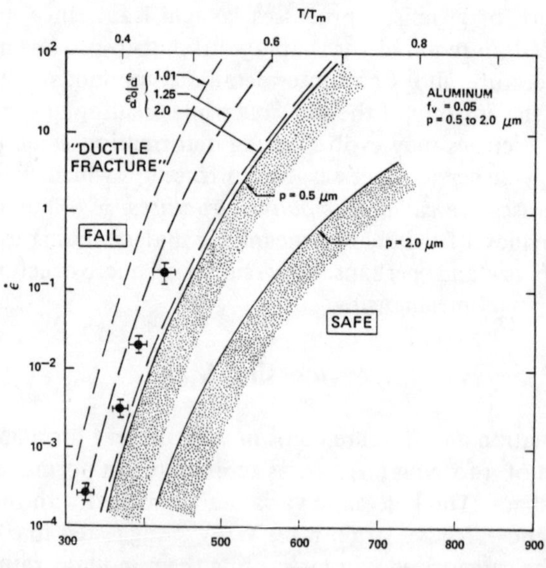

Fig. 2. Processing map for initiation of cavities at second-phase particles, as in ductile fracture.

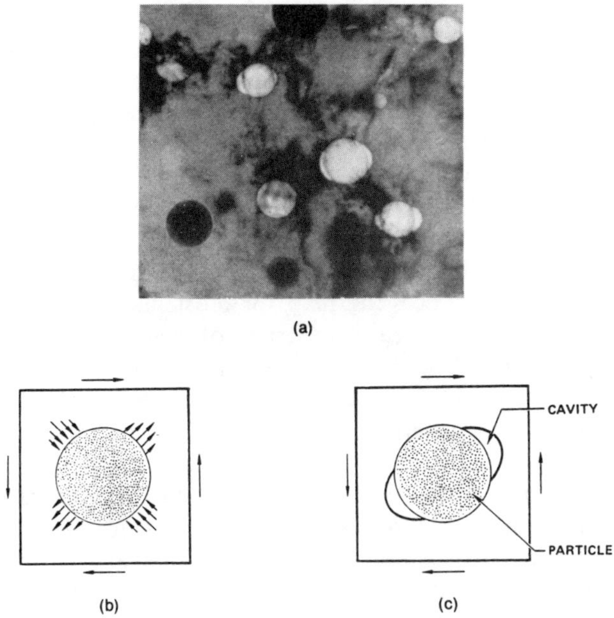

Fig. 3. Cavities initiating at hard particles in ductile fracture, as shown in (a), a micrograph by Palmer and Smith.[9] If the tensile stresses produced by inhomogeneous flow around particle, shown in (b), had been relaxed by diffusional transport, then cavity shown in (c) would *not* have initiated. Micrograph magnification: 150,000×.

of T for various values of p and nucleation strain up to the nucleation-strain limit, ϵ_n/ϵ_0.

An important result of this cavitation-model prediction is that the nucleation of cavities at hard particles depends strongly on the f_v/p^3 ratio, indicating that the boundary position is very sensitive to particle size and, to a lesser extent, to f_v. For a given volume fraction of hard particles, the "safe" region will be shifted to higher temperatures with increasing particle size. This knowledge concerning cavity nucleation enables alloy developers to make an intelligent selection of the composition range and proper processing scheme for controlling f_v and p, if improvement of workability and product reliability is the objective.

Mechanistic modeling has certain limitations. The cavitation model, as formulated, does not include a particle-size-distribution function. A distribution in p will produce a spread in the critical strain ϵ_d for nucleation of a cavity. A stochastic equation should be used to obtain a more accurate position of the boundary. An accurate prediction of the boundary position is required to guide the successful selection of the proper process parameters in an industrial forge shop, where process control is of the utmost importance. In fact, it may be impossible to use the mechanistic modeling methodology for predicting industrial processing maps, since the prior thermomechanical history of the material strongly influences material behavior during processing; there appears to be no effective method for introducing processing history into a mechanistic prediction.

Wedge cracking at grain boundaries is another important high-temperature deformation phenomenon. Because grain-boundary sliding is always associated with superplastic forming, the onset of wedge cracking represents a critical boundary in the processing map for most materials at temperatures above 0.4 T(m). The classical micrograph of wedge cracking by Yim and Grant,[10] presented in Fig. 4, shows that grain-boundary sliding is a necessary condition for the initiation and growth of wedge cracks. Wedge cracking is very sensitive to the applied strain rate. If the matrix deforms at a rate which is faster than that of the grain-boundary movement, then wedge cracking will not occur. If the strain rate is slow, such as in superplastic forming, sufficient time is available to relax the stresses at triple junctions by forming a wedge crack. If the strain rate is low and the alloy can form a second phase which is of lower density than the continuous phase, as in the case of Ti-6242 alloy, then time is available to relax the stresses by forming the lower-density α phase. Figure 5 shows an α phase precipitated at a triple junction having the morphology of a wedge crack.[11] In this case the local strain rate was $\sim 5 \times 10^{-3}\ s^{-1}$, and the temperature was $\sim$1200 K.

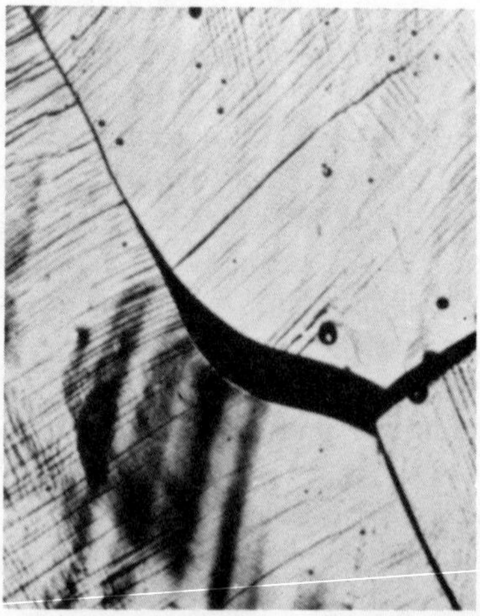

Fig. 4. Classical micrograph by Yim and Grant[10] showing sliding to be a necessary condition for initiation and propagation of wedge cracks.

For simple alloys the strain rate above which no wedge cracking occurs depends on the grain-boundary-sliding resistance k_b.[12] Assuming that the sliding rate is linear, the following equation for the shear stress, S, applied to the boundary holds:

$$S = k_b \dot{U} \tag{3}$$

where $\dot{U}$ is the rate of sliding. Since Eq 3 is applicable only for bicrystals, a correction must be made for polycrystalline materials to account for the increased resistance at triple junctions. If $\dot{U}$ is the maximum possible rate of sliding in a polycrystalline material, the maximum possible strain rate below which sliding is still possible is given by

$$\dot{\epsilon}_w = \frac{\dot{U}}{d} \tag{4}$$

where d is the grain size. In polycrystalline materials, wedge cracking will occur only if the applied strain rate $\dot{\epsilon}$ is less than $\dot{\epsilon}_w$. k_b can be calculated using the following equation:

$$k_b = \frac{kT}{8\Omega} \frac{f_b p^2}{\delta D_b} \tag{5}$$

where f_b is the area fraction of particles if the boundary contains a hard second phase, and δD_b is the interface width times the self-diffusivity along the interface. The similarity between the approaches for constructing "dam-

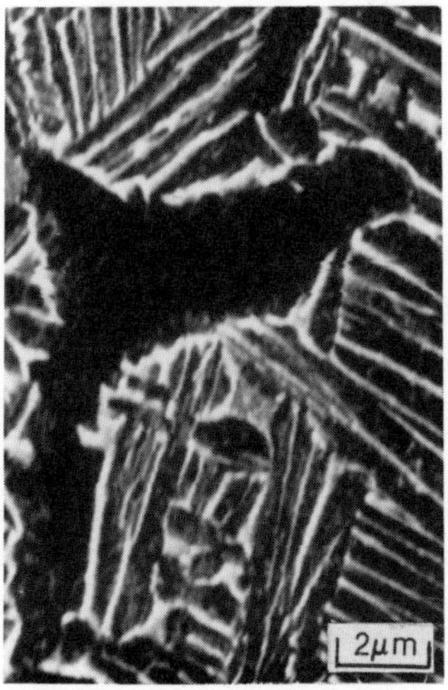

Fig. 5. Ti-6242 α phase precipitated at triple junction having morphology of wedge crack[11] ($\dot{\epsilon} \simeq 5 \times 10^{-3}$ s^{-1}, T $\simeq$ 1200 K).

age" maps and deformation/fracture maps is apparent. These approaches require a significant amount of basic data such as activation energies, frequency factors, diffusion coefficients, structural parameters, and size distributions. Vast physical knowledge is required concerning any combination of processes to enable their approximation by means of an equation. Some high-temperature deformation processes such as dynamic recrystallization are difficult to model because the processing conditions associated with their occurrence are not well understood.

Dynamic recrystallization is an important process because it provides a mechanism for refining the grain structure of a polycrystalline material and it is useful for eliminating the prior-particle boundaries of powder-metallurgical (P/M) alloys by hot deformation. Although much research has been conducted in attempts to understand this process, an effective model has not been developed.[13,14] Some investigators suggest that a critical strain is required to nucleate the dynamic recrystallization process, while others suggest that stored elastic energy is the driving force. The equation used by Raj[8] was based on the phenomenological observation that a critical strain ϵ_{cr} is required to nucleate the process, which yields an equation for the recrystallization strain rate of the form

$$\dot{\epsilon}_{cr} = Z(\epsilon_{cr})e^{-Q/kT} \tag{6}$$

The adiabatic-shearing process is classified as a continuum-effects mechanism and is characterized by ease of nucleation in many materials, particularly if they are characterized by relatively high density, low heat capacity, low thermal conductivity, large values of effective flow stress, low values of strain hardening $\partial\bar{\sigma}/\partial\bar{\epsilon}$, and strong temperature dependence of the flow stress $\partial\bar{\sigma}/\partial T$. When a polycrystalline material is slowly deformed at strain rates in the neighborhood of 0.001 s^{-1}, the deformation process is essentially isothermal. As the rate of deformation is increased, the idealization of isothermal behavior is untenable. The rate of temperature rise $\dot{T}$ for a material can be calculated by means of

$$\dot{T} = \frac{\bar{\sigma}\dot{\bar{\epsilon}}}{nC_p\rho}\,\frac{1}{1+m} \tag{7}$$

where n is the mechanical equivalent of heat, $\bar{\sigma}$ is the effective flow stress, $\dot{\bar{\epsilon}}$ is the effective strain rate, and m is the strain-rate-sensitivity parameter. A critical effective strain rate can be derived using an appropriate criterion for catastrophic shear such that

$$\dot{\bar{\epsilon}}_c = 4\pi\rho\kappa C_p\bar{\epsilon}\,\frac{n}{L^2\bar{\sigma}^2}\left[\frac{(\partial\bar{\sigma}/\partial\bar{\epsilon})}{(\partial\bar{\sigma}/\partial T)}\right]^2 \tag{8}$$

where L is the shear band length, $\bar{\epsilon}$ is the effective strain, and κ is the thermal diffusivity.

These mechanistic models are sufficient to predict a processing map which delineates safe and unsafe deformation regions. A composite processing map by Raj[8] based on these various "damage" mechanisms is shown in Fig. 6. The boundaries are conservative predictions since the construction is based on process nucleation. This does not constitute a serious problem since it is always desirable to avoid the formation of defects in forged parts.

In summary, the construction of deformation, fracture, and processing maps based on atomic mechanisms and "structure" makes important contributions to the scientific understanding of material behavior encountered during deformation processing. These maps are limited in practical application as a design tool for the following reasons: (1) the net-strain-rate equations for these maps are for steady-state conditions; (2) the analyses are most valid for pure materials and simple alloys and are not applicable for complicated engineering materials whose strain-rate sensitivity is a function of temperature and strain rate; (3) the large number of material parameters which must be determined, such as diffusivity, makes use of a model difficult; (4) the location of the various boundaries is very sensitive to prior thermomechanical history and microstructure, making accurate determination of location

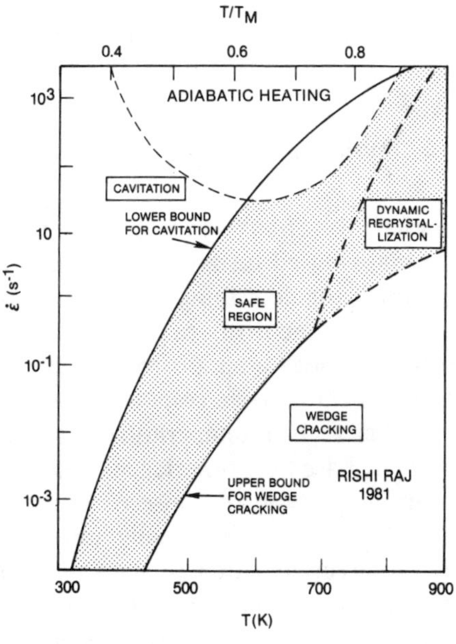

Fig. 6. Composite processing map by Raj[8] based on various "damage" mechanisms for aluminum.

almost impossible; (5) the stochastic nature of deformation and fracture processes is not taken into account by the atomic-mechanism approach, and engineering alloys are too complex for characterization by simple mechanistic models.

For the treatment of complex engineering alloys, a modeling approach is needed which is compatible with advanced numerical methods for modeling deformation processes. Such a model would satisfy the requirement for a near-optimal solution to a deformation problem. It is easy to obtain an admissible solution to a finite-element model of a given forging problem, but it is necessary to introduce the processing "personality" of the material into the solution in order to optimize the results in terms of the ability of the material to satisfy the demands of the process. Mechanistic models are not sufficiently accurate for achieving this goal.

A method which has been developed[(15)] for describing the behavior of the workpiece material under processing conditions is based on combining continuum mechanics and thermodynamics. The thermomechanics approach allows one to view the workability problem in a macroscopic sense without having to make a priori assumptions concerning the atomic mechanisms in operation under a given set of conditions. Although it is necessary to develop a constitutive equation which describes the relationship between the flow stress of the material and the kinematic variables of temperature T, effective strain rate $\dot{\bar{\epsilon}}$, and effective strain $\bar{\epsilon}$, the amount of testing is reduced by the modeling. The modeling approach is called Dynamic Material Modeling (DMM).

4. Dynamic Material Modeling

Recognizing the practical difficulty in making use of deformation, fracture, and processing maps based on atomic mechanisms, Prasad et al.[(15)] developed a new approach for describing the material behavior under processing conditions. The approach was called Dynamic Material Modeling (DMM) because the methodology attempts to describe the actual or dynamic path which a material element takes in response to an instantaneous change in effective strain rate $\dot{\bar{\epsilon}}$ at a given temperature T and effective strain $\bar{\epsilon}$.

It is a well-known fact, based on experience, that prior thermomechanical history strongly influences the response of the workpiece material to changes in process variables. Since it is virtually impossible to control the initial state of the material precisely—that is, the so-called as-received condition—one cannot predict a priori the atomic mechanism(s) in operation for a given instantaneous power input (P) to the system, which is defined as

$$P = \bar{\sigma}\dot{\bar{\epsilon}} = \sigma_1\dot{\epsilon}_2 + \sigma_2\dot{\epsilon}_2 + \sigma_3\dot{\epsilon}_3 \quad (9)$$

where σ_1, σ_2, and σ_3 are principal stresses and $\dot{\epsilon}_1$, $\dot{\epsilon}_2$, and $\dot{\epsilon}_3$ are principal strain rates. An important characteristic of any dynamic material or physical system is its stochastic nature. For example, it is well known that a bifurcation process may occur, but it is impossible to predict when it will occur. In a stochastic material system, an instantaneous response such as an effective stress $\bar{\sigma}$ to an input such as an instantaneous effective strain rate $\dot{\bar{\epsilon}}$ depends on the prior history of the material. Raj,[8] in his work on processing maps, discussed these characteristics in terms of microstructural effects on the position of a boundary. With the realization that a macroscopic approach was needed for simplifying the problem of defining desirable processing windows and that workability is both a continuum-mechanics and a thermodynamics problem, the DMM approach was developed based on combining these sciences. A framework for conceptualizing the DMM is given in Fig. 7, where the thermomechanics universe is given by

$$\text{Universe} = \text{E} + \text{M} + \text{C} \tag{10}$$

where E is the equipment system (that is, a forge press); M is the material system, including the dies, lubricating system, and workpiece material; and C is the control system. E channels energy into M, and M disperses it in terms of heat and microstructure evolution. M is a nonconservative system due to friction, and the workpiece material is assumed to be dissipative, dispersing energy as heat and structural changes.

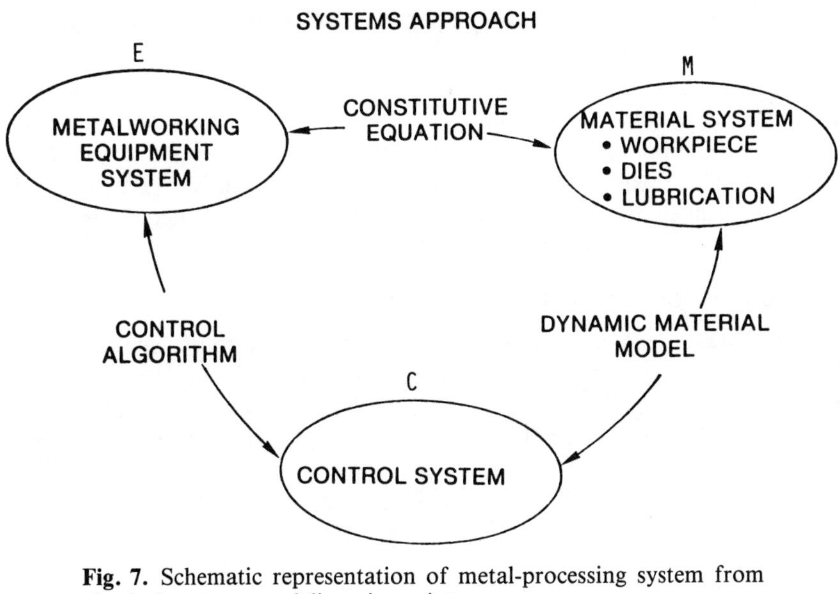

Fig. 7. Schematic representation of metal-processing system from physical-systems-modeling viewpoint.

The role of the constitutive equation—which is defined to be a mapping between the static variable $\bar{\sigma}$ and the kinematic variables $\dot{\bar{\epsilon}}$, t, and T—is to partition power between potential energy V and kinetic energy K. The kinetic energy is converted to heat, and the negative potential-energy gradient with respect to a generalized mechanical coordinate $-\partial V/\partial x_k$ is a force X_k^i which drives the irreversible atomic processes.

The DMM is a map in temperature-effective strain-rate-efficiency space which graphically describes dissipation of power by the material in stable and unstable ways. The boundaries on the map correspond to the loci of bifurcation points presented as boundaries between safe and unsafe regions in the Raj[(8)] processing maps. Hence, a correspondence exists between the mechanistic approach to mapping and the DMM. Since the boundaries in the DMM are determined by a Liapunov[(16)] stability analysis, they correspond to a narrow region where the material energy-dispersal processes are in a stationary state. If one could derive a constitutive equation describing the net effective strain rate in this region for a given temperature and effective strain $\bar{\epsilon}$ level, that equation could be used to model this particular line. A major difference between the DMM approach and the mechanistic approach is that in the DMM the mechanism(s) active in a particular region are determined after the map has been constructed. An example of a DMM with stability analysis for Ti-6242 is shown in Fig. 8.

The control system in the universe, C, functions to control the flow of energy into the material system M; therefore, the process always remains in the stable region, thereby avoiding the formation of defects or the occurrence of plastic-instability phenomena.

In summary, the thermomechanics approach allows one to view material behavior macroscopically, avoiding the problem of making a priori determinations of the operating atomic mechanisms. This represents a top-down approach which begins as a macroscopic analysis and ends as a microscopic evaluation aimed at understanding the evolution of microstructure and properties in a finished product—a forging.

4.1. Foundations of the Dynamic Material Model

The DMM approach is directed toward the dissipation of instantaneous power P (Eq 9) flowing into the material system M. P is dissipated by partitioning it into J, whose primitive is potential energy (V), and G, whose primitive is kinetic energy (K). The partitioning is given by the relation

$$\bar{\sigma}\dot{\bar{\epsilon}} = \int_0^{\dot{\bar{\epsilon}}} \bar{\sigma}\, d\dot{\bar{\epsilon}} + \int_0^{\bar{\sigma}} \dot{\bar{\epsilon}}\, d\bar{\sigma} \qquad (11)$$

or

$$P = G + J \qquad (12)$$

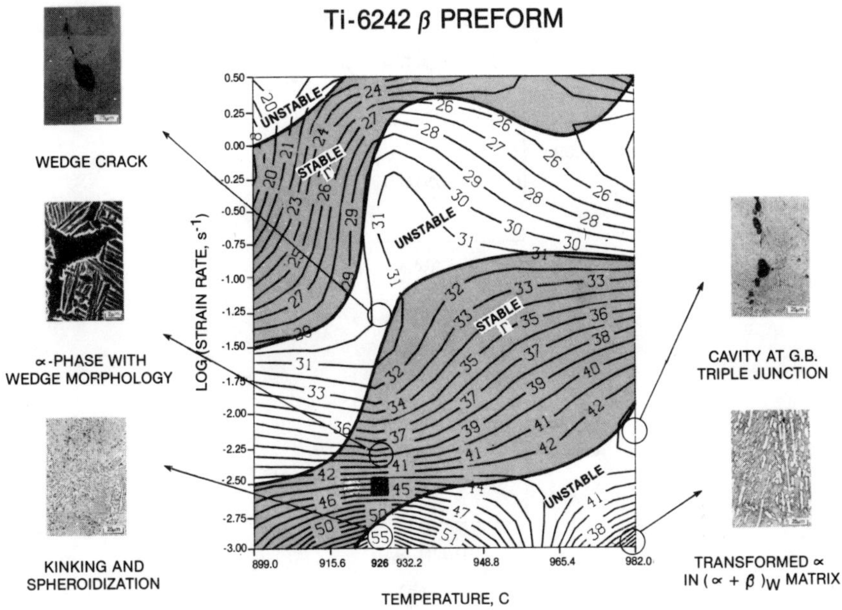

Fig. 8. Processing map for Ti-6242 β microstructure, with stable regions identified.

Power dissipated by plastic work is denoted by the area G under the curve which is illustrated in Fig. 9(a) and designated the dissipator content. The area J above the curve is termed the dissipator co-content and is related to the structural changes which occur dynamically to disperse energy. Thus, the fundamental assumption in the continuum model is that J is related to structural changes and G to continuum effects. These functions are complementary and are related by a Legendre transformation.[17-19]

From Eq 11 it follows that at any given temperature and effective strain, the partitioning of power between J and G is given by

$$\left[\frac{\partial J}{\partial G}\right]_{T,\epsilon} = \left[\frac{\partial \ln \bar{\sigma}}{\partial \ln \dot{\bar{\epsilon}}}\right]_{T,\epsilon} \tag{13}$$

and

$$\frac{\partial \ln \bar{\sigma}}{\partial \ln \dot{\bar{\epsilon}}} = m \tag{14}$$

the well-known strain-rate-sensitivity parameter of the material. From Eq 14 it follows that the dynamic constitutive equation is of the type

$$\bar{\sigma} = A\dot{\bar{\epsilon}}^{m} \tag{15}$$

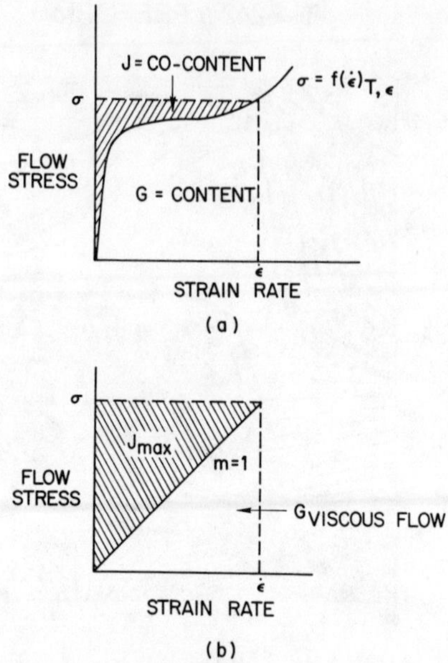

Fig. 9. Schematic representation of constitutive relation of material system as energy converter (dissipator). (a) Material system as nonlinear (general case) energy dissipator; (b) material system as linear (special case) dissipator.

Equation 15 is termed a dynamic constitutive equation because it is a result of integrating along the actual or dynamic path which an element of material takes when it changes instantaneously from an initial strain rate to another strain rate according to the Principle of Least Action,[20] where m is assumed to be constant. In the hot working range for pure metals,[8,21-22] m is temperature and strain-rate independent; but in complicated alloy systems, m has been shown[23] to vary with temperature and strain rate.

The power dissipated in changing the material structure is

$$J = \int_0^{\dot{\bar{\epsilon}}} \dot{\bar{\epsilon}}\, d\bar{\sigma} = \int_0^{\bar{\sigma}} (\bar{\sigma}/A)^{1/m}\, d\bar{\sigma} = \frac{\dot{\bar{\epsilon}}\bar{\sigma} m}{m+1} \tag{16}$$

and the power dissipated as heat is

$$G = \int_0^{\dot{\bar{\epsilon}}} \bar{\sigma}\, d\dot{\bar{\epsilon}} = \frac{1}{1+m}\, \bar{\sigma}\dot{\bar{\epsilon}} \tag{17}$$

Addition of J and G yields

$$\bar{\sigma}\dot{\bar{\epsilon}} = \frac{m}{1+m}\, \bar{\sigma}\bar{\epsilon} + \frac{1}{1+m}\, \bar{\sigma}\dot{\bar{\epsilon}} \tag{18}$$

which shows that m can be regarded as a *power-partitioning* index for distributing energy between J and G. For values of $m < 1$, the curve for $\bar{\sigma}$ versus $\dot{\bar{\epsilon}}$ appears to be similar to that shown in Fig. 9(a). When $m = 1$, the curve is a straight line, yielding the maximum dissipator co-content of area $J_{max} = \bar{\sigma}\dot{\bar{\epsilon}}/2$ as shown in Fig. 9(b). Prasad et al.[15] defined a parameter η by use of the ratio J/J_{max} given by

$$\eta = \frac{J}{J_{max}} = \frac{2m}{m + 1} \tag{19}$$

and called it efficiency. η is a dimensionless system parameter which describes how well the material is dissipating power by structural changes relative to J_{max}. It will be used later to derive a bifurcation line as a result of a stability analysis. Three-dimensional plots showing the variation of efficiency η with T and $\dot{\bar{\epsilon}}$ for Ti-6242 preform material at an effective strain of 0.6 are presented in Fig. 10 and 11. Figure 10 shows how the variation of η can be used to measure the dispersal of energy through metallurgical processes with forging temperature for a Ti-6242 $\alpha + \beta$ preform at an effective strain level of 0.6 for several effective strain rates. Typical microstructures

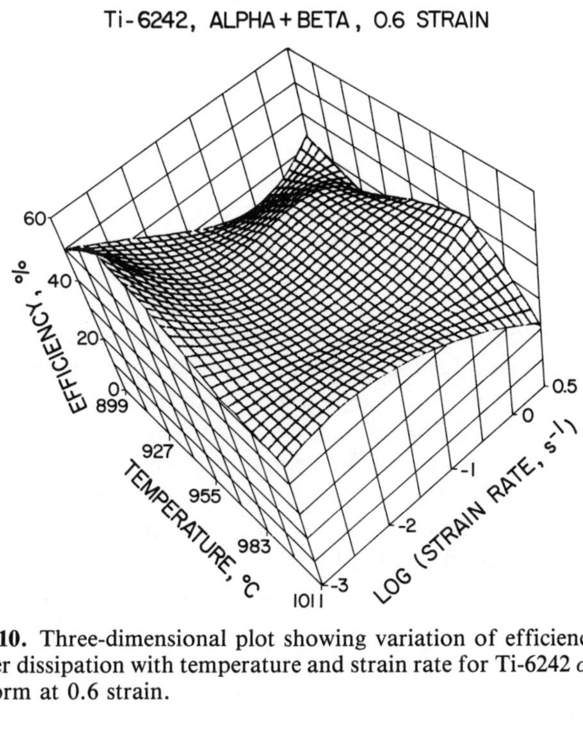

Fig. 10. Three-dimensional plot showing variation of efficiency of power dissipation with temperature and strain rate for Ti-6242 $\alpha + \beta$ preform at 0.6 strain.

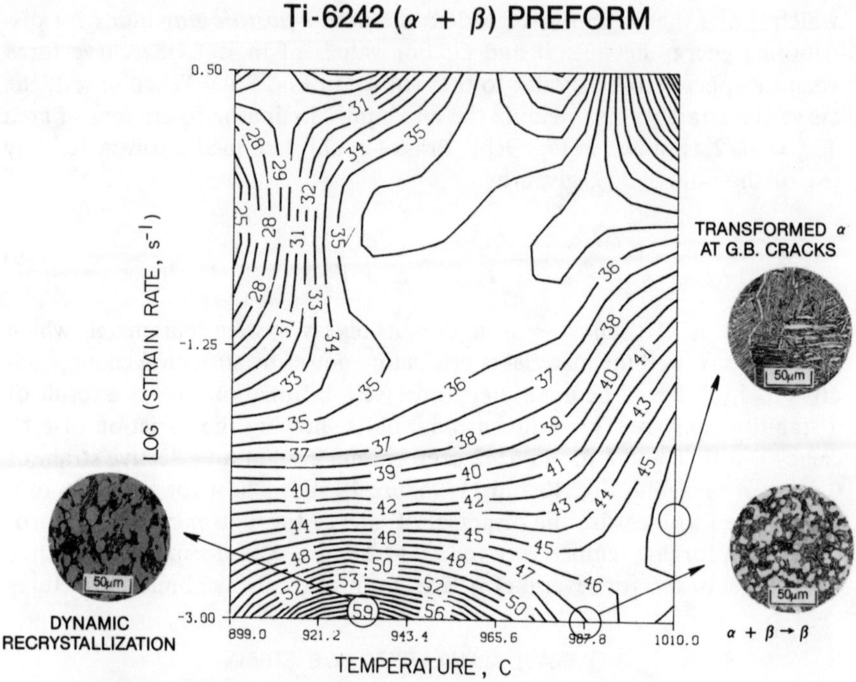

Fig. 11. Map showing constant efficiency contours in temperature-strain rate space for Ti-6242 $\alpha + \beta$ preform at 0.6 strain.

corresponding to particular processing temperatures and strain rates are also shown.

Several important features can be observed in the efficiency plot in Fig. 12: (1) at a low strain rate of 0.001/s, a peak occurs in the neighborhood of 1200 K (917°C); (2) a microscopic examination of the material at T < 1200 K revealed that the α phase recrystallized during hot deformation; (3) at T > 1200 K the α phase begins to dissolve, and this process continues until the β transus is reached; (4) at the 1200-K peak, steady state between the dynamic-recrystallization process in the α phase and the α-phase dissolution is reached; and (5) the 1200-K peak corresponds to a bifurcation process. The peak just described corresponds to a single point on a boundary which could be described by an equation of the type used by Raj[8] to predict the boundaries shown on the composite map and displayed in Fig. 8. Efficiency η will be used later as a parameter in a Liapunov function for axiomatically deriving the bifurcation lines from information contained in a constitutive equation.

The efficiency parameter is a system variable which can be used to create a three-dimensional map for displaying the dynamic behavior of the workpiece material. However, these maps require additional stability information

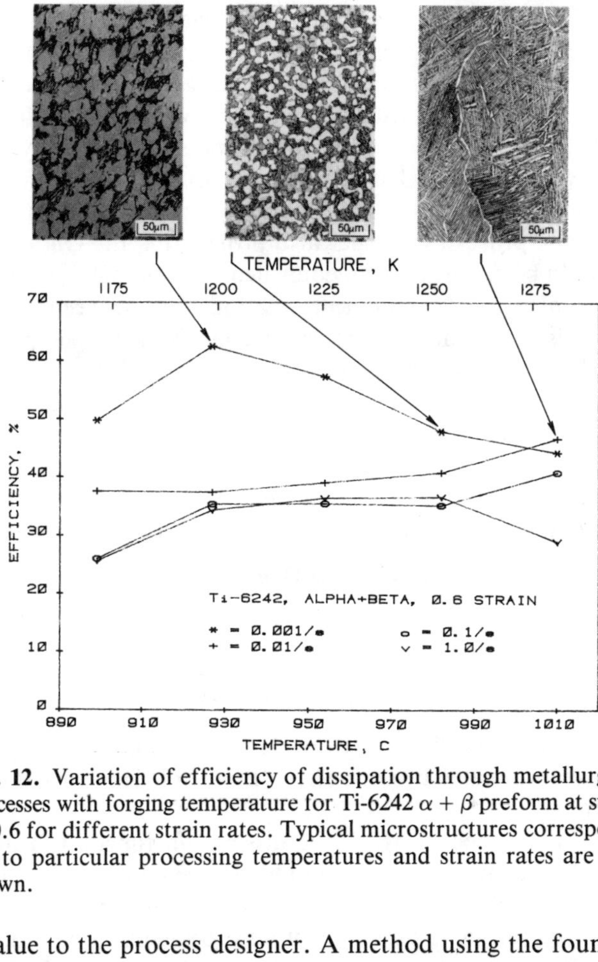

Fig. 12. Variation of efficiency of dissipation through metallurgical processes with forging temperature for Ti-6242 $\alpha + \beta$ preform at strain of 0.6 for different strain rates. Typical microstructures corresponding to particular processing temperatures and strain rates are also shown.

to be of value to the process designer. A method using the foundations of continuum mechanics, thermodynamics, and stability theory for delineating the boundaries which represent the loci of bifurcation points $b[(\eta, T, \dot{\bar{\epsilon}})]_{\bar{\epsilon}}$ will be described below.

4.2. Theoretical Background

In the early 1960s[24,25] considerable theoretical effort was expended to combine thermodynamics and continuum mechanics, and a successful synthesis of the two fields resulted. Studies in the border region between thermodynamics and mechanics have been impeded by lack of communication;[26] as a result the modeling of material behavior under processing conditions has suffered.

Plastic deformation from a thermodynamic viewpoint is an irreversible process because even the smallest increment of plastic deformation is irreversible. Onsager's reciprocity (symmetry) relations[27] have served as the basis for the linear and nonlinear theory of irreversible thermodynamics. Biot[28-30] recognized the implications of Onsager's work in continuum mechanics and used it to establish an analogy between elastic deformation and viscous flow. Ziegler,[24] modifying Gibb's statistical thermodynamic phase-space theory,[31] showed that Onsager's symmetrical relations could be derived by a simple extension of the theory, making it valid for arbitrary thermodynamic processes. Ziegler also proved that the Onsager symmetry relations could be replaced by an equivalent orthogonality statement which remains meaningful for nonlinear relationships between velocities and forces.[24] This represents an important contribution since an irreversible process is always characterized by certain velocities (fluxes) and corresponding forces which are connected by a dissipation function D.

The orthogonality condition referred to earlier represents the simplest invariant connection between D and the force vector. For the case of linear irreversible systems, the D surfaces are ellipsoids; but for nonlinear systems they are of a more general type. The orthogonality condition, expressed as

$$X_k^i = \lambda(\partial D/\partial \dot{x}_k) \tag{20}$$

still holds. In Eq 20, X_k^i is a generalized irreversible force and λ is a factor determined by the identity

$$X_k^i \dot{x}_k = D(\dot{x}_k) \tag{21}$$

The orthogonality condition is equivalent to various extremum principles such as those formulated by Onsager,[27] Biot,[30] Prigogine and DeGroot,[32] and Ziegler.[33-36]

An important property which the dissipation function D must possess from a physical standpoint to avoid indeterminant conditions between the fluxes and irreversible forces is that the D surfaces must be at least weakly convex; to avoid unstable situations, it must be assumed that the value of D increases sufficiently fast on any radius from the origin.[24,26] The principle of maximum dissipation rate is one theorem which is valid under these conditions. This theorem can also be called the principle of maximum rate of entropy production since $D(\dot{x}_k)/T = \dot{S}$. The extremum principles in this form will be used to develop a dynamic material model (DMM) since, for any closed system, the rate of entropy production under prescribed forces is expected to be a maximum and should approach its final state on the path of least action—that is, the dynamic path.

4.3. Thermodynamic Foundations

The energy balance for the system defined by Eq 10 can be expressed in terms of the time derivative as

$$\dot{K} + \dot{E} = P + Q \tag{22}$$

where the dot denotes the time rate; K is the kinetic energy; E is the internal energy; P is the power input; and Q is the working heat, which has the dimension of power.

The rate at which energy may increase in the workpiece has an upper bound B[37] which is related to the rate of entropy production $\dot{S}$. The upper limit to the rate of energy increase for a specific material is given by

$$\dot{E} - (P - \dot{K}) = Q \leq B \tag{23a}$$

B, an arbitrary constant which bounds Q, varies from body to body in the way in which each body (workpiece) varies in prior thermomechanical history.

The entropy for a given workpiece can be expressed in terms of the arbitrary constant B as

$$S = \int \frac{B}{T}\, dt \tag{23b}$$

or

$$B = T\dot{S} \tag{23c}$$

and Eq 23c assumes the form

$$Q \leq T\dot{S} \tag{23d}$$

Equation 23c should be interpreted as stating that a universal bound B can be assigned for all thermally energetic processes in a given material which has experienced some unique thermomechanical history, and Eq 23d states that the rate of entropy storage should be at least as fast as the ratio Q/T.

The difference between the power input P and the kinetic energy rate $\dot{K}$ is the excess of the mechanical work over the kinetic energy rate. It is termed the *work storage* and written as

$$W^* = P - \dot{K} \tag{24}$$

and, in terms of the balance of energy, Eq 22 appears as

$$\dot{E} = W^* + Q \tag{25}$$

A *reduced dissipation* inequality can be obtained if Q is eliminated between Eq 25 and 23d

$$\dot{E} - W^* \leq T\dot{S} \tag{26}$$

A process in which the equality of Eq 23d and 26 holds is called *reversible*; all other processes are called *irreversible.* Thus, $Q = T\dot{S}$ for reversible processes, and "adiabatic" and "isentropic" have the same meaning. $\dot{S}$ is greater than zero for irreversible adiabatic processes, while Q is less than zero for irreversible isentropic processes.

The Helmholtz *free energy* function is given as

$$F = E - TS \tag{27}$$

and, in terms of F, the reduced dissipation inequality becomes

$$\dot{F} + S\dot{T} - W^* \leq 0 \tag{28}$$

The *rate of dissipation work* D is given by

$$D(\dot{x}_k) = \frac{dW}{dt} = X_k^i \frac{dx_k}{dt} = X_k^i \dot{x}_k \tag{29}$$

where W is the irreversible work performed on an element in an infinitesimal plastic deformation, X_k^i is the irreversible component of the generalized force as indicated by the superscript i, and x_k is a generalized mechanical coordinate. dW is a scalar quantity, and X_k^i and dx_k are vectors. Summation over the index k is implied. $\dot{x}_k$ is a flux and can be taken as a strain-rate component. $D(\dot{x}_k)$ is always greater than zero and is a very important function; it is the only means of introducing *irreversibility* into statistical thermodynamics.[24] $D(\dot{x}_k)$ also allows one to evaluate the overall stability of the system. When other nonmechanical quantities such as heat flow and diffusion exist, this *dissipation work* function can be generalized and called the *dissipation rate* function.

4.4. Extremum Principles

$D(\dot{x}_k)$ is a continuous function with continuous partial derivatives $\partial D/\partial \dot{x}_k$, and the surfaces of constant D are smooth. The surface can be drawn for a fixed force X_k^i, as shown in Fig. 13, following the normality condition which is expressed as follows:

$$X_k^i = \lambda \nabla D = \frac{\partial D}{\partial \dot{x}_k} \quad (\lambda > 0) \tag{30}$$

As shown in Fig. 13, the point P lies on the hyperplane E and on the surface $D = M > 0$. If the dissipation rate M and the unit vector v_k of the

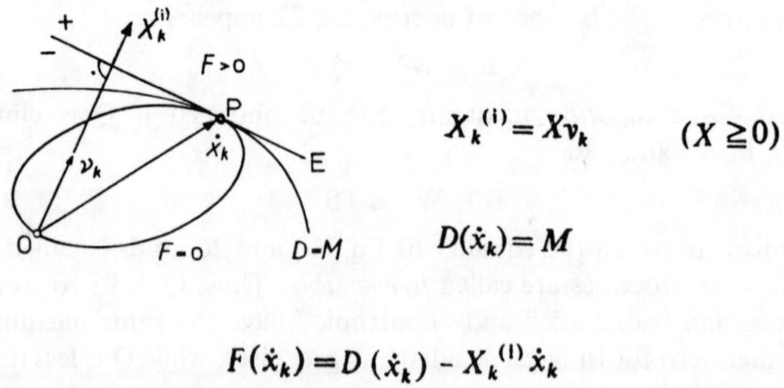

Fig. 13. Schematic representation of dissipation rate function in velocity space (from Ref 24).

force vector X_k^i are prescribed, as shown in the figure, then the actual flux vector $\dot{x}_k$ minimizes the magnitude X of X_k^i, according to the Principle of Least Force. Therefore, the value of M fixes a certain D surface; and the scalar product $v_k\dot{x}_k$ is maximum, and the value of the magnitude X is minimum. Thus, the actual flux vector $\dot{x}_k$ maximizes the scalar product $v_k\dot{x}_k$ and, thereby, minimizes X. The scalar product $v_k\dot{x}_k$ also maximizes the value of the rate function while minimizing the force. This leads to another extremum principle, namely, the Principle of Maximum Dissipation Rate. The properties of the D function can also be related to the Principle of Maximum Rate of Entropy Production $\dot{S}$ by defining the rate of entropy production as

$$\dot{S} = D(x_k)/T = M/T \tag{31}$$

where $T > 0$. The maximum rate of entropy production $\dot{S}$ differs from $D = M$ only by a factor of absolute temperature T and, hence, the Maximum Dissipation Rate Principle can also be stated as the Principle of Maximum Entropy Rate Production. Thus, the extremum principles are equivalent in the sense that the conditions of least force, maximum dissipation rate, and maximum rate of entropy production are established for the construction of $D(\dot{x}_k)$ surfaces in velocity space. These principles also give rise to the following relations:

$$\left.\frac{\partial D}{\partial \dot{x}_k}\right|_{X_k^i} = X_k^i \tag{32}$$

where X_k^i is the least irreversible force,

$$\left.\frac{\partial D}{\partial X_k^i}\right|_{\dot{x}_k} = \dot{x}_k \tag{33}$$

where $\dot{x}_k$ is the flux vector which maximizes the scalar product $v_k\dot{x}_k$, and

$$\left.\frac{\partial D}{\partial T}\right|_{\dot{S}} = \dot{S}_{max} \tag{34}$$

where $\dot{S}_{max}$ is the maximum rate of entropy production. These relations are very important and used in deriving stability criteria in terms of state variables of $\log \dot{\epsilon}$ and T.

The $D(\dot{x}_k)$ surface in velocity space can be incremented as shown in Fig. 14. The upper limit for this increment is bounded by $B = \dot{S}T$, and the step increment is controlled by the following relation for stability:

$$\frac{dD}{dR} > \frac{D}{R} \tag{35}$$

When $D(\dot{x}_k)$ is incremented, Eq 35 should be satisfied for stable conditions. Hence, in a stable system D grows faster than the scalar product $X_k^i\dot{x}_k$ cal-

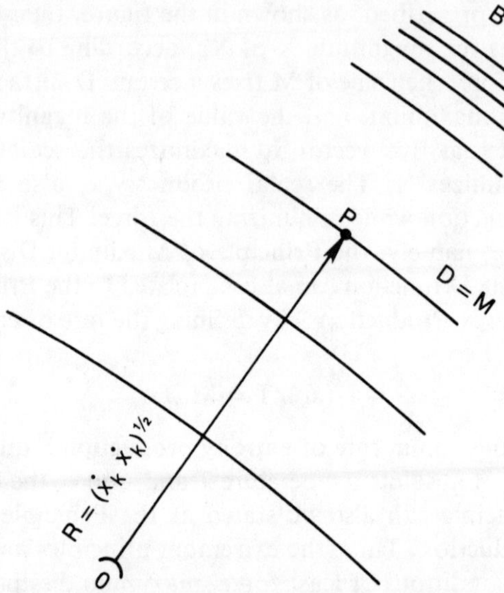

Fig. 14. Representation of distance R from origin 0 to point P on dissipation rate function D in velocity space. D is not defined at origin since it does not exist until an irreversible force becomes operative.

culated with a fixed value of X^i_k. It follows that in a stable system, the function

$$F(\dot{x}_k) = D(\dot{x}_k) - X^i_k \dot{x}_k \tag{36}$$

defined with a fixed least irreversible force is positive on the positive side of E. This result is shown in Fig. 13 and will be discussed further in a later section.

The entropy of the system, which can vary from body to body depending on the prior history of the material, determines the maximum value of the dissipative rate function. An example to illustrate this statement can be found in the starting microstructures of Ti-6242 alloy commonly used as a forging preform material. These microstructures are shown in Fig. 15(a) and (b). Figure 15(a) is an equiaxed $\alpha + \beta$ microstructure and Fig. 15(b) represents a classical Widmanstätten structure. The $\alpha + \beta$ microstructure results because of a series of thermomechanical treatments; and, according to Gegel,[11] it represents the lowest dissipated energy state for the system. The Widmanstätten structure is the starting microstructure for producing the equiaxed $\alpha + \beta$ microstructure by thermomechanical processing. Thomas and Dadras[23] have shown that these two starting microstructures for Ti-6242 have markedly different constitutive behavior. Thus, the entropy production produced by plastic strain rate can be expected to be significantly

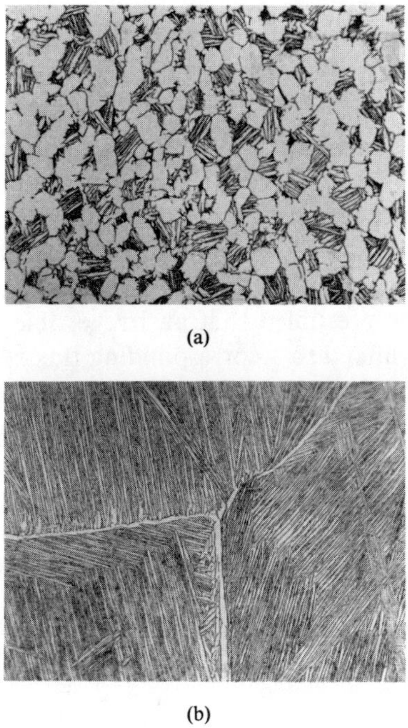

Fig. 15. 500× microstructures of as-received Ti-6242 mults. (a) $\alpha + \beta$ preform microstructure; (b) β preform microstructure.

different for the Ti-6242 equiaxed $\alpha + \beta$ and the Widmanstätten microstructures. The rate of entropy production was used in this research to determine the stable region required for a given process to operate in processing space.

4.5. Complementary Functions Between Force and Velocity Space

Equation 30 expresses the irreversible force X_k^i in terms of the generalized velocity coordinates $\dot{x}_k$ and, if the second derivative of a potential function $\psi(\dot{x}_k)$ exists,[17-19]

$$dX_k^i = -\sum_k^n \frac{\partial^2 \psi(\dot{x}_k)}{\partial \dot{x}_k \partial \dot{x}_j} = d\dot{x}_j \tag{37}$$

If the determinant of the second derivative does not vanish, the equations can be inverted and

$$\dot{x}_k = \dot{x}_k(X_k^i) \tag{38}$$

In view of Eq 38, the complementary function can be defined as

$$\Omega(X_k^i) \equiv \sum_k^n X_k^i \dot{x}_k + \psi \tag{39}$$

The partial derivative of Eq 39 follows:

$$\frac{\partial \Omega}{\partial X_j^i} = \dot{x}_j + \left(X_k^i + \frac{\partial \psi}{\partial x_k} \right) \frac{\partial \dot{x}_k}{\partial X_j^i} = \dot{x}_j \tag{40}$$

It can be seen that the variables $\dot{x}_k$ are related to the complementary function $\Omega(X_k^i)$ much as the variables X_k^i are related to the function $\psi(\dot{x}_k)$. The transformation which reverses the roles of the complementary functions ψ and Ω and the variables $\dot{x}_k$ and X_k^i is known as Legendre's transformation.

A graphical representation of the duality of force and velocity space is shown in Fig. 16. It is presumed that an irreversible force $X_k^{(i)}$ exists and that it subsequently is linked to a corresponding flux $\dot{x}_k$. A normal vector at point P′ in force space is constructed in the usual manner. A hyperplane E′ is made to be tangent at P′ where D′, the dissipative rate function on force space, is equal to M′ > 0. The resulting flux vector $\dot{x}_k$ in velocity space produced by the presumed irreversible force vector $X_k^{(i)}$ in force space is required by the normality condition to be orthogonal with respect to the hyperplane E′. Point P is a point on the hypersurface D = M > 0 in velocity space, and it is the terminus for the velocity vector $\dot{x}_k$ starting from the origin O in velocity space. The irreversible force vector $X_k^{(i)}$ has its origin at the point P′ in force space and its terminus at X_k^i.

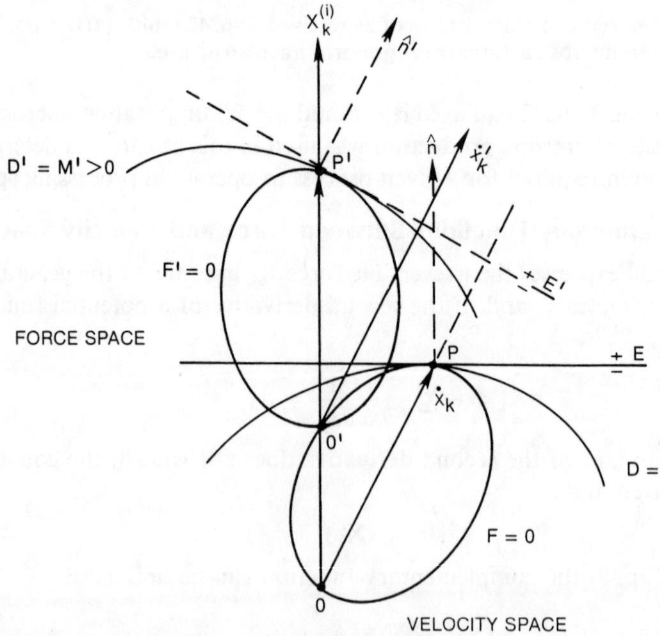

Fig. 16. Representation of duality existing between dissipation rate functions in velocity space and force space.

The surface F = 0 is defined as a plastic hardness surface. It is a potential surface which represents the onset of irreversible forces and material processes which are time dependent. Inside the surface F = 0, the material behaves as a linear elastic body with reversible properties. Thus, this graphical representation is for an imperfectly irreversible body. Examples of such a solid are viscoelastic and viscoplastic materials.

In the present work, the process is completely irreversible, and no elastic-strain components are assumed to be present. For example, the finite-element code ALPID used in this research is based on the theory of rigid-thermo-viscoplastic solids. If it were important to describe an elastic-thermoviscoplastic solid, then elastic strain, plastic strain, and temperature would need to be taken as the state variables. In the present work the rate of entropy production is strictly determined by the plastic strain rate. Hence, strain rate is also one of the state variables.

In summary, the graphical representation of force and velocity space shows the existence of two complementary functions, which, in fact, can also be related by a Legendre transformation.[17-19]

The dissipation rate functions in velocity and force space D and D′, respectively, are equal to the total instantaneous power $P = X_k^{(i)}\dot{x}_k$ being applied to the solid body. Ω and ψ are functions of D′ and D, respectively, and an inverse function exists between Ω and D′ and between ψ and D.

The complementary function Ω leads to the following definition:

$$\Omega = \int_{c'} \dot{x}_k \, dX_k^{(i)} \tag{41}$$

Similarly,

$$\psi = \int_{c} X_k^{(i)} \, d\dot{x}_k \tag{42}$$

where c is an arbitrary curve in a domain b and c′ is the projection of P in the b′ domain. These relationships are shown in Fig. 17. OP defines an arbitrary curve with coordinates $\dot{x}_k$ in b, and O′P′ defines the same arbitrary curve in b′ with coordinates $X_k^{(i)}$.

The instantaneous absorbed power in a dissipator body is given by

$$P = \int_{c} X_k^{(i)} \, d\dot{x}_k + \int_{c'} \dot{x}_k \, dX_k^{(i)} = X_k^{(i)}\dot{x}_k \tag{43}$$

4.6. Complementary Functions in Terms of Effective Stress and Effective Strain Rate

In a paper by Prasad et al.[15] Eq 43 was written in the following

$$\bar{\sigma}\dot{\bar{\epsilon}} = \int_0^{\dot{\bar{\epsilon}}} \bar{\sigma} \, d\dot{\bar{\epsilon}} + \int_0^{\bar{\sigma}} \dot{\bar{\epsilon}} \, d\bar{\sigma} \tag{44}$$

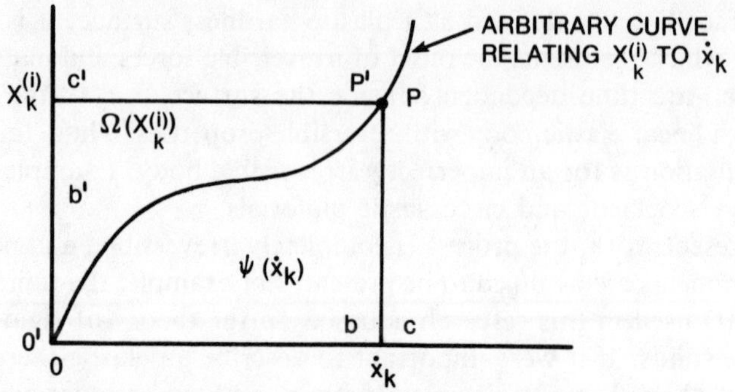

Fig. 17. Representation of complementary functions Ω and ψ and their summation which leads to absorbed power P.

$\bar{\sigma}$ and $\dot{\bar{\epsilon}}$ being equivalent stress and strain rate, making $P = \bar{\sigma}\dot{\bar{\epsilon}}$ in Eq 43. $\int_0^{\dot{\bar{\epsilon}}} \bar{\sigma}\, d\dot{\bar{\epsilon}}$ was termed the G-content and $\int_0^{\bar{\sigma}} \dot{\bar{\epsilon}}\, d\bar{\sigma}$ the J-co-content, indicating that J is the complementary dissipator function of the G-content.

4.7. Stability Analysis

The stability of the workpiece material can be assessed by making use of the D function and normality condition as described by Eq 30. Because practical workability is partially determined by the *mechanical stability* of the material, which can be regarded as a continuum effect, a criterion can be developed for this type of stability as follows.

According to Eq 30

$$X_k^i = \lambda \nabla D$$

where $\lambda > 0$. Therefore,

$$D = X_k^i \cdot \dot{x}_k = \lambda \nabla D \cdot \dot{x}_k \tag{45}$$

$$\nabla D = \nabla(\lambda \nabla D \cdot \dot{x}_k) \tag{46}$$

$$\nabla D = \lambda \nabla^2 D \dot{x}_k + \lambda \nabla D \tag{47}$$

Also, $R^2 = (\dot{x}_k \cdot \dot{x}_k) > 0$ as shown in Fig. 14, where R represents the distance from origin O to point P on the dissipation rate function D in velocity space. From this relation, the following equation is obtained:

$$2R\nabla R = 2\dot{x}_k \tag{48}$$

Hence,

$$\nabla R = \dot{x}_k / R \tag{49}$$

Substituting the above equation into

$$\nabla D = (dD/dR)\nabla R \tag{50}$$

yields

$$(dD/dR)(\dot{x}_k/R) = \lambda\nabla D + \lambda\nabla^2 D\dot{x}_k \tag{51}$$

Multiplying Eq 51 by $\dot{x}_k$ yields

$$(dD/dR)(\dot{x}_k \cdot \dot{x}_k/R) = \lambda\nabla D\dot{x}_k + \lambda\nabla^2 D + \dot{x}_k \cdot \dot{x}_k \tag{52}$$

$$(dD/dR)R = D + \lambda\nabla^2 DR^2 \tag{53}$$

Therefore,

$$dD/dR = D/R + \lambda\nabla^2 DR \tag{54}$$

and

$$dD/dR > D/R \tag{55}$$

because $\nabla^2 D > 0$. It follows that

$$\frac{d(X_k^i \cdot \dot{x}_k)}{d\dot{x}_k} > \frac{X_k^i}{\dot{x}_k} \tag{56}$$

and

$$\frac{d \log X_k^i}{d \log \dot{x}_k} > 0 \tag{57}$$

Substituting state variables $\log \bar{\sigma}$ and $\log \dot{\bar{\epsilon}}$ into Eq 57 for $\log X_k^i$ and $\log \dot{x}_k$, respectively, yields a stability criterion for the material system

$$\frac{d \log \bar{\sigma}}{d \log \dot{\bar{\epsilon}}} > 0 \tag{58}$$

where Eq 58 is the well-known strain-rate-sensitivity parameter m. It usually appears as an exponent in a constitutive equation of the form $\bar{\sigma} = K\dot{\bar{\epsilon}}^m$.

4.8. m Range for Stable Flow

If $\log \bar{\sigma}$ and $\log \dot{\bar{\epsilon}}$ are substituted into Eq 57, it follows that $m > 0$ is a lower bound. If m were equal to zero, according to Eq 16, all of the power would be dissipated by heat, and this could lead to a plastic instability by a continuum process such as adiabatic shearing. Matson[38] has measured a value for m in a borosilicate glass which is equal to zero, and the test specimen was observed to exhibit a shape after compression testing which is typical of a specimen having undergone plastic instability. The deformed borosilicate specimen is shown in Fig. 18.

An upper bound for m can be deduced by making use of the dynamic constitutive equation, Eq 15, and substituting it into the following relation:

$$F' = D(\dot{x}_k) - X_k^i \dot{x}_k \tag{59}$$

Fig. 18. Deformed 7740 borosilicate (Pyrex) specimen; T = 650 °C, $\dot{\epsilon} = 5 \times 10^{-2}\ s^{-1}$.

such that

$$F' = D(\dot{x}_k) - C\dot{x}_k^{1+m} \tag{60}$$

where C is a material constant. Systems which have been observed to be stable have dissipative rate functions which increase faster than the power input.[(24)]

The properties of a dissipative rate function $D(\dot{x}_k)$ for nonlinear irreversible systems were discussed earlier; in summary, they are:

(a) $D(\dot{x}_k)$ is at least piecewise quadratic and weakly convex.
(b) The Orthogonality Principle applies.

These geometric properties of $D(\dot{x}_k)$ require m to be less than unity ($m < 1$). If $D(\dot{x}_k)$ were to have the geometric properties of an ideal ellipsoid, as is the case for linear irreversible processes, then m = 1 would be the upper bound. Thus, F′ can be written as

$$F' = D(\dot{x}_k) - C\dot{x}_k^{1+m} \geq 0 \tag{61}$$

$F' = 0$ would correspond to the case where the material would behave in a linear, irreversible way. $F' > 0$ would correspond to a nonlinear, irreversible thermodynamic process far away from equilibrium or steady state.

Another important system parameter can be derived by dividing F′ by $T > 0$, which yields

$$\frac{F'}{T} = \frac{D(\dot{x}_k)}{T} - \frac{C\dot{x}_k^{1+m}}{T} \geq 0 \tag{62}$$

F′/T can be interpreted as the entropy rate difference if $D(\dot{x}_k)/T$ is defined as the rate of storage of entropy and $C\dot{x}_k^{1+m}/T$ as the power input to temperature T ratio P/T. Equation 62 can be written as

$$\Delta\dot{S} = \dot{S}_{sys} - \dot{S}_{app} \tag{63}$$

where $D(\dot{x}_k)/T = \dot{S}_{sys}$ and $C\dot{x}_k^{1+m}/T = \dot{S}_{app}$. Equation 63 can be normalized by dividing both sides by $\dot{S}_{app}$ to obtain the normalized expression

$$\frac{\Delta\dot{S}}{\dot{S}_{app}} = s - 1 \tag{64}$$

The system parameter s is, thus, the ratio $D(\dot{x}_k)/C\dot{x}_k^{1+m}$; in other words, it is the ratio of the power dissipated to the power supplied. According to the Second Law of Thermodynamics, the system is stable when Eq 62 and 63 are positive, unstable when they are less than zero, and marginally stable when they are equal to zero. Thus, for stability, s must be greater than or equal to 1. Because s is a dimensionless system parameter and related to the entropy through the Second Law of Thermodynamics, it can be assumed to be related to *material stability*. Furthermore, it is reasonable to assume that s is a system parameter which may be useful for describing the evolution of "structure" as a function of time t or strain rate. It will be shown later that the parameter m can be associated with continuum effects or mechanical stability and that s is related to the material or structural stability. The allowable ranges for these system parameters are

$$0 < m \leq 1 \tag{65}$$

and

$$s \geq 1 \tag{66}$$

4.9. Evaluation of m and s

Evaluation of m. The two methods most commonly used to evaluate m in the literature are shown in Fig. 19. Continuous flow curves measured at a constant strain rate are used for experimentally determining the values of m for different temperatures and effective strain levels since it is neither necessary nor desirable to assume that "structure" will remain constant during a strain-rate change. Experience clearly shows that microstructural changes frequently occur when the strain rate is changed. For example, processes such as wedge cracking and cavity formation at second-phase hard particles are theoretically expected and observed to be very strain-rate sensitive.[8]

Evaluation of s. s can be evaluated easily by making use of Eq 31 to obtain

$$\Delta = \left.\frac{\partial D}{\partial T}\right|_{\dot{\epsilon}} = \left.\frac{\partial P}{\partial T}\right|_{\dot{\epsilon}} = \frac{\partial}{\partial T}\,[\dot{S}T] \tag{67}$$

Therefore,

$$\Delta = \dot{S}_{sys} \tag{68}$$

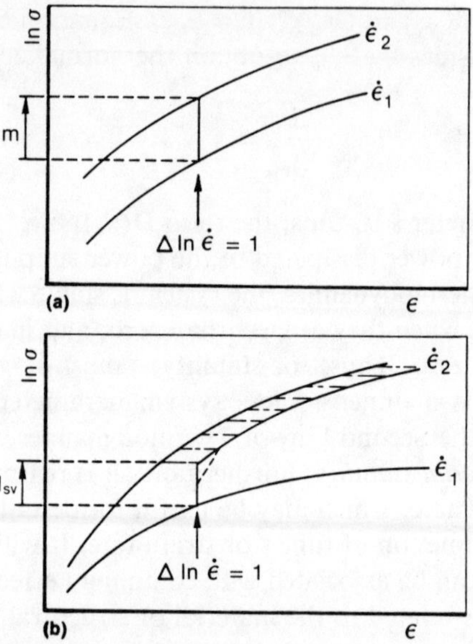

Fig. 19. Schematic representations of definitions of strain-rate-sensitivity index. Difference between definitions is due to strain not being a state variable. (a) m, based on continuous flow curves measured at constant true strain rates; (b) M_{SV}, state variable description based on strain-rate-change tests.

where $\dot{S}_{sys}$ is the rate of entropy production by the system. Furthermore,

$$\Delta = \left.\frac{\partial P}{\partial T}\right|_{\dot{\epsilon}} = -\frac{P d(\log P)}{T^2 \partial(1/T)} \tag{69}$$

or

$$\Delta = \frac{P}{T}\left[-\frac{1}{T}\,\frac{\partial \log P}{T\partial(1/T)}\right] \tag{70}$$

where

$$\left.\frac{\partial \log P}{T\partial(1/T)}\right|_{\dot{\epsilon}} = \left.\frac{\partial \log \bar{\sigma}}{T\partial(1/T)}\right|_{\dot{\epsilon}} \tag{71}$$

Define a new coefficient s such that

$$s = -\frac{1}{T}\left.\frac{\partial \log \bar{\sigma}}{\partial(1/T)}\right|_{\dot{\bar{\epsilon}}} \tag{72}$$

It follows that the rate of entropy production by the system is given by

$$\Delta = \frac{P}{T}\,s \tag{73}$$

where P/T can be thought of as the applied rate of entropy input ($\dot{S}_{app}$).

According to the Second Law of Thermodynamics, s should be greater than one for stable material flow. This implies that the workpiece should store entropy at least as fast as the heat working for stable flow.

4.10. Intrinsic Workability

Materials being forged undergo large, nonlinear irreversible deformation, which is made possible by linking together several atomic processes. The string of processes, in addition to providing the degrees of freedom needed for producing the large deformation, is also responsible for the evolution of "structure." Each of these processes may be thought of as a channel for dispersing energy supplied by the forging process, and each may give rise to entropy production. Forging, as a rule, is a shape-changing process where the deformation is generally inhomogeneous and transient over a large volume of the deforming continuum. Steady-state phenomena often represent a limiting condition which can only be achieved under unique combinations of temperature T and effective strain rate $\dot{\bar{\epsilon}}$. It is, therefore, highly desirable to be able to define stable regions in terms of the kinematic variables T, $\dot{\bar{\epsilon}}$, and t.

A solution to this problem would require the limiting conditions to be identified. These limiting conditions would be loci of bifurcation points where two atomic mechanisms are operating simultaneously to produce a maximum in the energy-dispersal rate at a unique combination of temperature T and effective strain rate $\dot{\bar{\epsilon}}$. Under these unique conditions the deformation process would be steady state. One branch of the deformation process would be stable and the other, unstable. Processes such as grain-boundary cavitation, wedge cracking, and cleavage are considered to cause the deforming continuum to become unstable because the free surfaces which form increase the free energy of the system rather than decrease it. A solution for this complex problem can be found by recognizing that the atomic processes associated with stable plastic flow and microstructure evolution lead to Liapunov functions.[5,39] In order to find a suitable Liapunov function, one must understand how the entropy is split for a closed system.[39-40]

The change of entropy dS can be split into two parts: the entropy production d_iS due to changes within the system, and the flow of entropy d_eS due to interaction with the outside world.[39-40] It follows that

$$dS = d_iS + d_eS \tag{74}$$

The entropy production within the system is always positive; thus,

$$d_iS \geq 0 \tag{75}$$

The entropy production d_iS vanishes when the system is subjected to reversible changes only and is always positive when the system undergoes irreversible thermodynamic changes. The workpiece material is considered to be a *closed* system because it may exchange energy with the surroundings but not with matter. The rate of generation of entropy per unit time is

$$\frac{d_iS}{dt} \geq 0 \tag{76}$$

and

$$\frac{d_iS}{dt} = \frac{dS}{dt} - \frac{dq/dt}{T} \geq 0 \tag{77}$$

where dq/dt is equal to the heat working Q. Thus, only irreversible processes in the material contribute to the production of entropy.

Prigogine[39-40] has shown that the state variable S can be expanded such that

$$S = S_0 + \delta S + 1/2(\delta^2 S) \tag{78}$$

where S_0 is the nominal entropy, δS is the first-order variation of entropy, and $\delta^2 S$ is the second-order variation of entropy. Furthermore, because of the principle of maximum rate of entropy production, it can be shown[37] that

$$\dot{S} = \frac{Q}{T} + \Delta \tag{79}$$

where Q is the working heat of the body and Δ is $\partial D/\partial T|_{\dot{x}_k}$. When the system is reversible, $D(\dot{x}_k)$ is equal to zero and, thus, $\Delta(\dot{x}_k)$ is the heat gain of the system arising from the irreversible contribution to entropy production. Therefore, following Truesdell[37] and Prigogine,[38] one can describe the state variable $\dot{S}$ as

$$\dot{S} = \frac{Q}{T} + \Delta = \dot{S}_0 + \delta\dot{S} + 1/2\delta^2\dot{S} \tag{80}$$

The linear terms in Eq 80 are equal to Q/T when the system is reversible; it is only the higher-order term which contributes to entropy production when the system is being plastically deformed and is irreversible. Thus, stability is given by the second-order term $1/2\delta^2\dot{S}$.

Glansdorff and Prigogine[39] have shown that $1/2\delta^2 S$ is a Liapunov function for linear irreversible systems, independent of boundary conditions. In this case, Prigogine assumed that *thermodynamic equilibrium* is the attractor for irreversibility. However, when plastic flow is involved, it is assumed that *steady state* is the attractor for irreversibility, making $\delta^2\dot{S}$ the Liapunov function rather than $\delta^2 S$.

It is necessary to invoke Ziegler's orthogonality principle and assert that steady state is the attractor for irreversibility since plastic deformation, no matter how small the increment of plastic flow, is highly irreversible. Thus, in an analogous way, $1/2\delta^2\dot{S}$ is a Liapunov function for nonlinear deforming systems as is $1/2\delta^2 S$ for linear irreversible systems. The Liapunov function is related to the dissipative rate function $D(\dot{x}_k)$ through Eq. 80, where

$$\Delta = 1/2\delta^2\dot{S} \tag{81}$$

Equation 81 can be set equal to

$$\Delta = 1/2\delta^2\dot{S} = \sum \delta X_k^i \, \delta\dot{x}_k \tag{82}$$

where k refers to the kth process, which may be combined with others to produce a net shear strain rate which will approach steady state as a limiting (critical) condition.

The forcing function of Eq 82 cannot be specified a priori because:

(a) The initial state cannot be precisely specified.
(b) The deformation process involves bifurcations which are stochastic.[39]

Therefore, it is not possible to solve the Liapunov function directly; it must be solved axiomatically by finding a sufficient number of dimensionless system variables which describe the evolution of the system by trial and error.

Thus, to obtain a suitable Liapunov function, one must find variables Y_i which will describe the evolution of the system beginning at some initial condition where the Y_i values are nonvanishing. The Liapunov function allows one to determine whether the deforming system, which is controlled by selected atomic processes, is approaching steady state. Steady state is defined as follows:

$$\frac{dY_i}{dt} = 0 \tag{83}$$

Consider a Liapunov function L having dimensionless system variables

$$L = L(Y_1, \ldots Y_n) \tag{84}$$

and suppose that it is positive definite (it may be also negative definite) throughout the region of interest and vanishes at $(dY_i/dt) = 0$. The time derivative of L as the system variables evolve with time (or, equivalently, strain rate) is

$$\frac{dL}{dt} = \sum \frac{\partial L}{\partial Y_i} \frac{dY_i}{dt} \tag{85}$$

The Liapunov theorem asserts that steady state will be an attractor if dL/dt has the sign opposite that of L.

To form the Liapunov function, the system variables m or, equivalently, η and s will be used. Thus,

$$L = L(\eta, s) \tag{86}$$

Because the kinematic variables of the material system include T, $\dot{\epsilon}$, and t, it is convenient to evaluate the Liapunov function with respect to log $\dot{\epsilon}$, which is regarded as a state variable. Since an inverse relationship exists between strain rate and time, the Liapunov function is inverted and assumes the opposite sign with respect to log $\dot{\epsilon}$. The conclusions are the same.

5. Interpreting and Understanding the Dynamic Material Model

The Dynamic Material Model (DMM) is a tool for assessing the intrinsic workability of a material during inelastic deformation and the evolution of microstructure. Metallurgical microstructures are relatively simple examples of a broad class of material pattern-forming phenomena. Theoretical studies to aid the understanding of such phenomena clearly show that steady-state theory and knowledge of stability conditions provide insufficient information for determining unique patterns.[41-42] Unresolved issues include how various microstructures form and are selected and controlled within a given system and whether sharp selection mechanisms exist during processing for creating structure.

The basic tenet of the DMM is that nonequilibrium systems may operate in such a way as to optimize some quantity—in this particular model, *the rate of entropy production*. The extensive experience of the forging industry indicates unequivocally that the prior thermomechanical history of a workpiece material frequently influences the response of that material to the demands of a forging process and evolution of the microstructure as a function of time. Although a family of microstructural stationary states may exist, only one of these states may be accessible from the initial state of the material. The term "stationary state" refers to a point in processing space where the derivative of the functional vanishes; however, this is not necessarily an extremum point. A microstructural stationary state is one in which the microstructure is stable for all $t > t_0$ until the boundary conditions are changed. When no sharp selection rules exist, different initial states will produce different steady-state patterns. Utility of the DMM will be demonstrated using Ti-6242 alloy; a processing map will be presented which shows the various steady-state processes which evolve during inelastic deformation.

The solution of $L = L(\eta,s)$ is represented in Fig. 8 for Ti-6242, whose initial microstructure has been characterized as a Widmanstätten pattern. The regions in the processing map where stability is observed are called Γ regions. These regions represent the upper and lower limits of the plastic strain rate and temperature ranges whose boundaries are the loci of the various bifurcation points. These processes coexist in steady state along a particular boundary. The terms "stable" and "unstable" are used in the sense of the

Second Law of Thermodynamics. Situations may occur where a particular microstructure forms and is labeled unstable. This labeling may be open to question because intuition dictates that the evolution of a particular structure should be regarded as natural. A specific case where the evolution of a microstructure would be considered unstable involves dynamic grain growth which occurs at elevated temperatures in some materials in competition with dynamic recrystallization. An example[14] is given in Fig. 20. This would be a bifurcation process, for example, in which the dynamic recrystallization process would be operating in the stable Γ region and the dynamic-grain-growth process would be operating in the unstable region.

The thermodynamic interpretation which should be used to understand this situation is that the dynamic-grain-growth process would be decreasing the rate at which entropy would be generated, making this thermodynamic process unstable. In contrast, dynamic recrystallization would be contributing to excess entropy production as new and smaller grains formed until steady state was reached. The Liapunov analysis would indicate that the system was reducing its total potential energy with time in this case because the sign of dL/dt would be opposite that of L(η,s). Dynamic grain growth, unlike dynamic recrystallization, is a process which increases the order of the system and reduces the efficiency of dispersing energy by microstructural evolution. It utilizes some of the supplied energy to create order, and order is always associated with a decrease in entropy production. Because a reduction in the efficiency factor accompanies dynamic grain growth, most of the

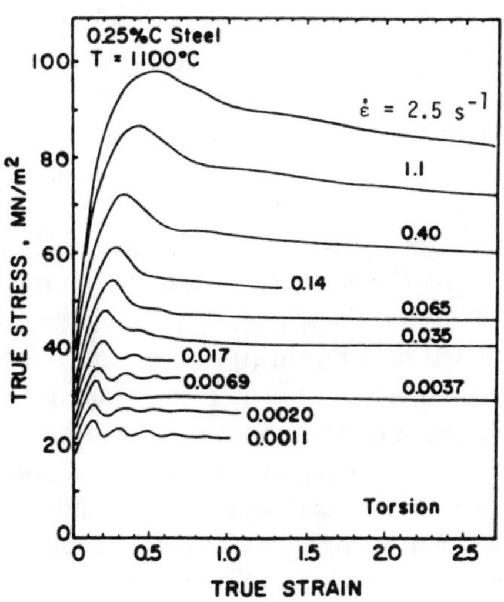

Fig. 20. Influence of strain rate upon flow curves derived from hot torsion data at 1100 °C for 0.25% C steel (from Ref 14).

supplied power will be incoherently dispersed as heat. The net entropy production for the universe would also be positive. Thus, in the DMM, when the system is far away from equilibrium, it is the *excess rate of entropy production* $\delta^2\dot{S}$, rather than internal entropy production $\dot{S}_i$, which creates structure and determines stability.

Dynamic grain growth might also be interpreted as an unstable process from the standpoint of workability. If superplastic flow were the mechanism for achieving enhanced flow, then grain growth during hot working could potentially eliminate the superplastic flow process and initiate another nonstationary state which could lead to defect formation such as grain-boundary cavitation.

Interpretation of the information presented in the DMM depends on both a practical understanding of workability and a firm knowledge of atomic mechanisms. Once the steady-state conditions have been identified in terms of the chosen dimensionless system variable and the process parameters, the mechanistic approach can be successfully merged with continuum modeling. The boundaries on the processing map are the loci of the points where the derivatives vanish or, more specifically, where $\partial\eta/\partial \log\dot{\bar{\epsilon}}$ and $\partial s/\partial \log\dot{\bar{\epsilon}}$ are equal to zero. Thus, when the boundary has been defined, the system has evolved to a stationary state, which is the attractor for irreversible processes.[40] The Liapunov function $L(\eta,s)$ is equal to the right-hand side of Eq 82, and Gladsdorff and Prigogine[39] have called this function the excess entropy production rate. If $\sum \delta X_k^i \delta\dot{x}_k \geq 0$ for all t starting at the time of perturbation, stability is ensured. This perturbation is testing the stability of a nonstationary state. The stable Γ regions should be interpreted as a field where the excess entropy production rate is greater than zero and a region where dissipative structures may form as the system proceeds along a trajectory toward a stationary boundary. Therefore, it follows that when the system is away from a steady-state boundary, the classical dynamic and kinetic equations of deformation break down.

The Liapunov analysis, in effect, seeks out the stationary states of the system until a steady state is reached. Although the boundaries on the processing map are represented as a single line, they should be thought of as a band. The bandwidth is related to the geometry of the functional where the stationary state is identified. If an attempt is being made to identify this type of boundary by metallographic techniques such as optical microscopy, SEM, and TEM, difficulty may be encountered. In the vicinity of the boundary, the microstructure will be bimodal because the steady state is achieved by a bifurcation. Experience has shown that if it is desirable to investigate the microstructure, one must process the material at a point within the stable and unstable fields at some distance from the steady-state boundary.

The method of boundary identification in the Ti-6242 processing map is illustrated in Fig. 21 (a and b) and 22 (a and b). The relationships between η and s as functions of $\log \dot{\epsilon}$ are shown in Fig. 21(a) and 22(a), and their respective partial derivatives are given in Fig. 21(b) and 22(b). The system variables and their partial derivatives were evaluated in this case at a true strain of 0.60. The stable regions shown in Fig. 21(a) and 22(b) were identified using the Liapunov axiom which states that for stability to be achieved, the sign of the partial derivatives must be opposite that of $L(\eta,s)$. This condition asserts that the excess entropy production rate is greater than zero.

The Ti-6242 processing map is a composite map showing both mechanical and material stability/instability regions.[43] The partial derivative $\partial\eta/\partial \log \dot{\epsilon}$ indicates the mechanical stability of the system, and $\partial s/\partial \log \dot{\epsilon}$ yields the material characteristics. The efficiency contours are plotted to provide an indication of the amount of power which is partitioned between J, the dissipator co-content, and G, the dissipator content. Because the processing map represents both material and mechanical stability limits, it is said to represent the *intrinsic workability* of a given material.

It is useful to refer to the bifurcation discussed previously and shown in Fig. 12 for illustration of another point about microstructural evolution. This point can be understood more easily by referring to Fig. 23, which shows a set of stress-strain curves for Ti-6242, with each curve having a different initial microstructure. The transformed-β or Widmanstätten structure clearly exhibits flow softening until the flow curve asymptotically approaches steady state at 0.60 plastic strain. The $\alpha + \beta$ equiaxed structure also asymptotically approaches steady state at this strain level, and the two curves eventually become superimposed at high strains. Coincident with the superpositioning of the flow curves, the microstructure of the formerly Widmanstätten specimen now becomes nearly identical to that of the test specimen which was originally equiaxed $\alpha + \beta$. The test specimens continue to deform plastically by identical mechanisms after reaching 0.60 true strain. The processing maps for these two different starting microstructures become identical at 0.60 true plastic strain.

Close inspection of the two flow curves reveals that even the so-called equiaxed $\alpha + \beta$ microstructure undergoes a softening process until steady-state flow is approached. The flow softening leads to the identification of a bifurcation at ~1200 K and a strain rate of 0.001 s^{-1}, which was associated with α-phase dynamic recrystallization in the stable Γ region and α-phase dissolution in the unstable region. The power-partitioning parameter m increases in the Γ region until the stationary point is reached and decreases in the unstable region as the volume fraction of β phase begins to increase. The formation of the β phase may be interpreted as a process which increases the order of the material system and reduces its entropy. Because the efficiency

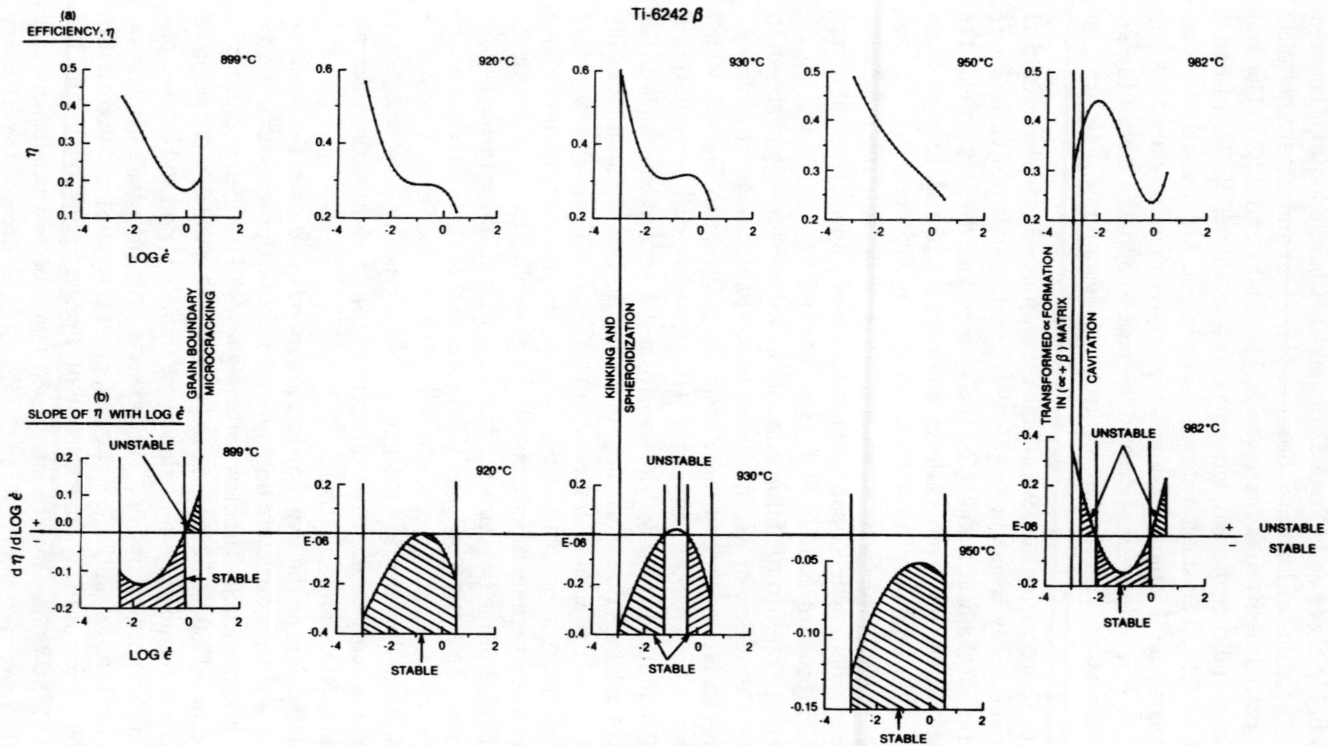

Fig. 21. Plots of (a) Liapunov function, efficiency η, vs. log strain rate (per sec.) and (b) $(\partial\eta/\partial \log \dot{\epsilon})|_T$ vs. log strain rate (per sec.) at true strain of 0.6 and at different temperatures for Ti-6242 β microstructure.

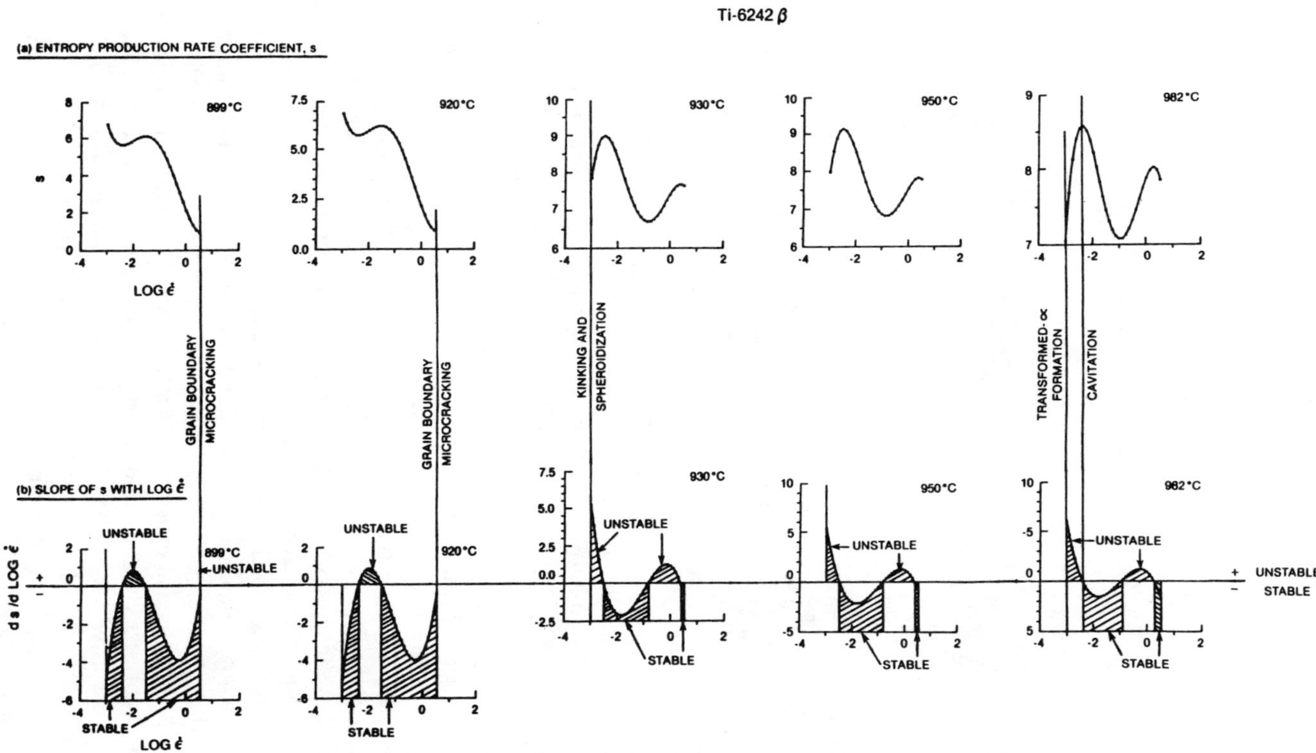

Fig. 22. Plots of (a) Liapunov function, entropy-production-rate coefficient s, vs. log strain rate (per second) and (b) $(\partial s/\partial \log \dot{\epsilon})|_T$ vs. log strain rate (per second) at true strain of 0.6 and at different temperatures for Ti-6242 β microstructure.

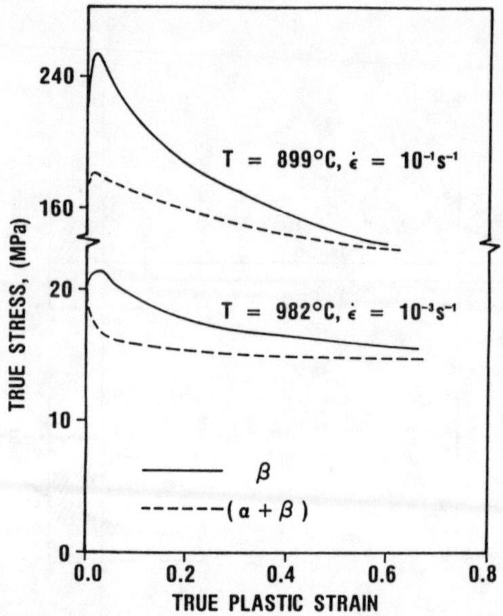

Fig. 23. Dependence of flow stress on preform microstructure for Ti-6242.

factor is also decreased, there is an increase in the amount of supplied energy which is dispersed incoherently as heat.

Prasad et al.[15] also showed that the transformed-β preform material produced a bifurcation at 1200 K and a strain rate of 0.001 s^{-1} by a different pair of mechanisms. The stress-strain curve for this material exhibited considerably more flow softening than that for equiaxed $\alpha + \beta$ preform material. After noting the wide range of complex microstructures which form in the transformed-β preform material as a function of deformation,[23] one can conclude that the Ti-6242 material approaches the same steady-state point of $\eta = 60\%$, T = 1200 K, and $\dot{\bar{\epsilon}} = 0.001$ s^{-1} for a fixed strain of 0.60 as the equiaxed $\alpha + \beta$ Ti-6242 material *but by an entirely different nonstationary trajectory.*

The microstructures which form during inelastic material flow are called *dissipative structures* because they arise from the exchange of energy with the outside and the rate of excess entropy production. The microstructural characteristics of the dissipative structures are a function of the particular trajectory in processing space and its prior thermomechanical history. The string of atomic processes which produce the dissipative structures in the nonequilibrium state is difficult—if not impossible—to define conclusively. Thus, the power of thermodynamics is demonstrated by its ability to reduce complexity to a few important state parameters.

Not all dissipative structures which form in the stable Γ region are desirable. Of course, *desirability* always depends on the forged-product specifications. If there is no need to be concerned with second-tier properties such as low-cycle fatigue and fracture toughness, then it is not necessary to reflect this concern in the judgment of microstructural qualities. However, assuming that these second-tier mechanical properties are important to reliability and performance, the α structure shown in the Γ region, which has the morphology of a wedge crack, might be considered undesirable. This microstructure formed at a temperature of about 1200 K and a plastic strain rate of $10^{-2.25}$ s^{-1}. Wedge cracking is known to be strain-rate sensitive, and wedge cracks tend to form when the grain-boundary shear strain rate is too fast to allow the tensile stresses which arise from grain-boundary sliding to be relaxed by diffusion or an accompanying phase transformation. In this example, wedge cracking was suppressed by the nucleation of α titanium, which has a lower density than β titanium.[44] The ability to relax tensile stresses in $\alpha + \beta$ titanium alloys by α-phase formation is a beneficial attribute because it explains why interphase cavitation is not observed for titanium systems during superplastic forming. The α titanium formed in this way is often loosely referred to as transformed α.

Dilational studies on the transformed-β forging-preform microstructure,[11] in addition to mechanical-property tests, indicate that 1189 K (916 °C) is a critical temperature for this material. Deforming this preform material at strain rates which are greater than ~2.0 s^{-1} at 1185 K (913 °C) produced two classical plastic instabilities. Figure 24 shows a wedge crack at a triple junction and evidence of grain-boundary cavitation. Referring to the processing map shown in Fig. 8, it can easily be seen that this material was deformed at a combination of temperature and strain rate which lies within the unstable region of the map. As shown in Fig. 25, increasing the strain rate to 10 s^{-1} at the same temperature causes the plastic instability to become more marked. According to Ref 11, the deformation temperature is too low for diffusional stability to be established, increasing the affinity for mechanical instability—particularly when the strain rates are relatively high. Associated with these observed phenomena are a low-efficiency η and a decreased entropy production rate.

It can be seen, with reference to Fig. 22(b), that the rate of entropy production increases with increasing time or decreasing strain rate for any given temperature. This increase in entropy production rate is entirely consistent with the principle of maximum entropy production rate. This point can be illustrated by examining the calculated entropy-production-rate parameter for each stationary state shown in the Ti-6242 processing map in Fig. 8 and observing that it increases with decreasing strain rate (or, equivalently, increasing time).

Fig. 24. Triple-point cracking in Ti-6242 β compression specimen deformed at 913 °C (1675 °F), $\dot{\bar{\epsilon}} \approx 2\ s^{-1}$.

The optimum domain for isothermal forging of Ti-6242 is shown on the processing map in Fig. 8 as a cross-hatched square. This so-called sweet zone was determined by the forging industry by trial-and-error processing. It was determined by varying the forging conditions and correlating the property-microstructure relationships with each condition. An interesting rule of thumb, which has evolved as a result of extensive practical experience and is used in designing an $\alpha + \beta$ titanium alloy forging, follows:

> If the plastic deformation is less than 0.60 effective strain in any region throughout the forging, then redesign the forging preform to redistribute the material.

It is more than fortuitous that the 0.60 effective strain rule of thumb exists from the standpoint of microstructural stability under dynamic conditions. Irreversible thermodynamic principles assert that steady state is an attractor for irreversible processes. The system, when subjected to an applied stress, will attempt to lower its potential energy state. Research on developing Ti-6242 constitutive equations for process modeling has shown that at the

Fig. 25. Ti-6242 β compression specimen deformed at 913 °C (1675 °F), $\dot{\varepsilon} \approx 10\ s^{-1}$.

0.60 effective strain level, the microstructure changes very little when subjected to further deformation during hot working. This technical knowledge leads directly to the particular heuristic design rule, bypassing the costly trial-and-error testing on the shop floor.

6. Application of DMM for Near-Optimal Process Design and Control

The Dynamic Material Model (DMM) is directly related to many of the accepted practices of the forging industry—at least where near-precision and forging quality are required. The cross-hatched region in Fig. 8 is an example of the correlation of control variables with forging quality and performance. Many examples exist for other engineering materials. A rule of thumb for Ti-6242 and other $\alpha + \beta$ titanium alloys follows:

> If microstructure, geometry, and tolerance control is the goal, then forge in a "processing window" near the steady-state boundary.

A corollary to this rule based on heuristics is as follows:

> If defects are to be avoided in the finished product, then always forge within the bandwidth of the steady-state boundary.

Another corollary to the basic rule follows:

> If the part is a precision forging and if geometry cannot be easily inspected by ultrasonic NDE, then forge in a "processing window" situated only in the stable region.

Other forging design rules can be written which are based on process modeling, dynamic material modeling, and experience; however, the objective here is not to develop the rules but to demonstrate that it is possible to combine process modeling with material-behavior modeling to produce a model dual-property Ti-6242 jet-engine compressor disk by using only the knowledge contained in the constitutive equations for Ti-6242 and the DMM. A detailed design of such a disk can be achieved by incorporating the material-behavior information in finite-element programs which are capable of analyzing large plastic incremental deformations.

The term "near-optimal design" indicates that criteria used for controlling the numerical solution are based on knowledge of how a given forging preform material will respond to the demands of the process. The initial object of process modeling should be to validate the choice of preform shape and to develop a process control algorithm (PCA) for regulating the forging process in real time. Process modeling provides a road map for process control. In forging the model must be able to predict the future due to the irreversible nature of the process; this can be accomplished using current finite-element programs if the material behavior under processing conditions can be specified as a nonholonomic constraint.

Malas[45] developed a method for achieving near-optimal design by incorporating the concept of a nonholonomic constraint in the finite-element program ALPID (Analysis of Large Plastic Incremental Deformation) – a rigid-thermoviscoplastic program which has been validated by the forging industry for accurate prediction of die filling and associated field quantities such as temperature, velocity, stress, and strain.

An isothermal closed-die forging process was simulated for producing a dual-property Ti-6242 compressor disk by incorporation of a desired strain-rate range as a nonholonomic constraint. The strain-rate range was selected after consideration of the practical control limitation of isothermal forging presses and use of design rules such as those described above for Ti-6242. Another consideration in the selection of temperature and strain-rate range is the rules which require the process variables to remain on the stable side of the steady-state boundary. The control target for the simulation was the cross-hatched area of the processing map. This analysis is regarded as conservative because of the narrow temperature and strain-rate range selected to prevent defect formation. Furthermore, these particular process variables were selected because they would fix the steady-state boundary, which would be the attractor for microstructural evolution.

The preform geometry was designed based on a tradeoff between the requirement for die filling and microstructure control throughout the forging. Initially, the transformed-β microstructure was chosen for the forging preform since material modeling showed that this structure would transform to an equiaxed $\alpha + \beta$ microstructure at strain levels where steady-state flow is approached. Material modeling also showed that the transformed-β microstructure could be retained as long as the strain level in the deforming preform did not exceed the effective strain range of 0.30–0.50. This strain range does not allow the steady-state condition to be reached in the region where the equiaxed microstructure becomes stable. These conditions were determined easily from the constitutive equation and DMM.

The design goal for the dual-property disk involved obtaining optimum low-cycle fatigue and burst-strength properties in the bore region and optimum creep, stress-rupture, and fracture-toughness properties in the rim region. The gradient in mechanical properties is determined by the performance requirements of the engine disk as specified by the design engineer. Once the design specifications have been determined, the process-engineering team of the forging vendor proceeds to design a forging preform which will ultimately produce an equiaxed $\alpha + \beta$ microstructure in the bore region and retain the transformed-β microstructure in the rim. Based on the results of the DMM, a strain rate of 0.003 s^{-1} and a temperature of 1200 K have been determined to be the optimum processing conditions for this material and application.

These processing conditions will provide the necessary microstructure and property gradients throughout the disk, while ensuring enhanced metal flow for die filling. The process modeling not only produces the correct preform shape for achieving the dual-property disk but also provides a control algorithm for maintaining the desired strain-rate range in critical locations of the deforming workpiece material as a function of time. The predicted final finite-element grid distortion and strain distribution are shown in Fig. 26 for the subscale disk. It should be noted that the effective strain level in the bore region exceeds unity, which is a requirement for producing the equiaxed microstructure, and it does not exceed the upper limit of 0.50 in the region where the transformed-β microstructure is required.

A subscale disk was forged to validate the near-optimal design concept. The microstructure gradients are shown in Fig. 27, and typical mechanical properties after final heat treatment are shown in Table 1. These results validate the design concepts based on material behavior modeling under processing conditions. These results demonstrate the ability to synthesize a set of prescribed microstructures and mechanical properties in a finished near-net-shape forging.

7. Summary

The goal of the studies described herein was to effect a synthesis of mechanistic and continuum modeling for describing the evolution of metallurgical

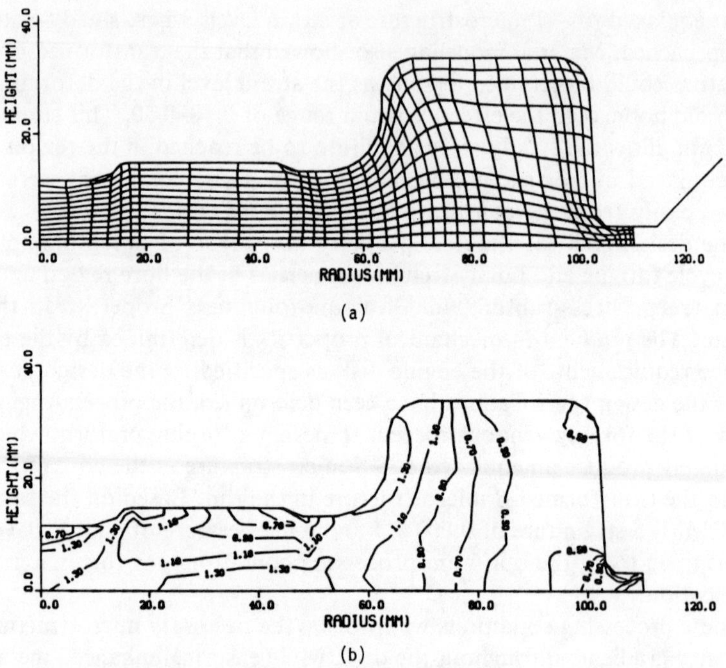

Fig. 26. Predicted grid distortion and final strain distribution from ALPID simulation of dual-property disk forging.

microstructures. These microstructures are relatively simple examples of a broad class of pattern-forming phenomena, in comparison with the complexity of structures formed in biological systems. Despite this relative simplicity, little is known about the evolution of microstructures during processing when the material is exchanging energy with the outside, and about the selection and control of microstructures within a given alloy system.

Dynamic material modeling has begun to suggest new directions of research which should provide answers to some of these theoretical questions, enabling metallurgical synthesis to become a science rather than an art.

Metal alloys can be thought of as dynamic systems. Their response to an effort variable such as stress is known to depend on prior thermomechanical history; it is also recognized that these systems have a memory. Dynamic material modeling has shown that when these alloy systems exchange energy with the outside, such as in forging, they may undergo a series of bifurcations (stationary states) which allows the alloy system to acquire a "historical" dimension.

The acquired history of a given material is recorded in constitutive equations which provide the values of arbitrary state variables at any given time.

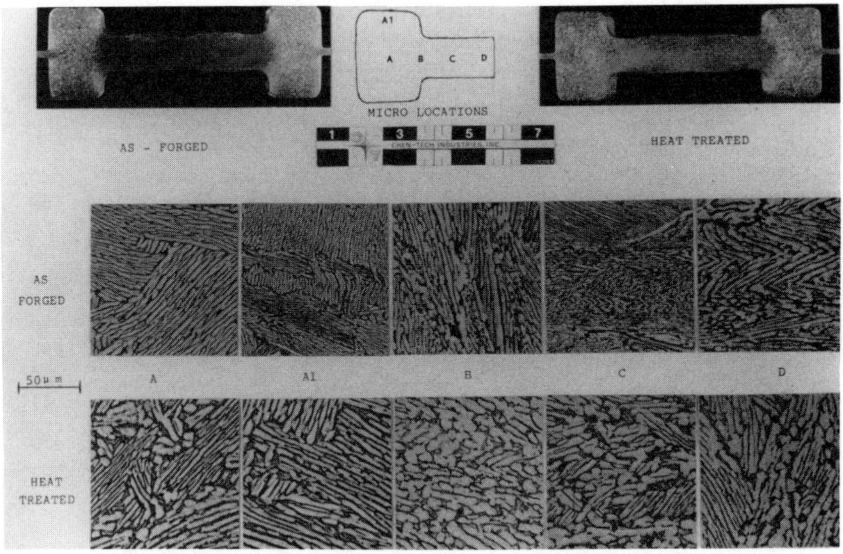

Fig. 27. Microstructural features of dual-property disk produced by selected forging and heat-treatment combinations.

These equations represent the new initial state from which microstructures may evolve, depending on certain selection rules. Because it is virtually impossible to describe a priori how the various atomic processes may be strung together in complex engineering materials, the constitutive equations must be experimentally determined. These equations contain the history which allows the future to be predicted. The DMM represents a method for extracting information concerning the evolution of structure which is contained in the constitutive equation. The difficulty of describing microstructure evolution by a mechanistic approach arises from the fact that the problem has deterministic, statistical, and indeterministic aspects.

One of the most attractive features of thermomechanics is its power of unification and its ability to reduce a large variety of phenomena to a few basic ideas. The DMM was developed from the extremum principles and the extension of Onsager's reciprocity relations for linear irreversible thermodynamic processes to nonlinear, highly irreversible processes far away from equilibrium. The extremum principles were applied based on the assumptions that nonequilibrium systems operate in such a way as to optimize the rate of production of entropy and that the rate of production of excess entropy may provide a theory for the evolution of microstructure. The dynamic material model (DMM) resulted from these hypotheses, and it has been tested against a wide variety of material systems such as cast metallic alloys, P/M alloys,

Table 1. Mechanical properties of Ti-6242 disks forged at 926 °C (1200 K) and solution heat treated at 955 °C (1228 K)(a)

Properties	Center region	Outer region
Room-temperature tensile properties		
Yield strength, MPa	652	636
Ultimate tensile strength, MPa	705	668
Elongation, %	7	4
Reduction in area, %	15	11
565 °C tensile properties		
Yield strength, MPa	363	353
Ultimate tensile strength, MPa	438	436
Elongation, %	20	22
Reduction in area, %	49	43
Time to 0.1% creep strain(b), hr	188	330
Fracture toughness, MPa $\sqrt{m}$	52	70

(a) Followed by aging at 595 °C for 8 hr and air cooling. (b) Test conditions: 510 °C (783 K)/240 MPa.

ceramics, and intermetallic-compound materials. Ti-6242, which was used herein as an example, represents a typical multiphase, multicomponent alloy system which exhibits a wide range of flow and fracture processes; its behavior during processing is strongly influenced by prior thermomechanical history.

Some scientific issues which require further analysis are:

1. The theory of DMM suggests that microstructure selection is related to prior thermomechanical history, rate of energy input, and energy dispersal by either a coherent or an incoherent process. It is thought that microstructure formation is related to coherent energy-dispersal mechanisms while heat storage is related to the incoherent dispersal of energy.
2. The stability of a structure is determined by a complex interplay of kinetic and thermodynamic quantities, and a multiplicity of possibilities of atomic processes seems necessary to describe the far-from-equilibrium situation. Flow of energy into the system, such as during the occurrence of dynamic grain growth, may organize the system and decrease its entropy; in other cases, such as dynamic recrystallization, the entropy may be increased.
3. Dissipative structures may form during nonstationary flow as the result of the excess entropy production rate which appears as either a positive definite or a negative definite function; as a consequence, the sta-

bility problem can be solved without reference to the properties of the stationary state.

4. Dissipative structures may have a "low" entropy value because they are believed to be associated with the dispersal of energy by a coherent mechanism. Each of the atomic mechanisms associated with plastic flow may be thought of as a channel for the dispersal of energy; these mechanisms may also contribute to excess entropy production which drives structure evolution.
5. The microstructure tends to become stable at stationary states, and the stability of the system under this condition is ensured by the maximum rate of internal entropy production rather than by the rate of excess entropy production. Optimized forging processes with respect to hot workability—and, in some cases, mechanical properties—have been observed to lie near the steady-state boundary in temperature-strain rate space in support of this conclusion.

Acknowledgments

The author wishes to acknowledge the Air Force Office of Scientific Research (AFOSR) for supporting this in-house work, and Dr. H. M. Burte, Chief Scientist of the Materials Laboratory, Air Force Wright Aeronautical Laboratories, Wright-Patterson AFB, Ohio, for his sustained efforts to stimulate basic research on metallurgical synthesis. The author is also grateful to his colleagues, K. A. Lark, J. C. Malas, L. M. Matson, J. T. Morgan, and W. T. O'Hara of the Processing and High Temperature Materials Branch for their scientific contributions and discussions. Special appreciation is expressed to Professor Y. V. R. K. Prasad, Indian Institute of Science, Bangalore, India, for the many scientific contributions which he made during his tenure as an NRC-AFSC Research Associate. The author wishes to expresses his gratitude to the Cameron Iron Works (Houston, Texas), Wyman-Gordon (Worcester, Massachusetts), Ladish (Cudahy, Wisconsin), and to Mr. R. Sprague and Dr. S. C. Jain of General Electric-AEBG for their encouragement and technical remarks and for sharing the results of their practical shop-floor experiences. The information which they shared during their efforts to implement advanced material and process-modeling methods developed by the Air Force as part of their technical modernization programs was extremely useful. The invaluable guidance and constant encouragement throughout the course of this research by Professor John Alexander, FRS and Stocker Visiting Professor of Engineering, and Professor J. S. Gunasekera, Department of Mechanical Engineering of Ohio University at Athens, Ohio, and Dr. S. M. Doraivelu, Universal Energy Systems, Dayton, Ohio, is greatly appreciated. Thanks are also due to Mrs. M. Whitaker for editorial assistance and preparation of the manuscript.

References

1. S. I. Oh, "Finite Element Analysis of Metal Forming Processes with Arbitrarily Shaped Dies," Int. J. Mech. Sci., Vol. 24, 1982, p 479.
2. G. E. Dieter, "Bulk Workability Testing," *Metals Handbook*, Ninth Edition, Vol 8, *Mechanical Testing*, American Society for Metals, Metals Park, OH, 1985, p 573-597.
3. M. F. Ashby and H. J. Frost, "The Kinetics of Inelastic Deformation Above 0 K," in *Constitutive Equations in Plasticity*, edited by A. S. Argon, MIT Press, Cambridge, MA, 1975, p 117.
4. A. Kelly, *Strong Solids*, Oxford University Press, Oxford, 1966.
5. I. Prigogine, *From Being to Becoming: Time and Complexity in the Physical Sciences*, W. H. Freeman and Company, New York, 1980.
6. P. W. Atkins, *The Second Law*, Scientific American Books, H. W. Freeman and Company, New York, 1984.
7. P. J. Wray, J. Appl. Phys., Vol 40, 1969, p 4018.
8. R. Raj, "Development of a Processing Map for Use in Warm Forming and Hot Forming Process," Metall. Trans. A, Vol 12, 1981, p 1089.
9. I. G. Palmer and G. C. Smith, *Oxide Dispersion Strengthening* (Second Bolton Landing Conference), Gordon and Breach, New York, 1960, p 252.
10. W. M. Yim and N. J. Grant, *Ductility*, American Society for Metals, Metals Park, OH, 1967, p 101.
11. H. L. Gegel, "Material Behavior Modeling—An Overview," in *Experimental Verification of Process Models*, edited by C. C. Chen, American Society for Metals, Metals Park, OH, 1983, p 3-32.
12. D. R. Mosher and R. Raj, Acta Met., Vol 22, 1974, p 1469.
13. J. J. Jonas and M. J. Luton, *Advances in Deformation Processing*, edited by J. J. Burke and V. Weiss, Plenum Press, New York, 1978, p 215.
14. J. J. Jonas and T. Sakai, "A New Approach to Dynamic Recrystallization," *Deformation, Processing and Structure*, edited by G. Kraus, American Society for Metals, Metals Park, OH, 1982, p 185-228; H. J. McQueen, "A Discussion of Paper by J. J. Jonas and T. Sakai," *Deformation, Processing and Structure*, edited by G. Kraus, American Society for Metals, Metals Park, OH, 1982, p 231-237.
15. Y. V. R. K. Prasad, H. L. Gegel, S. M. Doraivelu, J. C. Malas, J. T. Morgan, K. A. Lark, and D. R. Barker, "Modeling of Dynamic Material Behavior in Hot Deformation–Forging of Ti-6242," Metall. Trans. A, Vol 15, 1984, p 1883.
16. M. W. Hirsch and S. Smale, *Differential Equations, Dynamical Systems and Linear Algebra*, Academic Press, New York, 1974, p 180-193.
17. P. E. Wellstead, *Introduction to Physical System Modeling*, Academic Press, New York, 1979, p 9-144.
18. H. Ziegler, "Some Extremum Principles in Irreversible Thermodynamics with Application to Continuum Mechanics," *Progress in Solid Mechanics*, Vol IV, John Wiley & Sons, Inc., New York, 1963, p 93-191.
19. R. Courant and D. Hilbert, *Methoden der Mathematischen Physik*, Springer-Verlag, Berlin, 1937, p 26.

20. F. W. Constant, *Theoretical Physics: Mechanics of Particles, Rigid and Elastic Bodies, Fluids, Heat Flow*, Addison-Wesley Publishing Co., Inc., New York, 1954, p 189.
21. J. J. Jonas, C. M. Sellars, and W. J. McG. Tegart, Metall. Rev., Vol 14, 1969, p 1.
22. H. J. McQueen and J. J. Jonas, *Treatise on Materials Science and Technology*, edited by R. J. Arsenault, Academic Press, New York, 1975, p 393.
23. P. Dadras and J. F. Thomas, Jr., Metall. Trans. A, Vol 12, 1981, p 1867.
24. H. Ziegler and P. M. Naghdi, *Progress in Solid Mechanics*, Vol IV, edited by J. N. Sneddon and R. Hill, North-Holland, Amsterdam, 1963.
25. *Irreversible Aspects of Continuum Mechanics and Transfer of Physical Characteristics in Moving Fluids* [International Union of Theoretical and Applied Mechanics (I.U.T.A.M.) Symposia, Vienna, June 22-28, 1966], edited by H. Parkus and L. I. Sedov, Springer-Verlag, New York, 1966.
26. H. Ziegler, "Thermomechanics," in Special Issue: Symposium on the Future of Applied Mathematics, J. Quart. Appl. Math., April 1972, p 93.
27. L. Onsager, "Reciprocal Relations in Irreversible Processes," Phys. Rev., Vol 37 (II), 1931, p 405, and Vol 38 (II), 1931, p 2265.
28. M. A. Biot, "Theory of Stress-Strain Relations in Anisotropic Viscoelasticity and Relaxation," J. Appl. Phys., Vol 25, 1954, p 1385.
29. M. A. Biot, "Variational Principles in Irreversible Thermodynamics With Application to Viscoelasticity," Phys. Rev., Vol 97, 1955, p 1463.
30. M. A. Biot, "Thermoelasticity and Irreversible Thermodynamics," J. Appl. Phys., Vol 27, 1956, p 240.
31. J. W. Gibbs, *Elementary Principles in Statistical Mechanics, Vol II The Collected Works*, Yale University Press, New Haven, CT, 1948.
32. S. R. Degroot, *Thermodynamics of Irreversible Processes*, North-Holland, Amsterdam, 1952, p 196.
33. H. Ziegler, "Zwei Extremalprinzipien der irreversiblen Thermodynamik," Ing. Arch., Vol 30, 1961, p 410.
34. H. Ziegler, "Die statistischen Grundlagen der irreversiblen Thermodynamik," Ing. Arch., Vol 31, 1962, p 317.
35. H. Ziegler, "Some Extremum Principles in Irreversible Thermodynamics, With Application to Continuum Mechanics," Progr. Solid Mech. Vol 4, 1963, p 91.
36. H. Ziegler, "Thermodynamic Considerations in Continuum Mechanics," The 1964 Minta Martin Lecture, MIT and AIAA, 1964.
37. C. Truesdell, "Thermodynamics for Beginners," in *Irreversible Aspects of Continuum Mechanics and Transfer of Physical Characteristics in Moving Fluids* [International Union of Theoretical and Applied Mechanics (I.U.T.A.M.) Symposia, Vienna, June 22-28, 1966], edited by H. Parkus and L. I. Sedov, Springer-Verlag, New York, 1966, p 373-389.
38. L. Matson, unpublished work, 1986.
39. P. Glandsdorff and I. Prigogine, *Thermodynamics of Structure, Stability and Fluctuations*, Wiley-Interscience, New York, 1971, p 63.
40. I. Prigogine, "Time, Structure, and Fluctuations," Sci., Vol 201, 1978, p 777.
41. J. S. Langer, "Pattern Selection in Solidification," Metall. Trans. A, Vol 15, 1984, p 961.
42. J. S. Langer and H. Muller-Krumbhaar, "Mode Selection in a Dendrite-Like Nonlinear System," Phys. Rev. A, Vol 27(1), 1983, p 499.
43. P. Germain, "The Role of Thermodynamics in Continuum Mechanics," in *Foun-*

dations of Continuum Thermodynamics, edited by J. J. Delgado Domingos, M. N. R. Nina, and J. H. Whitelaw, Halstead Press, John Wiley & Sons, Inc., New York, 1973, p 317.

44. J. C. Malas, H. L. Gegel, S. I. Oh, and G. D. Lahoti, "Metallurgical Validation of a Finite-Element Program for the Modeling of Isothermal Forging Processes," in *Experimental Verification of Process Models*, edited by C. C. Chen, American Society for Metals, Metals Park, OH, 1983, p 358.
45. J. C. Malas, *A Thermodynamic and Continuum Approach to the Design and Control of Precision Forging Processes*, M. S. Thesis, Wright State University, Dayton, OH, 1985.

Index

NOTE: The symbol (F) or (T) following a page number or page span indicates that at least some of the information on that page or within that page span is presented in a figure or table, respectively.